Hitler

A Study in Tyranny

Hitler

A Study in Tyranny

by ALAN BULLOCK

Abridged Edition

HarperPerennial
A Division of HarperCollins*Publishers*

To My Mother and Father

First Perennial Library edition published 1971. First HarperPerennial edition published 1991.

LIBRARY OF CONGRESS CATALOG CARD NUMBER 90-54030

ISBN 0-06-092020-3

91 92 93 94 95 MPC 10 9 8 7 6 5 4 3 2 1

CONTENTS

Preface vii

BOOK I
PARTY LEADER, 1889–1933

1 *The Formative Years, 1889–1918* 1
2 *The Years of Struggle, 1919–24* 30
3 *The Years of Waiting, 1924–31* 64
4 *The Months of Opportunity, October 1931–30 January 1933* 95

BOOK II
CHANCELLOR, 1933–9

5 *Revolution after Power, 30 January 1933–August 1934* 137
6 *The Counterfeit Peace, 1933–7* 172
7 *The Dictator* 207
8 *From Vienna to Prague, 1938–9* 233
9 *Hitler's War, 1939* 284

BOOK III
WAR-LORD, 1939–45

10 *The Inconclusive Victory, 1939–40* 321
11 *'The World Will Hold Its Breath,' 1940–41* 353
12 *The Unachieved Empire, 1941–3* 381
13 *Two Julys, 1943–4* 411
14 *The Emperor Without His Clothes* 448

EPILOGUE 485

PREFACE TO THE ABRIDGED EDITION

I am greatly indebted to Mrs. Mildred Owen for the skill she has shown in reducing my original text to its present form. She has done this without sacrificing any of the essential parts or the over-all balance of the book, and I am very well satisfied to see it published in this abridged edition.

When I first published my study of Hitler, I said in the preface that I had started with two questions in mind. The first, suggested by much that was said at the Nuremberg Trials, was to discover how great a part Hitler played in the history of the Third Reich and whether Goring and the other defendants were exaggerating when they claimed that under the Nazi régime the will of one man, and of one man alone, was decisive. This led to the second and larger question: if the picture of Hitler given at Nuremberg was substantially accurate, what were the gifts Hitler possessed which enabled him first to secure and then to maintain such power. I determined to reconstruct, so far as I was able, the course of his life from his birth in 1889 to his death in 1945, in the hope that this would enable me to offer an account of one of the most puzzling and remarkable careers in modern history.

The book is cast, therefore, in the form of a historical narrative, interrupted only at one point by a chapter in which I have tried to present a portrait of Hitler on the eve of his greatest triumphs (Chapter 7). I have not attempted to write a history of Germany, nor a study of government and society under the Nazi régime. My theme is not dictatorship, but the dictator, the personal power of one man, although it may be added that for most of the years between 1933 and 1945 this is identical with the most important part of the history of the Third Reich. Up to

1934 the interest lies in the means by which Hitler secured power in Germany. After 1934 the emphasis shifts to foreign policy and ultimately to war, the means by which Hitler sought to extend his power outside Germany. If at times, especially between 1938 and 1945, the figure of the man is submerged beneath the complicated narrative of politics and war, this corresponds to Hitler's own sacrifice of his private life (which was meagre and uninteresting at the best of times) to the demands of the position he had created for himself. In the last year of his life, however, as his empire begins to crumble, the true nature of the man is revealed again in all its harshness.

No man can sit down to write about the history of his own times—or perhaps of any time—without bringing to the task the preconceptions which spring out of his own character and experience. This is the inescapable condition of the historian's work, and the present study is no more exempt from these limitations than any other account of the events of the recent past. Nevertheless, I wrote this book without any particular axe to grind or case to argue. I have no simple formula to offer in explanation of the events I have described; few major historical events appear to me to be susceptible of simple explanations. Nor has it been my purpose either to rehabilitate or to indict Adolf Hitler. If I cannot claim the impartiality of a judge, I have not cast myself for the role of prosecuting counsel, still less for that of counsel for the defence. However disputable some of my interpretations may be, there is a solid substratum of fact —and the facts are eloquent enough.

In 1962 I made a thorough revision of the text in the light of the new material which had been published on the history of the Nazi Party and the Third Reich. After much thought, I found no reason to alter substantially the picture I drew of Hitler when the book was first published, although I have not hesitated to change the emphasis where it no longer seemed right. It was in the account of the events leading up to the Second World War that I made the most complete revision, partly because of the large number of new diplomatic documents that have been published, partly because it is here that my own views have been most affected by the longer perspective in which we are now able to see these events. I am indebted to Mr. A. J. P. Taylor's *Origins of the Second World War* for stimulating me to re-read the whole of the documentary evidence for Hitler's foreign policy in the years 1933–9. The fact that I disagree with Mr. Taylor in his view of Hitler and his foreign policy —more than ever, now that I have re-read the documents—does not reduce my debt to him for stirring me up to take a critical look at my own account.

When the revised edition of my book was published, I took the opportunity to revise the bibliography as well. Any reader who wishes to consult this can find it in the unabridged edition, together with the original list of acknowledgments and documentation in the form of footnotes for all the quotations I have made from original sources. In view of the number of publications which would now have to be included in the bibliography, I have decided to omit it from the present abridged version.

I have already expressed my thanks to Mrs. Owen. My debt to my wife remains the greatest of all, not only for the support she gave me in first undertaking this study twenty years ago, but for her encouragement and help in facing the task of its revision and present abridgment.

ALAN BULLOCK

St. Catherine's College
Oxford
January 1971

Men do not become tyrants in order to keep out the cold.

ARISTOTLE, *Politics*

Hitler

A Study in Tyranny

BOOK I

PARTY LEADER
1889–1933

CHAPTER ONE

THE FORMATIVE YEARS
1889–1918

Adolf Hitler was born at half past six on the evening of 20 April 1889, in the Gasthof zum Pommer, an inn in the small town of Braunau on the River Inn which forms the frontier between Austria and Bavaria.

In the summer of this same year, 1889, Lenin, a student of nineteen in trouble with the authorities, moved with his mother from Kazan to Samara. Stalin was a poor cobbler's son in Tiflis, Mussolini the six-year-old child of a blacksmith in the bleak Romagna. The three republics Hitler was to destroy, the Austria of the Treaty of St Germain, Czechoslovakia, and Poland, were not yet in existence. Four great empires—the Hapsburg, the Hohenzollern, the Romanov, and the Ottoman—ruled over Central and Eastern Europe. The Bolshevik Revolution and the Soviet Union were not yet imagined: Russia was still the Holy Russia of the Tsars.

Hitler's family, on both sides, came from the Waldviertel, a poor, remote country district, lying on the north side of the Danube, some fifty miles north-west of Vienna. In this countryside of hills and woods, with few towns or railways, lived a peasant population cut off from the main arteries of Austrian life.

It was from this country stock, with its frequent intermarriages, that Hitler sprang. The family name, possibly Czech in origin and spelled in a variety of ways, first appears in the Waldviertel in the first half of the fifteenth century.

Johann Georg Hiedler, the presumed grandfather of the future chancellor, seems to have been a wanderer who never settled down, but followed the trade of a miller in several places in Lower Austria. In the course of these wanderings he picked up with a peasant girl from the village of Strones, Maria Anna Schicklgruber, whom he married at Döllersheim in May 1842.

Five years earlier, in 1837, Maria had given birth to an illegitimate child, who was known by the name of Alois. According to the accepted tradition the father of this child was Johann Georg Hiedler, but when he married Maria, he did not bother to legitimize Alois, who continued to be known by his mother's maiden name of Schicklgruber until he was nearly forty. Alois was brought up at Spital in the house of his father's brother, Johann Nepomuk Hiedler.

In 1876 Johann Nepomuk took steps to legitimize the young man who had grown up in his house. He called on the parish priest at Döllersheim and persuaded him to cross out the word 'illegitimate' in the register and to append a statement signed by three witnesses that his brother Johann Georg Hiedler had accepted the paternity of the child Alois. In all probability, we shall never know for certain who Adolf Hitler's grandfather, the father of Alois, really was, or whether he was Jewish, as had been suggested. From the beginning of 1877, twelve years before Adolf was born, his father called himself Hitler, and his son was never known by any other name until his opponents dug up this long-forgotten village scandal and tried, without justification, to label him with his grandmother's name of Schicklgruber.

Alois left his uncle's home at the age of thirteen to serve as a cobbler's apprentice in Vienna, but he did not take to a trade, and by the time he was eighteen he had joined the Imperial Customs Service. From 1855 to 1895 Alois served as a customs officer in the towns of Upper Austria. He earned the normal promotion and as a minor state official he had certainly moved up in the social scale from his peasant origins. In the resplendent imperial uniform of the Hapsburg service Alois Hitler appeared the image of respectability. But his private life belied appearances.

In 1864 he married Anna Glass, the adopted daughter of another customs collector. There were no children and, after a separation, Alois's wife, who was considerably older and had long been ailing, died in 1883. A month later Alois married a young hotel servant, Franziska Matzelberger, who had already borne him a son out of wedlock and who gave birth to a daughter, Angela, three months after their marriage. Within a year of her daughter's birth, Franziska was dead of tuberculosis.

This time he waited half a year before marrying again. His third wife, Klara Pölzl, twenty-three years younger than himself, came from the village of Spital, where the Hitlers had originated. The two families were already related by marriage, and Klara herself was the granddaughter of Johann Nepomuk Hiedler, in whose house Alois had been brought up as a child. She had even lived with Alois and his first wife for a time at Braunau, but at the age of twenty had gone off to Vienna to earn her living as a domestic servant. An episcopal dispensation had to be secured for such a marriage between second cousins, but finally, on 7 January 1885, Alois Hitler married his third wife, and on 17 May of the same year their first child, Gustav, was born at Braunau.

Adolf was the third child of Alois Hitler's third marriage. Gustav and Ida, both born before him, died in infancy; his younger brother, Edward, died when he was six; only his younger sister, Paula, born in 1896, lived to grow up. There were also, however, the two children of the second marriage with Franziska, Adolf Hitler's half-brother Alois, and his half-sister Angela. Angela was the only one of his relations with whom Hitler maintained any sort of friendship. She kept house for him at Berchtesgaden for a time, and it was her daughter, Geli Raubal, with whom Hitler fell in love.

When Adolf was born his father was over fifty and his mother was under thirty. Alois Hitler was a hard, unsympathetic, and short-tempered man, and his domestic life suggests a difficult and passionate temperament. Towards the end of his life he seems to have become bitter over some disappointment, perhaps connected with an inheritance. When he retired in 1895 at the age of fifty-eight, he stayed in Upper Austria. The family finally settled at Leonding, a village just outside Linz, where they lived in a small house with a garden.

Hitler attempted to represent himself in *Mein Kampf* as the child of poverty and privation. In fact, his father had a perfectly adequate pension and gave the boy the chance of a good education. After five years in primary schools, in September 1900 the eleven-year-old Adolf entered the Linz Realschule, a secondary school designed to train boys for a technical or commercial career. At the beginning of 1903 Alois Hitler died, but his widow continued to draw a pension and was not left in need. Adolf left the Linz Realschule in 1904 not because his mother was too poor to pay the fees, but because his record at school was so indifferent that he had to accept a transfer to another school at Steyr, where he finished his education at sixteen.

In *Mein Kampf* Hitler makes much of a dramatic conflict between himself and his father over his ambition to become an artist. There is no doubt that he did not get on well with his father, but it is unlikely that his ambition to become an artist (he was not fourteen when his father died) had much to do with it. A more probable explanation is that his father was dissatisfied with his school reports and made his dissatisfaction plain. Hitler glossed over his poor performance at school, which he left in 1905 without securing the customary Leaving Certificate. He found every possible excuse for himself, from illness and his father's tyranny to artistic ambition and political prejudice. It was a failure which rankled and found frequent expression in sneers at the 'educated gentlemen' with their diplomas and doctorates.

There is no doubt that Hitler was fond of his mother, but she had little control over her self-willed son who refused to earn his living and spent the next two years indulging in dreams of becoming an artist or architect, living at home, filling his sketch book with entirely unoriginal drawings and elaborating grandiose plans for the rebuilding of Linz. His one friend was August Kubizek, the son of a Linz upholsterer. Together they visited the theatre, where Hitler acquired a life-long passion for Wagner's opera. Wagnerian romanticism and vast dreams of his own success as an artist and Kubizek's as a musician filled his mind. He lived in a world of his own, content to let his mother provide for his needs, scornfully refusing to concern himself with such petty mundane affairs as money or a job.

A visit to Vienna in May and June 1906 fired him with enthusiasm for the splendour of its buildings, its art galleries and Opera. His ambition now was to enter the Academy of Fine Arts in Vienna. His mother was anxious and uneasy but finally capitulated.

His first attempt to enter the Academy in October 1907 was unsuccessful. The Academy's Classification List contains the entry: 'The following took the test with insufficient results or were not admitted. . . . Adolf Hitler, Braunau a. Inn, 20 April 1889. German. Catholic. Father, civil servant. 4 classes in *Realschule*. Few heads. Test drawings unsatisfactory.'

The result, he says in *Mein Kampf*, came as a bitter shock. The Director advised him to try his talents in the direction of architecture: he was not cut out to be a painter. But Hitler refused to admit defeat. Even though his mother was dying of cancer, he did not return to Linz until after her death (21 December 1907), in time for the funeral, and in February 1908 he went back to Vienna to resume his life as an 'art student.'

He was entitled to draw an orphan's pension and had the small savings left by his mother to fall back on. He was soon joined by his friend Kubizek, whom he had prevailed upon to follow his example and seek a place at the Vienna Conservatoire. The two shared a room in which there was hardly space for Kubizek's piano and Hitler's table.

Apart from Kubizek, Hitler lived a solitary life. Women were attracted to him, but he showed complete indifference to them. Much of the time he spent dreaming or brooding. His moods alternated between abstracted preoccupation and outbursts of excited talk. He wandered for hours through the streets and parks, staring at buildings which he admired, or suddenly disappearing into the public library in pursuit of some new enthusiasm. Again and again, the two young men visited the Opera and the Burgtheater. But while Kubizek pursued his studies at the Conservatoire, Hitler was incapable of any disciplined or systematic work. He drew little, wrote more and even attempted to compose a music drama on the theme of Wieland the Smith.

In July 1908, Kubizek went back to Linz for the summer. A month later Hitler set out to visit two of his aunts in Spital. They had expected to meet again in Vienna in the autumn, but when Kubizek returned to the capital, he could find no trace of his friend.

In mid-September Hitler had again applied for admission to the Academy of Art. This time, he was not even admitted to the examination. Perhaps it was wounded pride that led him to avoid Kubizek. Whatever the reason, for the next five years he chose to bury himself in obscurity.

Vienna, at the beginning of 1909, was still an imperial city, capital of an Empire of fifty million souls stretching from the Rhine to the Dniester, from Saxony to Montenegro. The massive, monumental buildings erected on the Ringstrasse in the last quarter of the nineteenth century reflected the prosperity and self-confidence of the Viennese middle class; the factories and poorer streets of the outer districts the rise of an industrial working class. Vienna was no place to be without money or a job. Hitler himself says that the years he spent there, from 1909 to 1913, were the unhappiest of his life. They were also in many ways the most important, the formative years in which his character and opinions were given definite shape.

In *Mein Kampf* Hitler speaks of his stay in Vienna as 'five years in which I had to earn my daily bread, first as a casual labourer, then as a painter of little trifles,' but a little further on, he gives another picture of those days: 'In the years 1909–10 I had so far improved my position that I no longer had to earn my daily bread as a manual labourer. I was now working independently as a draughtsman and painter in water-colours.' Hitler explains that he made very little money at this, but that he was master of his own time and felt that he was getting nearer to the profession he wanted to take up, that of an architect.

According to Konrad Heiden, who was the first man to piece together the scraps of independent evidence, in 1909, Hitler was obliged, for lack of funds, to give up the furnished room in which he had been living. In the summer he could sleep out, but with the coming of autumn he found a bed in a doss-house, and at the end of the year he moved to a hostel for men started by a charitable foundation at 27 Meldemannstrasse. Here he lived for the remaining three years of his stay in Vienna.

A few others who knew Hitler at this time have been traced and questioned, amongst them Reinhold Hanisch, a tramp from German Bohemia, who for a time knew Hitler well. Hanisch's testimony is partly confirmed by one of the few pieces of documentary evidence which have been discovered for the early years. For in 1910, after a quarrel, Hitler sued Hanisch for cheating him of a small sum of money, and the records of the Vienna police court have been published.

Hanisch describes his first meeting with Hitler in the doss-house in 1909. 'On the very first day there sat next to the bed that had been allotted to me a man who had nothing on except an old torn pair of trousers—Hitler. His clothes were being cleaned of lice, since for days he had been wandering about without a roof and in a terribly neglected condition.'

Hanisch and Hitler joined forces in looking for work; they beat carpets, carried bags outside the West Station, and did casual labouring jobs, sometimes shovelling snow off the streets. Hitler had no overcoat, and he felt the cold badly. Hanisch asked Hitler one day what trade he had learned. ' "I am a painter," was the answer. Thinking that he was a house decorator, I said that it would surely be easy to make money at this trade. He was offended and answered that he was not that sort of painter, but an academician and an artist.' When the two moved to the Meldemannstrasse, 'we had to think out better ways of making money. Hitler proposed that we should fake pictures. He told me that already in Linz he had painted small landscapes in oil, had roasted them in an oven until they had become quite brown and had several times been successful in selling these pictures to traders as valuable old masters.' This sounds highly improbable, but in any case Hanisch, who had registered as Walter Fritz, was afraid of the police. 'So I suggested to Hitler that it would be better to stay in an honest trade and paint postcards. I myself was to sell the painted cards, we decided to work together and share the money we earned.'

Hitler bought a few cards, ink and paints, and produced little copies of views of Vienna, which Hanisch peddled. They made enough to keep them until, in the summer of 1910, Hanisch sold a copy which Hitler had made of a drawing of the Vienna Parliament for ten crowns. Hitler, who was sure it was worth far more—he valued it at fifty in his statement to the police—was convinced he had been cheated. When Hanisch failed to return to the hostel, Hitler brought a lawsuit against him which ended in Hanisch spending a week in prison and the break-up of their partnership.

This was in August 1910. For the remaining four years before the First World War, first in Vienna, later in Munich, Hitler continued to eke out a living in the same way. Some of Hitler's drawings, mostly stiff, lifeless copies of buildings in which his attempts to add human figures are a failure, were still to be found in Vienna in the 1930s, when they had acquired the value of collectors' pieces. More often he drew posters and crude advertisements for small shops—Teddy Perspiration Powder, Santa Claus selling coloured candles, or St Stefan's spire rising over a mountain of soap, with the signature 'A. Hitler' in the corner.

After their quarrel Hanisch lost sight of Hitler, but he gives a description of Hitler as he knew him in 1910 at the age of twenty-one. He wore an ancient black overcoat, which had been given him by an old-clothes dealer in the hostel, a Hungarian Jew named Neumann, and which reached down over his knees. From under a greasy, black derby hat, his hair hung long over his coat collar. His thin and hungry face was covered with a black beard above which his large staring eyes were the one prominent feature. Altogether, Hanisch adds, 'an apparition such as rarely occurs among Christians.'

From time to time Hitler had received financial help from his aunt in Linz, Johanna Pölzl and, when she died in March 1911, it seems likely that he was left some small legacy. In May of that year his orphan's pension was stopped, but he still avoided any regular work.

Hanisch depicts him as lazy and moody, two characteristics which were often to reappear. He disliked regular work. If he earned a few crowns, he refused to draw for days and went off to a café to eat cream cakes and read newspapers. He neither smoked nor drank and, according to Hanisch, was too shy and awkward to have any success with women. His passions were reading newspapers and talking politics. 'Over and over again,' Hanisch recalls, 'there were days on which he simply refused to work. Then he would hang around night shelters, living on the bread and soup that he got there, and discussing politics, often getting involved in heated controversies.'

When he became excited in argument he would shout and wave his arms until the others in the room cursed him for disturbing them, or the porter came in to stop the noise. These outbursts of violent argument and denunciation alternated with moods of despondency.

Everyone who knew him was struck by the combination of ambition, energy, and indolence in Hitler. He was not only desperately anxious to impress people but was full of clever ideas for making his fortune and fame—from water-divining to designing an aeroplane. In this mood he would talk exuberantly and begin to spend the fortune he was to make in anticipation, but he was incapable of the application and hard work needed to carry out his projects. He would relapse into moodiness and disappear until he began to hare off after some new trick or short cut to success. His intellectual interests followed the same pattern. He spent much time in the public library, but his reading was indiscriminate and unsystematic. Ancient Rome, the Eastern religions, Yoga, Occultism, Hypnotism, Astrology, Protestantism, each in turn excited his interest for a moment. He started a score of jobs but failed to make anything of them and relapsed into the old hand-to-mouth existence, living by expedients and little spurts of activity.

As time passed these habits became ingrained, and he became more eccentric, more turned in on himself. He struck people as unbalanced. He gave rein to his hatreds—against the Jews, the priests, the Social Democrats, the Hapsburgs—without restraint. The few people with whom he had been friendly became tired of him, of his strange behaviour and wild talk. Yet these Vienna days stamped an indelible impression on his character and mind. 'During these years a view of life and a definite outlook on the world took shape in my mind. These became the granite basis of my conduct at that time. Since then I have extended that foundation very little, I have changed nothing in it . . . Vienna was a hard school for me, but it taught me the most profound lessons of my life.' However pretentiously expressed, this is true. It is time to examine what these lessons were.

Hitler wrote in *Mein Kampf*, 'The idea of struggle is as old as life itself, for life is only preserved because other living things perish through struggle. . . . In this struggle, the stronger, the more able, win, while the less able, the weak, lose. Struggle is the father of all things. . . . It is not by the principles of humanity that man lives or is able to preserve himself above the animal world, but solely by means of the most brutal struggle. . . . If you do not fight for life, then life will never be won.'

This is the natural philosophy of the doss-house. In this struggle any trick or ruse, however unscrupulous, the use of any weapon or opportunity, however treacherous, are permissible. To quote another typical sentence from Hitler's speeches: 'Whatever goal man has reached is due to his originality plus his brutality.' Astuteness; the ability to lie, twist, cheat and flatter; the elimination of sentimentality or loyalty in favour of ruthlessness, these were the qualities which enabled men to rise; above all, strength of will. Such were the principles which Hitler drew from his years in Vienna. He never trusted anyone; he never committed himself to anyone, never admitted any loyalty. His lack of scruple later took by surprise even those who prided themselves on their unscrupulousness. He learned to lie with conviction and dissemble with candour. To the end he refused to admit defeat and still held to the belief that by the power of will alone he could transform events.

Distrust was matched by contempt. Men were moved by fear, greed, lust for power, envy, often by mean and petty motives. Politics, Hitler was later to conclude, is the art of knowing how to use these weaknesses for one's own ends. Already in Vienna Hitler admired Karl Lueger, the famous Burgomaster of Vienna and leader of the Christian Social Party, because 'he had a rare gift of insight into human nature and was very careful not to take men as something better than they were in reality.' Hitler felt particular contempt for the masses—'everybody who properly estimates the political intelligence of the masses can easily see that this is not sufficiently developed to enable them to form general political judgements on their own account.' Here again was material to be manipulated by a skilful politician, although as yet Hitler had no idea of making a political career.

Hitler writes in *Mein Kampf* of the misery in which the Vienna working class lived at this time, but it is evident from every line of his account that these conditions produced no feeling of sympathy in him. 'I do not know which appalled me most at that time: the economic misery of those who were then my companions, their crude customs and morals, or the low level of their intellectual culture.' Least of all did he feel any sympathy with the attempts of the poor and the exploited to improve their position by their own efforts. Hitler's hatred was directed not so much against the rogues, beggars, bankrupt business men, and *déclassé* 'gentlemen' who were the flotsam and jetsam drifting in and out of the hostel in the Meldemannstrasse, as against the working men who belonged to organizations like the Social Democratic Party and the trade unions and who preached equality and the solidarity of the working classes. It was these, much more than the former, who threatened his claim to superiority. He passionately refused to join a trade union, or in any way to accept the status of a working man.

The whole ideology of the working-class movement was alien and hateful to him. They disparaged the nation, the Fatherland, the law, religion, and morality. 'There was nothing they did not drag in the mud,' he said, and he struggled with the question whether such men 'were worthy to belong to a great people.'

Hitler found the solution of his dilemma in the 'discovery' that the working men were the victims of a deliberate system for poisoning the popular mind, organized by the Social Democratic Party's leaders, who cynically exploited the distress of the masses for their own ends. Then came the crowning revelation:

'I discovered the relations existing between this destructive teaching and the specific character of a people, who up to that time had been almost unknown to me. Knowledge of the Jews is the only key whereby one may understand the inner nature and the real aims of Social Democracy.'

There was nothing new in Hitler's anti-Semitism; it was endemic in Vienna, and everything he ever said or wrote about the Jews is only a reflection of the anti-Semitic literature he read in Vienna before 1914. In Linz there had been very few Jews—'I do not remember ever having heard the word at home during my father's lifetime.' Even in Vienna Hitler had at first been repelled by the violence of the anti-Semitic Press. Then, 'one day, when passing through the Inner City, I suddenly encountered a phenomenon in a long caftan and wearing black sidelocks. My first thought was: is this a Jew? . . . I watched the man stealthily and cautiously, but the longer I gazed at this strange countenance and examined it section by section, the more the question shaped itself in my brain: is this a German? I turned to books for help in removing my doubts. For the first time in my life I bought myself some anti-Semitic pamphlets for a few pence.'

The language in which Hitler describes his discovery has the obscene taint to be found in most anti-Semitic literature: 'Was there any shady undertaking, any form of foulness, especially in cultural life, in which at least one Jew did not participate? On putting the probing knife carefully to that kind of abscess one immediately discovered, like a maggot in a putrescent body, a little Jew who was often blinded by the sudden light.'

Characteristic of Viennese anti-Semitism was its sexuality. 'The black-haired Jewish youth lies in wait for hours on end, satanically glaring at and spying on the unsuspicious girl whom he plans to seduce, adulterating her blood and removing her from the bosom of her own people. . . . The Jews were responsible for bringing negroes into the Rhineland with the ultimate idea of bastardizing the white race which they hate and thus lowering its cultural and political level so that the Jew might dominate.'

Hitler's wild assertions about Jews in *Mein Kampf* are pure fantasy. The Jew is no longer a human being; he has become a mythical figure, a grimacing, leering devil invested with infernal powers, the incarnation of evil, into which Hitler projects all that he hates and fears—and desires. Like all obsessions, the Jew is not a partial, but a total explanation. The Jew is everywhere, responsible for everything—the Modernism in art and music Hitler disliked; pornography and prostitution; the anti-national criticism of the Press; the exploitation of the masses by Capitalism, and its reverse, the exploitation of the masses by Socialism; not least for his own failure to get on.

Behind all this, Hitler soon convinced himself, lay a Jewish world conspiracy to destroy and subdue the Aryan peoples. Their purpose was to weaken the nation by fomenting social divisions and class conflict, and by attacking the values of race, heroism, struggle, and authoritarian rule in favour of the false internationalist, humanitarian, pacifist, materialist ideals of democracy. 'The Jewish doctrine of Marxism repudiates the aristocratic principle of nature and substitutes for it and the eternal privilege of force and energy, numerical mass and its dead weight. Thus it denies the individual worth of the human personality, impugns the teaching that nationhood and race have a primary significance, and by doing this takes away the very foundations of human existence and human civilization.'

In Hitler's eyes the inequality of individuals and of races was one of the laws of Nature. This poor wretch, often half-starved, without a job, family, or home, clung obstinately to any belief that would bolster up the claim of his own superiority over the labourers, the tramps, the Jews, and the Slavs with whom he rubbed shoulders in the streets.

Hitler had no use for any democratic institution: free speech, free press, or parliament. During the earlier part of his time in Vienna he had sometimes attended the sessions of the Reichsrat, the representative assembly of the Austrian half of the Empire, and he devotes fifteen pages of *Mein Kampf* to expressing his scorn for what he saw. 'The majority represents not only ignorance but cowardice. . . . The majority can never replace the man.'

Belief in equality between races was an even greater offence in Hitler's eyes than belief in equality between individuals. He had already become a passionate German nationalist. In Austria-Hungary this meant even more than it meant in Germany, and the fanatical quality of Hitler's nationalism reflects his Austrian origin.

For several hundred years the Germans of Austria played the leading part in the politics and cultural life of Central Europe. Until 1871 there had been no single unified German state. Germans had lived under the rule of a score of different states loosely grouped together in the Holy Roman Empire, and then, after 1815, in the German Federation. In the middle of the nineteenth century it was still Vienna, not Berlin, which ranked as the first of German cities. Moreover, the Hapsburgs not only enjoyed a pre-eminent position among the German states, but also ruled over wide lands inhabited by many different peoples.

On both counts the Germans of Vienna and the Austrian lands, who identified themselves with the Hapsburgs, looked on themselves as an imperial race with political privilege, boasting of a cultural tradition which few other peoples in Europe could equal. From the middle of the nineteenth century, however, this position was first challenged and then undermined.

Prussia defeated Austria at Sadowa in 1866, and thereafter the new German Empire with its capital at Berlin increasingly took the place hitherto occupied by Austria and Vienna as the premier German state. At the same time the pre-eminence of the Germans within the Hapsburg Empire itself was challenged, first by the Italians, who secured their independence in the 1860s; then by the Magyars of Hungary, to whom equality had to be conceded in 1867; finally by the Slav peoples. Especially in Bohemia and Moravia, where the most advanced of the Slav peoples, the Czechs, lived, the demand for equal rights was bitterly resented by the Germans and fiercely resisted. This conflict of the nationalities dominated Austrian politics from 1870 to the break-up of the Empire in 1918.

In this conflict Hitler had no patience with concessions. The Germans should rule the Empire, at least the Austrian half of it, with an authoritarian and centralized administration; there should be only one official language—German—and the schools and universities should be used 'to inculcate a feeling of common citizenship,' an ambiguous expression for Germanization. The representative assembly of the Reichsrat, in which the Germans (only 35 per cent of the population of Austria) were permanently outnumbered, should be suppressed. Here was a special reason for hatred of the Social Democratic Party, which refused to follow the nationalist lead of the Pan-Germans, and instead fostered class conflicts at the expense of national unity.

The political ideas and programme which Hitler picked up in Vienna were entirely unoriginal. They were the clichés of radical and Pan-German gutter politics, the stock-in-trade of the anti-Semitic and nationalist Press. The originality was to appear in Hitler's grasp of how to create a mass-movement and secure power on the basis of these ideas.

The three parties which interested Hitler were the Austrian Social Democrats, Georg von Schönerer's Pan-German Nationalists, and Karl Lueger's Christian Social Party.

From the Social Democrats Hitler derived the idea of a mass party and mass propaganda. Studying the Social Democratic Press and Party speeches, Hitler reached the conclusion that: 'the psyche of the broad masses is accessible only to what is strong and uncompromising. . . . The masses of the people prefer the ruler to the suppliant and are filled with a stronger sense of mental security by a teaching that brooks no rival than by a teaching which offers them a liberal choice. They have very little idea of how to make such a choice and thus are prone to feel that they have been abandoned. Whereas they feel very little shame at being terrorized intellectually and are scarcely conscious of the fact that their freedom as human beings is impudently abused. . . . I also came to understand that physical intimidation has its significance for the mass as well as the individual. . . . For the successes which are thus obtained are taken by the adherents as a triumphant symbol of the righteousness of their own cause; while the beaten opponent very often loses faith in the effectiveness of any further resistance.'

From Schönerer Hitler took his extreme German Nationalism, his anti-Socialism, his anti-Semitism, his hatred of the Hapsburgs and his programme of reunion with Germany. But he learned as much from what he believed were the three cardinal errors committed by Schönerer and the Nationalists in their political tactics: The Nationalists failed to grasp the importance of the social problem, directing their attention to the middle classes and neglecting the masses. They wasted their energy in a parliamentary struggle and failed to establish themselves as the leaders of a great movement. Finally they made the mistake of attacking the Catholic Church and split their forces instead of concentrating them. 'The art of leadership,' Hitler wrote, 'consists of consolidating the attention of the people against a single adversary and taking care that nothing will split up this attention. . . . The leader of genius must have the ability to make different opponents appear as if they belonged to one category.'

It was in the third party, the Christian Socialists, and their remarkable leader, Karl Lueger, that Hitler found brilliantly displayed that grasp of political tactics, the lack of which hampered the success of the Nationalists. Lueger had made himself Burgomaster of Vienna—in many ways the most important elective post in Austria—and by 1907 the Christian Socialists under his leadership had become the strongest party in the Austrian parliament. Hitler saw much to criticize in Lueger's programme. His anti-Semitism was based on religious and economic, not on racial, grounds ('I decide who is a Jew,' Lueger once said), and he rejected the intransigent nationalism of the Pan-Germans, seeking to preserve and strengthen the Hapsburg State with its mixture of nationalities. But Hitler was prepared to overlook even this in his admiration for Lueger's leadership.

The strength of Lueger's following lay in the lower middle class of Vienna, the small shopkeepers, business men and artisans, the petty officials and municipal employees. 'He devoted the greatest part of his political activity,' Hitler noted, 'to the task of winning over those sections of the population whose existence was in danger.'

Finally, instead of quarrelling with the Church, Lueger made it his ally and used to the full the traditional loyalty to crown and altar. In a sentence which again points forward to his later career, Hitler remarks: 'He was quick to adopt all available means for winning the support of long-established institutions, so as to be able to derive the greatest possible advantage for his movement from those old sources of power.'

It would be an exaggeration to suppose that at this period Hitler had already formulated clearly the ideas he set out in *Mein Kampf* in the middle of the 1920s. None the less the greater part of the experience on which he drew was already complete when he left Vienna, and to the end Hitler bore the stamp of his Austrian origins.

In May of 1913 Hitler left Vienna for good and moved across the German frontier to Munich. He was then twenty-four years old, awkward, moody and reserved, yet nursing a passion of hatred and fanaticism which from time to time broke out in a torrent of excited words. Years of failure had laid up a deep store of resentment in him, but had failed to weaken the conviction of his own superiority.

In *Mein Kampf* Hitler speaks of leaving Vienna in the spring of 1912, but the Vienna police records report him as living there until May 1913. Hitler is so careless about dates and facts in his book that the later date seems more likely to be correct. Hitler is equally evasive about the reasons which led him to leave. He writes in general terms of his dislike of Vienna and the state of affairs in Austria, but gives no specific reason why, on one day rather than another, he decided to go to the station, buy a ticket and at last leave the city he had come to detest.

The most likely explanation is that Hitler was anxious to escape military service, for which he had failed to report each year since 1910. Inquiries were being made by the police, and he may have found it necessary to slip over the frontier. Eventually he was located in Munich and ordered to present himself for examination at Linz. Hitler denied that he had left Vienna to avoid conscription, and asked, on account of his lack of means, to be allowed to report at Salzburg, which was nearer to Munich than Linz. His request was agreed to, and he duly presented himself for examination at Salzburg on 5 February 1914. He was rejected for military or auxiliary service on the grounds of poor health.

In Munich he found lodgings with a tailor's family, by the name of Popp, which lived in a poor quarter near the barracks. In retrospect, Hitler described this as 'by far the happiest time of my life. . . . I came to love that city more than any other place known to me. A German city. How different from Vienna.'

It may be doubted if this represented Hitler's feelings at the time. His life followed much the same pattern as before. His dislike of hard work and regular employment had by now hardened into a habit. He made a precarious living by drawing advertisements and posters, or peddling sketches to dealers. He was perpetually short of money.

The shadowy picture that emerges from the reminiscences of the few people who knew him in Munich is once again of a man living in his own world of fantasy. He gives the same impression of eccentricity and lack of balance, brooding and muttering to himself over his extravagant theories of race, anti-Semitism, and anti-Marxism, then bursting out in wild, sarcastic diatribes. He spent much time in cafés and beer-cellars, devouring the newspapers and arguing about politics.

Frau Popp, his landlady, speaks of him as a voracious reader, an impression Hitler more than once tries to create in *Mein Kampf*. Yet nowhere is there any indication of the works he read. Hitler's own comment on reading is illuminating: 'Reading is not an end in itself, but a means to an end. . . . One who has cultivated the art of reading will instantly discern, in a book or journal or pamphlet, what ought to be remembered because it meets one's personal needs or is of value as general knowledge.'

This is a picture of a man with a closed mind, reading only to confirm what he already believes, ignoring what does not fit in with his preconceived scheme.

Hitler was indignant at the ignorance and indifference of people in Munich to the situation of the Germans in Austria. Since 1879 the two states, the German Empire and the Hapsburg Monarchy, had been bound together by a military alliance, which remained the foundation of German foreign policy up to the defeat of 1918. Hitler felt that this predisposed most Germans to refuse to listen to the exaggerated accounts he gave of the 'desperate' position of the German Austrians in the conflict of nationalities within the Monarchy.

Hitler's objection to the alliance of Germany and Austria was twofold. It crippled the Austrians in their resistance to what he regarded as the deliberate anti-German policy of the Hapsburgs. At the same time, for Germany herself it represented a dangerous commitment to the support of a state which, he was convinced, was on the verge of disintegration.

When Franz Ferdinand was assassinated by Serbian students, at Sarajevo on 28 June 1914, Hitler's first reaction was confused. For, in his eyes, it was Franz Ferdinand, the heir to the Hapsburg throne, who had been more responsible than anyone else for that policy of concessions to the Slav peoples which roused the anger of the German nationalists. But, as events moved towards the outbreak of a general European war, Hitler brushed aside his doubts. At least Austria would be compelled to fight, and could not, as he had always feared, betray her ally Germany. 'I believed that it was not a case of Austria fighting to get satisfaction from Serbia, but rather a case of Germany fighting for her own existence . . . for its freedom and for its future. . . . For me, as for every other German, the most memorable period of my life now began. Face to face with that mighty struggle all the past fell away into oblivion.'

There were other, deeper and more personal reasons for his satisfaction. War meant to Hitler the opportunity to slough off the frustration, failure, and resentment of the past six years. Here was an escape from the tension and dissatisfaction of a lonely individuality into the excitement and warmth of a close, disciplined, collective life, in which he could identify himself with the power and purpose of a great organization.

On 1 August Hitler was in the cheering, singing crowd which gathered on the Odeonsplatz to listen to the proclamation declaring war. In a chance photograph that has been preserved his face is clearly recognizable, his eyes excited and exultant; it is the face of a man who has come home at last. Two days later he addressed a formal petition to King Ludwig III of Bavaria, asking to be allowed to volunteer, although of Austrian nationality, for a Bavarian regiment. The reply granted his request. 'I opened the document with trembling hands; no words of mine can describe the satisfaction I felt. . . . Within a few days I was wearing that uniform which I was not to put off again for nearly six years.'

Together with a large number of other volunteers he was enrolled in the 1st Company of the 16th Bavarian Reserve Infantry Regiment, known from its original commander as the List Regiment. Another volunteer in the same regiment was Rudolf Hess; the regimental clerk was a Sergeant-major Max Amann, later to become business manager of the Nazi Party's paper and of the Party publishing house. After a period of initial training in Munich, they spent several weeks at Lechfeld, and then, on 21 October 1914, entrained for the Front.

At Lille they were sent up into the line as reinforcements for the 6th Bavarian Division of the Bavarian Crown Prince Rupprecht's VIth Army. Hitler's first experience of fighting was in one of the fiercest and most critical engagements of the war, the First Battle of Ypres, when the British succeeded in stemming an all-out effort by the Germans to burst through to the Channel coast. For four days and nights the List Regiment was in the thick of the bitter fighting with the British round Becelaere and Gheluvelt, which reduced the regiment from three thousand five hundred to six hundred men.

Throughout the war Hitler served as a *Meldegänger,* a runner whose job was to carry messages between Company and Regiment H.Q. Although he was not actually in the trenches, there is little doubt that his was a dangerous enough job, and for the greater part of four years he was at the Front or not far in the rear.

In 1916 the List Regiment took part in the heavy fighting on the Somme, and near Bapaume on 7 October Hitler was wounded in the leg. He was sent back to Germany for the first time in two years. After a period in hospital at Beelitz, near Berlin, and at Munich with the Reserve battalion of his regiment, he returned to the Front at the beginning of March 1917, now promoted to lance-corporal. He was in time to take part in the later stages of the Battle of Arras and in the Third Battle of Ypres in the summer. With the rest of the regiment Hitler went forward in the last great German offensive in the spring of 1918.

During the night of 13–14 October the British opened a gas attack. Hitler was caught on a hill south of Werwick and his eyes were affected. By the time he got back to Rear H.Q. he could no longer see. On the morning of 14 October he collapsed and was sent back to a military hospital at Pasewalk, not far from Stettin. He was still there, recovering from the injury to his eyes, when the war ended with the capitulation of 11 November.

What sort of a soldier was Hitler? As early as December 1914, he had been awarded the Iron Cross, Second Class, and when Hitler, in March 1932, brought a lawsuit against a newspaper which had accused him of cowardice, his former commanding officer, Lieutenant-Colonel Engelhardt, testified to his bravery in the fighting of November 1914, when the regiment had first gone into action. Much more interesting is the Iron Cross, First Class, an uncommon decoration for a corporal, which Hitler was awarded in 1918. The most varied and improbable accounts have been given of the action for which he won this. The official history of the List Regiment says nothing at all. Whatever the occasion, it was certainly a decoration of which Hitler was proud and which he habitually wore after he had become Chancellor.

In view of his long service and the shortage of officers in the German Army in the last months of the war, the fact that Hitler never rose above the rank of corporal aroused curiosity and was much discussed in the German Press before 1933. There is no evidence that Hitler ever applied or was eager for promotion to the rank of non-commissioned officer, leave alone a commission. Probably the impression of eccentricity which he continued to give was no recommendation.

While not unpopular with his comrades, they felt that he did not share their interests or attitude to the war. He did not care about leave or women. He was silent when the others grumbled about the time they had to spend in the trenches or the hardships. Konrad Heiden, in his book *Der Führer*, says, 'We all cursed him and found him intolerable. There was this white crow among us that didn't go along with us when we damned the war.'

The few photographs of this time seem to bear this out—a solemn pale face, prematurely old, with staring eyes. He took the war seriously, identifying himself with the failure or success of German arms.

Many years afterwards Hitler would still refer to 'the stupendous impression produced upon me by the war—the greatest of all experiences. For that individual interest—the interest of one's own ego—could be subordinated to the common interest—that the great, heroic struggle of our people demonstrated in overwhelming fashion.' Many Germans regarded the comradeship, discipline and excitement of life at the Front as vastly more attractive than the obscurity, aimlessness, and dull placidity of peace, but this was particularly true of Hitler, for he had neither family, wife, job, nor future to which to return: there was much greater warmth and friendliness in the orderlies' mess than he had known since he left Linz. In the years after the war it was from ex-servicemen like this who could never settle down into monotonous routine that the Freikorps (post-war illegal armed bands), the Nazis, and a score of extremist parties recruited their members. The war, and the impact of the war upon the individual lives of millions of Germans, were among the essential conditions for the rise of Hitler and the Nazi Party.

It is surprising, in view of his later pretensions as a strategist in the Second World War, that Hitler has nothing to say in *Mein Kampf* about the conduct of the military operations. At the time he wrote his book he was still too anxious to secure the favour of the Army leaders to indulge in the attitude of contempt he later adopted towards the generals. In any case, Hitler followed the conventional Nationalist line of argument: the German Army had never been defeated, the war had been lost by the treachery and cowardice of the leaders at home, the capitulation of November 1918 was a failure of political not military leadership.

At the time of his stay in hospital at Beelitz and his visit to Munich (October 1916–March 1917) Hitler became indignant at the contrast between the spirit of the Army at the Front and the poor morale and lack of discipline at home. There he encountered shirkers who boasted of dodging military service, grumbling, profiteering, the black market, and other familiar accompaniments of wartime civilian life; it was with relief that he returned to the Front. Hitler had no use for a government which tolerated political discussion, covert anti-war propaganda and strikes in time of war. In *Mein Kampf* his contempt for parliamentary deputies and journalists is lavish: 'All decent men who had anything to say, said it point-blank in the enemy's face; or, failing this, kept their mouths shut and did their duty elsewhere. Uncompromising military measures should have been adopted to root out the evil. Parties should have been abolished and the Reichstag brought to its senses at the point of the bayonet, if necessary. It would have been still better if the Reichstag had been dissolved immediately.'

This is no more than the common talk of any one of the ex-servicemen's (*Frontkämpfer*) associations which sprang up after the war and comforted their wounded pride by blaming Socialist agitators, Jews, profiteers, and democratic politicians for the 'shameful treachery' of the 'Stab in the Back.' But Hitler adds a characteristic twist which shows once more the originality of his ideas as soon as he was faced with a question of political leadership. It was not enough, he concluded, to use force to suppress the Socialist and anti-national agitation to which he attributed the sapping of Germany's will to go on fighting. 'If force be used to combat a spiritual power, that force remains a defensive measure only, so long as the wielders of it are not the standard bearers and apostles of a new spiritual doctrine . . . It is only in the struggle between *Weltanschauungen* [world views] that physical force, consistently and ruthlessly applied, will eventually turn the scale in its own favour.' This was the reason for the failure of every attempt to combat Marxism hitherto, including the failure of Bismarck's anti-socialist legislation —'it lacked the basis of a new *Weltanschauung.*'

Out of this grew the idea of creating a new movement which would fight Social Democracy with its own weapons. For power lay with the masses, and if the hold of the Jew-ridden Marxist parties on their allegiance was to be broken, a sustitute had to be found. The key, Hitler became convinced, lay in propaganda, and the lesson Hitler had already drawn from the Social Democrats and Lueger's Christian Socialists in Vienna was completed by his observation of the success of English propaganda during the war, by contrast with the failure of German attempts.

There were two themes on which Hitler constantly played in the years that followed the war: Man of the People, and Unknown Soldier of the First World War. When he spoke to the first Congress of German Workers in Berlin on 10 May 1933, he assured them: 'Fate, in a moment of caprice or perhaps fulfilling the designs of Providence, cast me into the great mass of the people, amongst common folk. I myself was a labouring man for years in the building trade and had to earn my own bread. And for a second time I took my place once again as an ordinary soldier amongst the masses.' These were the twin foundations of his demagogy and, in however garbled a fashion, they correspond to the two formative experiences of his life, the years in Vienna and Munich, and the years at the Front.

Those years between the end of 1908 and the end of 1918 had hardened him, taught him to be self-reliant, confirmed his belief in himself, toughened the power of his will. From them he emerged with a stock of fixed ideas and prejudices which were to alter little in the rest of his life: hatred of the Jews; contempt for the ideals of democracy, internationalism, equality, and peace; a preference for authoritarian forms of government; an intolerant nationalism; a rooted belief in the inequality of races and individuals; and faith in the heroic virtues of war. Most important of all, in the experiences of those years he had already hit upon a conception of how political power was to be secured and exercised which, when fully developed, was to open the way to a career without parallel in history. Much of what he had learned remained to be formulated even in his own mind, and had still to be crystallized into the decision to become a politician. It required only a sudden shock to precipitate it. That shock was supplied by the end of the war, the capitulation of Germany, and the overthrow of the Empire.

CHAPTER TWO

THE YEARS OF STRUGGLE

1919–24

The news that Germany had lost the war and was suing for peace came as a profound shock to the German people and the German Army. Some of the most spectacular German successes had occurred in the first half of 1918, and the swift reversal of this situation in August and September was concealed from the German people.

Although the initiative for ending the war had come from the High Command, from General Ludendorff himself, the civil government, hitherto denied any voice in the conduct of the war, was obliged to take the full responsibility for ending it. Here was the germ of the legend of the 'Stab in the Back.'

The end of the war brought the collapse of the Imperial régime and the reluctant assumption of power by the democratic parties in the Reichstag. The Republican Government had to bear the odium of signing, first the surrender and then the peace terms. It was easy for the embittered and unscrupulous to twist this into the lie that the Social Democrats and the Republican Parties had deliberately engineered the capitulation, betrayed Germany, and stabbed the German Army in the back, in order to hoist themselves into power. The fact that the Provisional Government, led by the Social Democrats, sacrificed party and class interests to the patriotic duty of holding Germany together in a crisis not of their making, was brushed aside. These were the 'November criminals,' the scapegoats who had to be found if the Army and the Nationalists were to rescue anything from the wreck of their hopes.

Throughout Central and Eastern Europe the end of the war was marked by a series of revolutionary changes. In Germany, where people now found themselves faced with new sacrifices demanded by the Peace Treaty and Reparations, the unrest, insecurity and fear lasted until the end of 1923.

The threat to the stability of the new Republican régime came both from the extremists of the Left who sought to carry out a social revolution on a Communist pattern, and from an intransigent Right, in whose eyes the Republic was damned from birth. It was associated with surrender, a shameful and deliberate act of treachery, as most of them soon came to regard it. In 1919 the Republican Government signed a Peace Treaty the terms of which were universally resented in Germany; the Government was henceforward branded as the agent of the Allies in despoiling and humiliating Germany. It was openly said that loyalty to the Fatherland required disloyalty to the Republic. This mood was not only to be found among the classes which had hitherto ruled Germany in their own interests; it was also characteristic of many wartime officers and ex-servicemen, who resented what they regarded as the ingratitude and treachery of the Home Front and the Republic towards the *Frontkämpfer*.

In this way the malaise which is the inevitable sequel to a long period of war found a political form. It was canalized into a campaign of agitation and conspiracy against the existing régime, a campaign in which free use was made of the habits of violence learned in the years of war.

When the war ended and the Republic was proclaimed, Hitler was still in hospital at Pasewalk. The acknowledgement of Germany's defeat and the establishment of a democratic Republic, in which the Social Democrats played the leading part, were both intolerable to him. Everything with which he had identified himself seemed to be defeated, swept aside in a torrent of events which had been released, as he had no doubt, by the same Jews who had always desired the defeat and humiliation of Germany.

Like many others among the mob of demobilized men who now found themselves flung on to the labour market at a time of widespread unemployment, he had little prospect of finding a job. The old problem of how to make a living, conveniently shelved for four years, reappeared. Characteristically, Hitler turned his back on it. After all, what had he to lose in the break-up of a world in which he had never found a place? Nothing. What had he to gain in the general unrest, confusion, and disorder? Everything, if only he knew how to turn events to his advantage. With a sure instinct, he saw in the distress of Germany the opportunity he had been looking for but had so far failed to find.

With considerable naïvety, he wrote in *Mein Kampf*: 'Generally speaking, a man should not take part in politics before he has reached the age of thirty.' Hitler was now in his thirtieth year, the time was ripe and the decision was taken: 'I resolved that I would take up political work.'

But how? Uncertain as yet of the answer, Hitler, after his discharge from hospital, but not yet demobilized, made his way through a disorganized country back to Munich, the capital of Bavaria.

Few towns in the Reich were as sensitive to the mood of unrest as Munich. Its political atmosphere was unstable and veered from one extreme to the other: revolution, political murder, a Soviet government, followed by a wave of suppression by the emerging Right-wing factions occurred in rapid succession through the year 1918.

In Bavaria ever since the unification of Germany there had been a strong dislike of government from Protestant, Prussian Berlin. In the disturbed and unstable condition of Germany between 1918 and 1923 the power of the central government in Berlin was weakened, and the Bavarian State Government, with a strong Right-wing bias, was able to exploit a situation in which the orders of the Reich Government were respected only if they were backed by the support of the authorities in Munich. Bavaria thus became a natural centre for all those who were eager to get rid of the Republican régime in Germany, and the Bavarian Government turned a blind eye to the treason and conspiracy against the legal government of the Reich which were being planned on its doorstep in Munich. It was in Bavaria that the irreconcilable elements of the Freikorps gathered, armed bands of volunteers formed by the Reichswehr at the end of the war to maintain order and protect the eastern frontiers of Germany but now just as willing to turn their guns against the Republic. The Freikorps were the training schools for the political murder and terrorism which disfigured German life up to 1924, and again after 1929.

Among the regular Army officers stationed in Munich were men like Major-General Ritter von Epp and his assistant, Captain Ernst Röhm. In the Freikorps and the innumerable defence leagues, patriotic unions, and ex-servicemen's associations which sprang up in Bavaria, they saw the nucleus of that future German Army which should one day revenge the humiliations of 1918. Until that day it was essential to keep together the men who had been the backbone of the old German Reichswehr, now reduced by the terms of the Treaty of Versailles to a mere hundred thousand.

Pöhner, the Police President of Munich, gave the famous reply, when asked if he knew there were political murder gangs in Bavaria: 'Yes, but not enough of them.' Wilhelm Frick, later Hitler's Minister of the Interior, was Pöhner's assistant; one of his colleagues in the Bavarian Ministry of Justice was Franz Gürtner, later Hitler's Minister of Justice.

At the back of the minds of all these men was the dream which bewitched the German Right for twenty years: overthrowing the Republic, restoring Germany to her rightful position as the greatest power of continental Europe and restoring the Army to its rightful position. Such was the promising political setting in which Hitler began his career.

Hitler lived through the exciting days of April and May 1919 in Munich itself. According to his account in *Mein Kampf*, he was to have been put under arrest by the Communist government then in power but drove off with his rifle the three men who came to arrest him. Once the Communists had been overthrown, he gave information before the Commission of Inquiry which tried and shot those reported to have been active on the other side. Hitler then got a job in the Press and News Bureau of the Army Political Department, a centre for the activities of such men as Röhm. After attending a course of 'political instruction' for the troops, Hitler was given the task of inoculating the men against contagion by socialist, pacifist, or democratic ideas. This was the first recognition that he had any political ability at all. In September he was instructed to investigate a small group, the German Workers' Party, which might possibly be of interest to the Army.

The German Workers' Party was originally set up by Anton Drexler, a Munich locksmith, whose idea was a party both working class and nationalist. It can scarcely have been a very impressive scene when, on the evening of 12 September 1919, Hitler attended his first meeting of the German Workers' Party in a Munich beer-cellar where about twenty-five people had gathered. One of them, a Bavarian separatist, urged secession from Germany and union with Austria. So vehement was Hitler's rebuttal of this that Drexler, impressed, gave Hitler a copy of his autobiographical pamphlet, and a few days later Hitler received a postcard inviting him to attend a committee meeting of the German Workers' Party.

After some hesitation Hitler went. The committee met in an obscure beer-house, the Alte Rosenbad, in the Herrnstrasse. 'I went through the badly lighted guest-room, where not a single guest was to be seen, and searched for the door which led to the side room; and there I was face to face with the Committee. Under the dim light shed by a grimy gas-lamp I could see four people sitting round a table. . . .' As Hitler frankly acknowledges, this very obscurity was an attraction. It was only in a party which, like himself, was beginning at the bottom that he had any prospect of playing a leading part and imposing his ideas. In the established parties there was no room for him, he would be a nobody. After two days' reflection he joined the Committee of the German Workers' Party as its seventh member.

Slowly and painfully he pushed the Party forward, and prodded his cautious and unimaginative colleagues on the committee into bolder methods of recruitment. Invitations were multigraphed, a small advertisement inserted in the local paper, a larger hall secured. When Hitler himself spoke for the first time, the result was only to confirm the chairman, Karl Harrer, in his belief that Hitler had no talent for public speaking. But Hitler persisted, and at the beginning of 1920 he was put in charge of the Party's propaganda. By the use of clever advertising he got nearly two thousand people into the Festsaal of the Hofbräuhaus on 24 February. The principal speaker was overshadowed by Hitler, who captured the audience's attention. Angered by the way in which Hitler was now forcing the pace, Harrer resigned as chairman. On 1 April 1920 Hitler left the Army and devoted all his time to the Party.

Hitler's and Drexler's group in Munich was not

the only National Socialist party. In Bavaria there were rival groups, and in Austria and the Sudetenland the pre-war German Social Workers' Party had been reorganized as the German National Socialists Workers' Party and had begun to use the swastika as its symbol. Up to August 1923 there were fairly frequent contacts between these different National Socialist groups, but little came of them. Hitler was too jealous of his independence to submit to interference from outside.

Much more important to Hitler was the support he received from Ernst Röhm, who exercised considerable influence in the shadowy world of the Freikorps. Röhm, a tough man with organizing ability, joined the German Workers' Party before Hitler. When Hitler began to build up the Party, Röhm pushed in ex-Freikorps men and ex-servicemen to swell the membership. From these elements the first 'strong-arm' squads were formed, the nucleus of the S.A. Röhm was the indispensable link in securing for Hitler the protection, or at least the tolerance, of the Army and the Bavarian Government. Without the unique power and influence of the Army, Hitler would never have been able to exercise with impunity his methods of incitement, violence and intimidation. At every step from 1914 to 1945 Hitler's varying relationship to the Army was of the greatest importance to him: never more so than in these early years in Munich.

Nevertheless, the foundation of Hitler's success was his own energy and ability as a political leader. Without this, the help would never have been forthcoming. Hitler's genius as a politician lay in his unequalled grasp of what could be done by propaganda, and his flair for seeing how to do it. He had to learn in a hard school, on his feet night after night, arguing his case in every kind of hall; often, in the early days, in the face of opposition, indifference or amused contempt; learning to hold his audience's attention, to win them over; most important of all, learning to read the minds of his audiences, finding the sensitive spots on which to hammer. Hitler came to know Germany and the German people at first hand as few of Germany's other leaders ever had. Over nearly all the politicians with whom he had to deal he had one great advantage: his immense practical experience of politics in the street, the level at which any politician must be effective if he is to carry a mass vote.

Hitler was the greatest demagogue in history. Those who add 'only a demagogue' fail to appreciate the nature of political power in an age of mass politics. As he himself said: 'To be a leader, means to be able to move masses.'

The lessons which Hitler drew from the activities of the Austrian Social Democrats and Lueger's Christian Socialists were now tried out in Munich. Success was far from being automatic. Hitler made mistakes and had much to learn before he could persuade people to take him seriously, but he learned from his mistakes, and by the time he came to write *Mein Kampf* in the middle of the 1920s he was able to set down the conditions of success. The pages in which he discusses the technique of mass propaganda stand out in brilliant contrast with the turgid attempts to explain his entirely unoriginal political ideas. He is quite open in explaining this procedure of political leadership: 'The receptive powers of the masses are very restricted, and their understanding is feeble. On the other hand, they quickly forget. Such being the case, all effective propaganda must be confined to a few bare necessities and then must be expressed in a few stereotyped formulas.' Furthermore, it is better to stick to a programme even when certain points in it become out of date: 'As soon as one point is removed from the sphere of dogmatic certainty, the discussion will not simply result in a new and better formulation, but may easily lead to endless debates and general confusion.'

When you lie, tell big lies. 'The grossly impudent lie always leaves traces behind it, even after it has been nailed down.'

Above all, never hesitate, never qualify what you say, never concede an inch to the other side, paint all your contrasts in black and white. This is the 'very first condition which has to be fulfilled in every kind of propaganda: a systematically one-sided attitude towards every problem that has to be dealt with. . . .'

Vehemence, passion, fanaticism, these are 'the great magnetic forces which alone attract the great masses; for these masses always respond to the compelling force which emanates from absolute faith in the ideas put forward, combined with an indomitable zest to fight for and defend them. . . . The doom of a nation can be averted only by a storm of glowing passion; but only those who are passionate themselves can arouse passion in others.'

Hitler showed a marked preference for the spoken over the written word. 'The broad masses of a population are more amenable to the appeal of rhetoric than to any other force.' The employment of verbal violence, the repetition of such words as 'smash', 'force', 'ruthless', 'hatred', was deliberate. Hitler's gestures and the emotional character of his speaking, lashing himself up to a pitch of near-hysteria in which he would scream and spit out his resentment, had the same effect on an audience. Men groaned or hissed and women sobbed involuntarily, caught up in the spell of powerful emotions of hatred and exaltation.

Propaganda was not confined to the spoken word. There were the red posters to provoke the Left; the flag, with its black swastika in a white circle on a red background, a design to which Hitler devoted the utmost care; the salute, the uniform, and the hierarchy of ranks. Mass meetings and demonstrations, another device which Hitler borrowed from the Austrian Social Democrats, were intended to create a sense of power, of belonging to a movement whose success was irresistible. Hitler here hit upon a psychological fact which was to prove of great importance in the history of the Nazi movement: that violence and terror have their own propaganda value, and that the display of physical force attracts as many as it repels. 'When our political meetings first started,' Hitler writes, 'I made it a special point to organize a suitable defence squad. . . . Some of them were comrades who had seen active service with me, others were young Party members who right from the start had been trained and brought up to realize that only terror is capable of smashing terror.' Defence is an ambiguous word to describe such activities, for, as Hitler adds, 'the best means of defence is attack, and the reputation of our hall-guard squads stamped us as a political fighting force and not as a debating society.'

From the first these men were used to provoke disturbance and to beat up political opponents as part of a deliberate campaign of intimidation. In September 1921 Hitler personally led his followers in storming the platform of a federalist meeting. When examined by the police commission which inquired into the incident, Hitler replied: 'It's all right. We got what we wanted. Ballerstedt did not speak.'

Far from using violence in a furtive underhand way, Hitler gave it the widest possible publicity so that people were forced to pay attention to what he was doing. No government of any determination would have tolerated such methods, but the Republican Government in Berlin had virtually no authority in Bavaria, and the Bavarian State Government showed remarkable complacence towards political terrorism, provided it was directed against the Left.

The 'strong-arm' squads were first formed in the

summer of 1920, but their definitive organization dates from August 1921, when a so-called Gymnastic and Sports Division was set up inside the Party. 'It is intended,' said the Party proclamation, 'to serve as a means for bringing our youthful members together in a powerful organization for the purpose of utilizing their strength as an offensive force at the disposal of the movement.' After 5 October, it changed its name to Sturmabteilung (the S.A., or Storm Section of the Party) and was largely composed of ex-Freikorps men.

In November 1921, the S.A. went into action in the so-called *Saalschlacht*, a fierce fight with the Reds in a Nazi meeting at the Hofbräuhaus, which was built up into a Party legend. In August 1922, S.A. formations paraded with swastika flags flying in a demonstration of the Patriotic Associations on the Munich Königsplatz, and a month later eight 'Hundreds' (*Hundertschaften*) were organized. In October 1922, Hitler took eight hundred of his stormtroopers to Coburg for a nationalist demonstration, defied the police ban on marching through the town and fought a pitched battle in the streets with the Socialists and Communists. To have been at Coburg was a much-prized distinction in the Nazi Party.

Inevitably, Hitler's propaganda methods, his attempt to turn the Party into a mass following for himself and to ride roughshod over the other members of the committee, produced resentment. In the early summer of 1921 Hitler spent some time in Berlin, where he got into touch with certain of the nationalist groups in the north. While he was away from Munich the other members of the Party committee tried to recapture the direction of the Party. Hitler returned immediately to Munich, and countered the move by offering his own resignation. There was no doubt who found the Party funds as well as the publicity, and the last thing the committee could afford was to let Hitler resign. Hitler, however, demanded dictatorial powers if he was to remain, together with the retirement of the committee and a ban on Party negotiations for six years. In a leaflet defending themselves, the members of the committee accused Hitler of 'a lust for power and personal ambition.' They had to repudiate the leaflet after Hitler brought a libel suit, and at two meetings on 26 and 29 July they capitulated, making Hitler president and giving him virtually unlimited powers. Drexler was kicked upstairs as Honorary President.

The split between Hitler and the committee went deeper than personal antipathy and mistrust. Drexler and Harrer had always thought of the Party as a workers' and lower-middle-class party, radical and anti-capitalist as well as nationalist. These ideas were expressed in the programme, with its Twenty-five Points (drawn up by Drexler, Hitler, and Gottfried Feder, and adopted in February 1920), as well as in the name of the German National Socialist Workers' Party. The programme was nationalist and anti-Semitic in character, and it came out strongly against Capitalism, the trusts, the big industrialists, and the big landowners.

There is no doubt that for Drexler and Feder this represented a genuine programme to which they always adhered. All programmes to Hitler, however, were means to an end, to be taken up or dropped as they were needed. 'Any idea,' he says in *Mein Kamp*, 'may be a source of danger if it be looked upon as an end in itself.' Hitler was as much interested in the working class as Drexler, but he was interested in them as material for political manipulation. Their grievances and discontents were the raw stuff of politics, a means, but never an end.

For the same reasons Hitler was not prepared to limit membership of the Party to any one class. All forms of discontent were grist to his mill; there was as much room in his Party for the unemployed ex-officer like Göring and Hess, or the embittered intellectual like Rosenberg and Goebbels, as for the working man who refused to join a trade union or the small shopkeeper who wanted to smash the windows of the big Jewish department stores. Ambition, resentment, envy, avidity for power and wealth—in every class—these were the powerful motive forces Hitler sought to harness. He was prepared to be all things to all men, because to him all men represented only one thing, a means to power.

The committee which had hitherto controlled the Party was now swept away—Hitler had long since ceased to attend its meetings. The new president put in Max Amann, the ex-sergeant-major of the List Regiment, to run the business side of the Party, and Dietrich Eckart as editor of the *Völkischer Beobachter*, the weekly Party paper. The power of making all big decisions Hitler kept in his own hands. He was working as he had never worked before. But it was work which suited him: his hours were irregular, he was his own master, his life was spent in talking, he lived in a whirl of self-dramatization, and the gap between his private dream-world and his outer life had been narrowed, however slightly.

Until the end of his life Hitler regarded these early years of the Nazi movement with pride. It was the heroic period of the Party's struggle, the *Kampfzeit*. 'It is truly a miracle to trace this development of our movement. To posterity it will appear like a fairy-tale. A people is shattered and then a small company of men arises and begins an Odyssey of wanderings, which begins in fanaticism, which in fanaticism pursues its course.'

Who were the men with whom Hitler began his 'Odyssey' in Munich? One of the most important was Ernst Röhm, a man for whom soldiering was his whole life and who had little but contempt for anything outside it. Röhm was too independent and had too much of the unruly temper of a condottiere to fit easily into the rigid pattern of the Reichswehr. None the less he provided an invaluable link with the Army authorities, even after his resignation, and more than any other man it was he who created the S.A.

Hess and Göring were two other ex-officers. Rudolf Hess, the son of a German merchant, was a solemn and stupid young man who took politics with great seriousness, conceived a deep admiration for Hitler and became his secretary and devoted follower. A very different figure from the humourless Hess was swaggering Hermann Göring, the last commander of the crack Richthofen Fighter Squadron and holder of Germany's highest decoration for bravery under fire. In the autumn of 1922 he heard Hitler speak, and shortly afterwards he became commander of the S.A.

Like Röhm, Gottfried Feder and Dietrich Eckart had joined the German Workers' Party before Hitler. Feder had unorthodox ideas about economics and made a great impression on Hitler, who is said to have clipped his moustache to the famous toothbrush in imitation of him. Eckart was well known as a journalist, poet, and playwright; he had violent nationalist, anti-democratic, and anti-clerical opinions, and an enthusiasm for Nordic folk-lore and Jew-baiting. Eckart had read widely, and he talked well. He lent Hitler books, corrected his style of expression in speaking and writing, and introduced him to influential people, including the Bechsteins, wealthy and famous piano manufacturers. Frau Hélène Bechstein took a great liking to the young man, gave parties for people to meet the new prophet, found money for the Party and later visited Hitler in prison.

Through Eckart, Hitler met Alfred Rosenberg, a refugee of German descent who had been trained as an architect in Moscow, but had fled to escape the

Revolution. Rosenberg introduced Hitler to a group of passionately anti-Bolshevik and anti-Semitic Russian émigrés.

Two years after the *Völkischer Beobachter* had been bought for him, Hitler made it into a daily. This required money. Some of it was provided by Frau Gertrud von Seidlitz, a Baltic lady who had shares in Finnish paper mills, while Putzi Hanfstängl, a son of the rich Munich family of art publishers, advanced a substantial loan. Hanfstängl, who had been educated at Harvard, took Hitler into his own home and introduced him to a number of other well-to-do Munich families.

Ill at ease on any formal social occasion, Hitler cleverly exploited his own awkwardness. He deliberately behaved in an exaggerated and eccentric fashion, arriving late and leaving unexpectedly, either sitting in ostentatious silence or forcing everyone to listen to him by shouting and making a speech.

There were other companions who came from the same lower middle class as Hitler himself and with whom he was more at home than with anyone else: Heinrich Hoffmann, who was to become the one man allowed to photograph Hitler; Max Amann, Hitler's tough, rude, but reliable business manager and publisher; Christian Weber, a former horse-trader, of great physical strength, who had worked as a 'chucker-out' at a disreputable Munich dive, and whose social life consisted in drinking endless seidels of beer; Ulrich Graf, Hitler's personal bodyguard, a butcher's apprentice and amateur wrestler, with a great taste for brawling. Röhm's own reputation—his homosexuality was later to become notorious—was none too good; nor was that of Hermann Esser, the only speaker in those early days who for a time rivalled Hitler. Esser was a young man to whom Hitler openly referred as a scoundrel. He boasted of sponging on his numerous mistresses and made a speciality of digging up Jewish scandals, the full stories of which in all their scabrous details were published in the *Völkischer Beobachter.* Esser's only competitor was Julius Streicher, an elementary-school teacher in Nuremberg, who excelled in a violent and crude anti-Semitism.

Such were the men with whom the 'miracle' of National Socialism was accomplished.

How Hitler managed to make a living at this time is far from clear. In the leaflet which was drawn up by the dissident members of the committee in July 1921, this was one of the principal points of accusation against Hitler: 'If any member asks him how he lives and what was his former profession, he always becomes angry and excited. Up to now no answer has been supplied to these questions. So his conscience cannot be clear, especially as his excessive intercourse with ladies, to whom he often describes himself as the King of Munich, costs a great deal of money.'

During the libel action to which this led, Hitler was asked to tell the court exactly how he lived. Did he, for instance, receive money for his speeches? 'If I speak for the National Socialist Party,' Hitler replied, 'I take no money for myself. But I also speak for other organizations, such as the German National Defence and Offensive League, and then, of course, I accept a fee. . . .' Hitler was obviously embarrassed by these inquiries, for Hess was put up to write an open letter to the *Völkischer Beobachter* assuring its readers that the leader was beyond reproach.

The probable answer is that Hitler was as careless about money as he had been in Vienna, that he lived from hand to mouth and bothered very little about who was going to pay for the next meal. At this time his home was at 41 Thierschstrasse, a poorish street near the river Isar, and his possessions were few. He habitually wore an old coat and troubled little about his appearance or comforts.

Undoubtedly Hitler received contributions from

those who sympathized with the aims of his Party, but their amount and importance have been exaggerated. Not until considerably later did he succeed in touching the big political funds of the German industrialists in the Ruhr and the Rhineland. In fact, the Nazi Party was launched on very slender resources.

Nazism was a phenomenon which throve only in conditions of disorder and insecurity. While these had been endemic in Germany ever since the defeat of 1918, two new factors made their appearance in 1923 which brought the most highly industrialized country of continental Europe to the verge of economic and political disintegration: the occupation of the Ruhr and the collapse of the mark.

By the autumn of 1922 the German Government professed itself unable to continue paying reparations and requested a moratorium. The French Government of Poincaré used the technical excuse of a German default in deliveries of timber to move French troops into the industrial district of the Ruhr on 11 January 1923. To cut off 80 per cent of Germany's steel and pig-iron production and more than 80 per cent of her coal from the rest of Germany, as the French proceeded to do, was to bring the economic life of the whole country to a standstill.

The result of the French occupation was to unite the German people as they had never been united since the early days of the war. The German Government called for a campaign of passive resistance, which, before long, became a state of undeclared war. The weapons on one side were strikes, sabotage, and guerrilla warfare, and on the other arrests, deportations, and economic blockade.

The occupation of the Ruhr gave the final touch to the deterioration of the mark. The savings of the middle classes and working classes were wiped out at a single blow with a ruthlessness which no revolution could ever equal. Even if a man worked till he dropped it was impossible to buy enough clothes for his family—and work, in any case, was not to be found.

The violence of Hitler's denunciations of the corrupt, Jew-ridden system which had allowed all this to happen, the bitterness of his attacks on the Versailles settlement and on the Republican Government which had accepted it, found an echo in the misery and despair of large classes of the German nation.

Hitler saw the opportunity clearly enough, but it was more difficult to see how to take advantage of it and turn the situation to his own profit. The National Socialists could overthrow the Republic only if Hitler succeeded in uniting all the nationalist and anti-republican groups in Bavaria, and if he succeeded in securing the patronage of more powerful forces—the most obvious being the Bavarian Government and the District Command of the Army—for a march on Berlin. He devoted his energies throughout 1923 to achieving these two objectives.

In the early months of 1923 he was afraid lest the French occupation of the Ruhr might unite Germany behind the Government. In the *Völkischer Beobachter* he wrote: 'So long as a nation does not do away with the assassins within its borders, no external successes can be possible. While written and spoken protests are directed against France, the real deadly enemy of the German people lurks within the walls of the nation. . . . Down with the November criminals, with all their nonsense about a United Front.'

With the tide of national feeling running high against the French, and in support of the Government's call for resistance, this was an unpopular line to take. To make people listen to him, Hitler summoned five thousand of the S.A. Stormtroopers to Munich for a demonstration at the end of January 1923. The authorities promptly banned it. Hitler began to rave: the S.A. would march, even if the police opened fire. The Bavarian Government retorted by issuing an additional ban on twelve meetings which Hitler was to address after the demonstration. Finally, only when General Otto von Lossow, commander of the Reichswehr in Bavaria, had satisfied himself that his officers could be relied on to fire on the National Socialists if necessary—a significant change of attitude—were Röhm and Epp able to secure his co-operation. The ban was lifted and Hitler held his demonstration.

In his speech at this first Party Day, 25 January 1923, Hitler made no secret of his hope that the Berlin Government would fail to unite the nation in resistance to the French. His emphasis again was on eliminating the 'betrayers of the German fatherland,' 'the perpetrators of the November crime,' whose participation in a new conflict in the field of foreign affairs could lead only to another stab in the back of Germany. Hitler could make no use of a united front in a new war of liberation which would strengthen the position of the Government. The time to deal with the French would come when the Republic had been overthrown. Here Hitler's essentially political outlook differed sharply from that of the Army and ex-Freikorps officers like Röhm, who thought of a war of revenge against France.

This conflict had been present from the beginning in the very different views Hitler and Röhm took of the S.A. For Röhm, and the other officers and ex-officers who helped to train the S.A., the first object was to build up in secret the armed forces forbidden by the Treaty of Versailles. The Army leaders believed that the outbreak of a state of undeclared war with France might well prove the prelude to a general war. In order to strengthen the Army it was planned to draw on the para-military formations like the S.A. Everything was to be done to bring them up to a high pitch of military efficiency, and Röhm flung himself into the task of expanding and training the S.A. with such effect that by the autumn of 1923 it numbered fifteen thousand men.

For Hitler, on the other hand, the S.A. was not just a disguised Army reserve; these were to be *political* troops used for political purposes. With shrewder insight than Röhm and his friends, Hitler saw that to rebuild Germany's national and military power, it was necessary to begin by capturing political power in the State, and the S.A. were to be used for that purpose.

In Hitler's speeches of early 1923, his hatred was still directed, not against the French, but against the Republic, which he depicted as a corrupt racket run by the Jews at the expense of the national interests. No accusation against the Jews was too wild for him, but his most bitter scorn was reserved for the 'respectable' parties of the Right who hesitated to act. But Hitler's speeches were not even reported in the Press.

The French occupation of the Ruhr still continued, but by fall, 1923, the initial mood of national unity on the German side had gone. Encouraged by the growing disorder, and the increasingly strained relations between Munich and Berlin, Hitler renewed his agitation in August. The fact that Gustav Stresemann, the new Chancellor, was known to be anxious to end the exhausting campaign of passive resistance in the Ruhr and Rhineland enabled Hitler to change front. He now adopted the more popular line of attacking the Berlin Government for the betrayal of the national resistance to the French, as well as for allowing the inflation to continue.

On 2 September, the anniversary of the German defeat of France at Sedan in 1870, a huge demonstration celebrated German Day at Nuremberg amidst scenes of great enthusiasm. All the Patriotic Associations took part. During the parade Hitler stood at the side of Ludendorff, and afterwards flayed the Government in a characteristically violent speech.

Ludendorff's presence was important. His reputation as the greatest military figure of the war and an unremitting opponent of the Republic made him the hero of the Right-wing extremists, while he still enjoyed considerable prestige in the Army. There was no one better placed to preside over a union of the quarrelsome and jealous patriotic leagues, and Hitler had carefully maintained close relations with the old man for some time past. Ludendorff was invincibly stupid as well as tactless in matters of politics. He disliked Bavarians, was on the worst possible terms with Crown Prince Rupprecht, the Bavarian Pretender, and constantly attacked the Church in the most Catholic part of Germany. But at least he was reliable on the question of Bavarian separatism, and his political stupidity was an asset from Hitler's point of view, for, skilfully managed, he could bring a great name to Hitler's support without entrenching on his control of policy.

The demonstration at Nuremberg had immediate practical consequences. The same day a new German Fighting Union (Deutscher Kampfbund) was set up and a manifesto issued over the old signatures of Friedrich Weber (Bund Oberland), Captain Heiss (Reichsflagge) and Adolf Hitler. The object of this renewed alliance between the Nazis and the Patriotic Leagues in Bavaria was declared to be the overthrow of the November Republic and of the *Diktat* of Versailles.

The crisis came to a head on 26 September when Stresemann announced the decision of the Reich Government to call off the campaign of passive resistance in the Ruhr unconditionally and to lift the ban on reparation deliveries to France and Belgium. This was a courageous and wise decision, intended as the preliminary to negotiations for a peaceful settlement. But it was also the signal the Nationalists had been waiting for to stir up a renewed agitation against the Government. 'The Republic, by God,' Hitler had declared on 12 September, 'is worthy of its fathers. . . . Subserviency towards the enemy, surrender of the human dignity of the German, pacifist cowardice, tolerance of every indignity, readiness to agree to everything until nothing remains.'

On 25 September the leaders of the Kampfbund—Hitler, Göring, Röhm, Kriebel, Heiss, and Weber—had already met. Hitler put his point of view and asked for the political leadership of the alliance. So strong was the impression he made that both Heiss and Weber agreed, while Röhm, convinced that they were on the edge of big events, next day resigned his commission and finally threw in his lot with Hitler.

Hitler's first step was to put his own fifteen thousand S.A. men in a state of readiness and announce fourteen immediate mass meetings in Munich alone. Whether he intended to try a *coup d'état* is not clear, but Knilling, the Bavarian Minister-President, was thoroughly alarmed. On 26 September the Bavarian Cabinet proclaimed a state of emergency and appointed Gustav von Kahr, one of the best-known Bavarian Right-wing politicians with strong monarchist and particularist leanings, as State Commissioner with dictatorial powers. Kahr promptly banned Hitler's fourteen meetings and refused to give way when Hitler, beside himself with rage, screamed that he would answer him with bloody revolution.

In the confused events that followed 26 September and led up to the unsuccessful putsch of 8–9 November, the position of two of the three parties is tolerably clear. Hitler consistently demanded a move on Berlin to be backed by the political and military authorities in Bavaria, but aiming at the substitution of a new régime for the whole of Germany. The twists and hesitations in Hitler's conduct arose from his recognition that he could not carry such a plan through without persuading Kahr and Lossow, the commanding officer in Bavaria, to join him.

The Central Government in Berlin had to face the threat of civil war from several directions: from Bavaria, where Hitler was openly calling for revolt, and where Kahr began to pursue an independent course of action which ran counter to the policy of Berlin; from Saxony, where the State Government came increasingly under the influence of the Communists, who were also aiming at a seizure of power; from the industrial centres, like Hamburg and the Ruhr, where Communist influence was strong; from the Rhineland, where the Separatists were still active, and from the nationalist extremists of the north.

The Stresemann Government's chances of mastering this critical situation depended upon the attitude of the Army. The High Command could be relied upon to use force to suppress any attempt at revolution from the Left, but its attitude towards a similar move from the Right might well be uncertain. In the years since the war the protection of the Army had been invoked again and again by those like Hitler who were scheming to overthrow the Republic.

Nothing could more clearly illustrate the unique position of the Army in German politics, a position fully appreciated by General Hans von Seeckt and the Army High Command. Seeckt, one of the most remarkable men in the long history of the German Army, had the insight to see that it was in the long-term interests of Germany, and of the German Army he served, to uphold the authority even of a Republican government, and so to preserve the unity of the Reich, rather than allow the country to be plunged into civil war for the momentary satisfaction of Party rancour and class resentment.

Seeckt's attitude allowed the political and military authorities in Berlin to speak with one voice, and on 26 September President Friedrich Ebert conferred emergency powers upon the Minister of Defence, Otto Gessler. This meant that the Army assumed the executive functions of the government and undertook the responsibility of safeguarding both the security of the Reich and the inviolability of the Republican Constitution. An attempt by Hitler—or anyone else—to carry out a march on Berlin would be met by force, with the Army on the side of the Government.

But there was a third party to be taken into account, the civil and military authority in Bavaria represented by Kahr and Lossow. The existence of this third factor, and the uncertainty of the policy Kahr and Lossow would adopt, gave Hitler a chance of success. Although the Bavarian Government had appointed Kahr to keep Hitler in check, relations between Munich and Berlin were strained. It was the action of the Bavarian Government in conferring dictatorial powers on Kahr which had led the Reich Government to declare a state of emergency itself, and Kahr's intentions were suspect in Berlin.

Kahr's aims are still far from clear. He was attracted by the idea of overthrowing the Republican régime and installing a conservative government which would give Bavaria back her old monarchy. At other times he played with the possibility of breaking away from the Reich altogether and establishing an independent South German State. If Kahr could only be persuaded to help overthrow the Republican régime in Berlin, Hitler had every hope of double-crossing his Bavarian allies once he was in power. It was equally possible for Kahr to use Hitler and the forces of the Kampfbund. An uneasy alliance developed between Kahr and the Nazis, each trying to exploit the other's support. Once again the critical decision lay with the Army, this time with the local commander in Bavaria, General von Lossow. Like Kahr, however, Lossow never quite succeeded in making up his mind until events decided for him.

In October 1923, the quarrel between Munich and

Berlin flared up, under direct provocation from Hitler. When the *Völkischer Beobachter* printed scurrilous attacks on Seeckt, Stresemann, and Gessler, the Minister of Defence in Berlin used his emergency powers to demand the suppression of the paper. Kahr refused to take orders from Berlin. When Gessler went over his head and ordered Lossow to execute the ban, Kahr persuaded him to disobey. When Berlin removed Lossow from his post, Kahr announced that Lossow would remain in command of the Army in Bavaria. He then demanded the resignation of the Reich Government and ordered the armed bands which supported him to concentrate on the borders of Bavaria and Thuringia.

All this suited Hitler admirably. Power in Bavaria was concentrated in the hands of a triumvirate consisting of Kahr, Lossow, and Colonel Hans von Seisser, the head of the State police. It was now, Hitler argued, only a question of whether Berlin marched on Munich, or Munich on Berlin. The situation in Saxony and Thuringia, on the northern borders of Bavaria, offered a splendid pretext for Kahr and Lossow to act. For there the Social Democratic cabinets of the two State Governments had been broadened to bring the Communists into power as partners of the Social Democrats, thereby providing the Communists with a spring-board for their own seizure of power. Action by the Bavarian Government to suppress this threat of a Left-wing revolution would undoubtedly command wide support, and, once at Dresden, Hitler reckoned, it would not be long before they were in Berlin.

Lossow and Kahr were full of smooth assurances that they would move as soon as the situation was ripe, yet Hitler and Röhm were mistrustful. They suspected that behind the façade of German Nationalism, with its cry of '*Auf nach Berlin*' (On to Berlin!), which Kahr kept up to satisfy the Kampfbund, he was playing with Bavarian separatist ideas under the very different banner of *'Los von Berlin'* (Away from Berlin!).

Meanwhile the Government in Berlin was beginning to master its difficulties. By the end of October the threat of a Communist revolution had been broken, and the offending governments in Saxony and Thuringia had been thrown out, thus depriving the Bavarian conspirators of their best pretext for intervention outside their own frontiers. Seisser was sent to Berlin to size up the situation. He reported that conditions were less favourable to the dissidents. Kahr and Lossow, who had no wish to become involved in an enterprise that was bound to fail, insisted that they alone should decide the time to act and that they should not be hustled. Hitler was by now convinced that the only way to get them to do what he wanted was to present them with a *fait accompli*. Otherwise, he feared, they might carry out their own *coup* without him.

His final plan was put into operation on 8 November. Kahr had arranged a big political meeting to be held in the Bürgerbräukeller, at which Lossow, Seisser and most of the other Bavarian political leaders were expected to be present. Kahr refused to see Hitler on the morning of the 8th, and Hitler was soon convinced that this meeting was to be the prelude to the proclamation of Bavarian independence and the restoration of the monarchy. On the spur of the moment Hitler decided to forestall Kahr.

Twenty minutes after Kahr had begun to speak to the assembly, Göring, with twenty-five armed brownshirts, burst into the hall. Hitler leaped on to a chair and fired a shot at the ceiling, then jumped down and began to push his way on to the platform. 'The National Revolution,' he shouted, 'has begun. This hall is occupied by six hundred heavily armed men. No one may leave the hall. The Bavarian and Reich Governments have been removed and a provisional National Government formed. The Army and police barracks have been occupied, troops and police are marching on the city under the swastika banner.' No one could be certain how far Hitler was only bluffing. There were six hundred S.A. men outside, and a machine gun in the vestibule. Leaving Göring to keep order in the hall, Hitler pushed Kahr, Lossow, and Seisser into a side room. Meanwhile Scheubner-Richter, Ludendorff's liaison man with Hitler, was driving through the night to fetch General Ludendorff, whom Hitler wanted as the figurehead of his revolution.

Hitler, who was wildly excited, began the interview with Kahr and his companions in melodramatic style: 'No one leaves this room alive without my permission.' He announced that he had formed a new government with Ludendorff (who actually knew nothing of what was happening). Waving his gun, Hitler shouted: 'I have four shots in my pistol. Three for my collaborators if they abandon me. The last is for myself.' Setting the revolver to his head, he declared: 'If I am not victorious by tomorrow afternoon, I shall be a dead man.'

The three men were less impressed than they should have been. They found it difficult to take Hitler's raving at all seriously, despite the gun and the armed guards at the windows. Hitler was making little progress. Now, leaving the room without a word, he dashed into the hall and announced that the three men had agreed to join him in forming a new German government.

His speech was another piece of bluff, but it worked. The announcement that agreement had been reached completely changed the mood of the crowd in the hall, which shouted its approval: the sound of cheering impressed the three men who were still held under guard in the side room.

As Hitler returned to them, Ludendorff appeared. He was thoroughly angry at Hitler for springing this surprise on him, and furious that Hitler had made himself the dictator of Germany, leaving Ludendorff with the command of an army which did not exist. But he kept himself under control: this was a great national event, he said, and he could only advise the others to collaborate. Hitler added: 'We can no longer turn back; our action is already inscribed on the pages of world history.'

In apparent unity they all filed back into the hall. While the audience cheered in enthusiasm, each made a brief speech, swore loyalty and shook hands.

Barely had this touching scene of reconciliation been completed than Hitler was called out to settle a quarrel which had started when Stormtroopers of the Bund Oberland tried to occupy the Engineers' barracks. By a bad error of judgement he left the hall without taking proper precautions. As the audience began to pour out of the exits, Lossow excused himself on the grounds that he must go to his office to issue orders, and left unobtrusively, followed by Kahr and Seisser. It was the last that was seen of Lossow or Kahr that night.

By morning Hitler's several hundred Stormtroopers of the S.A. and Kampfbund had grown to some three thousand men, for considerable forces continued to come in from the countryside during the night. While his own bodyguard occupied the offices of the Social Democratic *Münchener Post* and smashed the machines, the Reichskriegsflagge, under Röhm's leadership, seized the War Ministry and set up barbedwire and machine guns.

Hitler came to Röhm before midnight and held a council of war with Ludendorff, Kriebel and Weber. As time passed, however, they became concerned at the absence of any news from Lossow or Kahr, and were at a loss what to do next. Messages to Lossow at the 19th Infantry Regiment's barracks produced no answer; nor did the messengers return. The night was allowed to pass without the seizure of a single key position, apart from Röhm's occupation of the

Army headquarters. Finally, between six and seven o'clock in the morning, Pöhner and Major Hühnlein were dispatched to occupy the police headquarters, but were promptly arrested instead, together with Frick.

As General von Lossow returned from the Bürgerbräukeller, Seeckt telegraphed from Berlin that if the Army in Bavaria did not suppress the putsch, he would do it himself. There was considerable sympathy with Hitler and Röhm among the junior officers from the rank of major downwards, but the senior officers were indignant at the insolence of this ex-corporal, and in the end discipline held. Orders were sent out to bring in reinforcements from outlying garrisons. Kahr issued a proclamation denouncing the promises extorted in the Bürgerbräukeller and dissolving the Nazi Party and the Kampfbund. From Crown Prince Rupprecht came a brief put pointed recommendation to crush the putsch at all costs, using force if necessary. Rupprecht had no use for a movement which had Ludendorff as one of its leaders.

By the morning of 9 November it was clear that the attempt had miscarried. At dawn Hitler returned with Ludendorff to the Bürgerbräu, leaving Röhm to hold out in the War Ministry. Ludendorff, convinced that the Army would never fire on the legendary figure of the First World War, had persuaded Hitler that they must take the offensive and try to restore the position by marching on Lossow's headquarters. Once he stood face to face with the officers and men of the Army, Ludendorff was convinced that they would obey him and not Lossow. Meanwhile troops of the Regular Army had surrounded Röhm and his men in the centre of the city, but both sides were reluctant to open fire.

Shortly after eleven o'clock on the morning of 9 November, a column of two or three thousand men left the Bürgerbräukeller and headed for the centre of the city. At the head of the column fluttered the swastika flag and the banner of the Bund Oberland. In the first row marched Hitler, between Ludendorff, Scheubner-Richter, and Ulrich Graf on one side, Dr Weber, Feder, and Kriebel on the other. Most of the men carried arms, and Hitler himself had a pistol in his hand. Crowds thronged the streets and there was an atmosphere of excitement and expectation. Julius Streicher, who had been haranguing the crowd in the Marienplatz, climbed down to take his place in the second rank. Rosenberg and Albrecht von Graefe, the sole representative of the North German Nationalists, who had arrived that morning at Ludendorff's urgent summons, trudged unhappily along with the rest. The column swung down the narrow Residenzstrasse, singing as it went. Beyond lay the old War Ministry, where Röhm was besieged. The time was half past twelve.

The police, armed with carbines, were drawn up in a cordon across the end of the street to prevent the column debouching on to the broad Odeonsplatz. The Stormtroopers completely outnumbered the police, but the narrowness of the street prevented them from bringing their superior numbers to bear. Who fired first has never been settled. Ulrich Graf ran forward and shouted to the police officer: 'Don't fire, Ludendorff and Hitler are coming,' while Hitler cried out: 'Surrender!' At this moment a shot rang out and a hail of bullets swept the street. The first man to fall was Scheubner-Richter, with whom Hitler had been marching arm-in-arm. Hitler fell, either pulled down or seeking cover. The shooting lasted only a minute, but sixteen Nazis and three police already lay dead or dying. Göring, badly wounded, was carried into a house. Weber, the leader of the Bund Oberland, stood against the wall weeping hysterically. All was confusion, neither side being at all sure what to do next. One man alone kept his head. Erect and unperturbed,

General Ludendorff, with his adjutant, Major Streck, by his side, marched steadily on, pushed through the line of police and reached the Odeonsplatz beyond.

The situation might still have been saved, but not a single man followed him. Hitler at the critical moment lost his nerve. He scrambled to his feet, stumbled back towards the end of the procession, and allowed himself to be pushed into a yellow motor-car which was waiting on the Max Josef Platz. He was undoubtedly in great pain from a dislocated shoulder, and probably believed himself to have been wounded. But there was no denying that under fire the Nazi leaders broke and fled. All but two of the killed and wounded were in the following ranks, exposed to the fire by the action of their leaders in taking cover.

Two hours later Röhm was persuaded to capitulate and was taken into custody. Göring, badly wounded, was smuggled across the Austrian frontier by his wife. On 11 November Hitler was arrested at Uffing, where he was being nursed by Hanfstängl's wife.

In many ways the attempt of 8–9 November was a remarkable achievement for a man like Hitler who had started from nothing only a few years before. In less than a couple of hours on the night of 8 November he had transformed the political situation in Bavaria and made a revolution by sheer bluff. It evidenced political talent of an unusual kind.

But the mistakes had been gross. Worst of all, from Hitler's point of view, was the contrast between his own behaviour under fire and that of Ludendorff, who, in the sight of all, had marched steadily forward and brushed aside the police carbines with contemptuous ease.

The truth is, however, that Hitler's plans had miscarried long before the column set out for the Odeonsplatz. He had never intended to use force; from the beginning his conception had been that of a revolution in agreement with the political and military authorities. 'We never thought to carry through a revolt against the Army: it was *with it* that we believed we should succeed.' Therefore no adequate preparations had been made for a seizure of power by arms. The *coup* was to be limited to forcing Kahr and Lossow into acting with him, in the belief that it was only hesitation, not opposition, that held them back. Again and again Hitler had told his men that when the moment came they need not worry, neither the Army nor the police would fire on them. The shots on the Odeonsplatz represented the final collapse of the premises upon which the whole attempt had been constructed. From the moment it became certain that Lossow and Kahr had taken sides against him, Hitler knew that the attempt had failed. There was a slender chance that a show of force might still swing the Army back to his side, and so he agreed to march. But it was to be a demonstration; the last thing Hitler wanted, or was prepared for, was to shoot it out with the Army.

Never was Hitler's political ability more clearly shown than in the way he recovered from this setback. The trial was held in Munich, and it was a trial for a conspiracy in which the chief witnesses for the prosecution—Kahr, Lossow, and Seisser—had been almost as deeply involved as the accused. The full story was one which most of the political leaders of Bavaria were only too anxious to avoid being made public. Hitler exploited this situation to the full.

The trial began before a special court on 26 February 1924. During the twenty-four days it lasted, it was front-page news in every German newspaper, and many foreign correspondents attended the trial. For the first time Hitler had an audience outside the frontiers of Bavaria.

From the first day Hitler's object was to recover the political initiative and virtually put the chief witnesses for the prosecution in the dock. He did this by the simple device of assuming full responsibility for the attempt to overthrow the Republic, and, instead of apologizing or trying to belittle the seriousness of this crime, indignantly reproaching Lossow, Kahr, and Seisser with the responsibility for its failure. This was a highly effective way of appealing to nationalist opinion, and turning the tables on the prosecution. Hitler maintained that the prosecution members were his collaborators in his enterprise to get rid of the Reich Government, and if it was treasonable they were equally guilty of treason. 'I alone bear the responsibility,' he concluded, 'but I am not a criminal because of that. If today I stand here as a revolutionary, it is as a revolutionary against the Revolution. There is no such thing as high treason against the traitors of 1918.'

Neither Kahr nor Seisser had the skill to withstand such tactics, while the judges sat placidly through Hitler's mounting attack on the Republic whose authority they represented. One man alone stood up to Hitler, and this surprisingly enough was General von Lossow.

Lossow was an angry man. His career had ended abruptly as a result of the November affair, and he had to listen in silence while his reputation was torn to shreds in the court, and he was represented as a coward. Now he had his chance to reply, and he expressed all the contempt of the officer caste for this jumped-up, ill-educated, loud-mouthed agitator who had never risen above the rank of corporal and now tried to dictate to the Army the policy it should pursue. He dealt bluntly with Hitler's own ambitions and lack of political qualifications.

But Hitler had the last word. In cross-examination he made Lossow lose his temper, and in his final speech he established a complete mastery over the court. He went out of his way to avoid recrimination and to renew the old offer of alliance with the Army. The failure of 1923 was the failure of individuals, of a Lossow and a Kahr; the most powerful and the most permanent of German institutions, the Army, was not involved. 'I believe that the hour will come when the masses, who today stand in the street with our swastika banner, will unite with those who fired upon them. . . . When I learned that it was the police who fired, I was happy that it was not the Reichswehr which had stained its record; the Reichswehr stands as untarnished as before. One day the hour will come when the Reichswehr will stand at our side, officers and men.' It took Hitler nine years to convince the Army that he was right.

In face of all the evidence Ludendorff was acquitted, and Hitler was given the minimum sentence of five years' imprisonment. Despite the attempts of the police to get him deported, Hitler was in fact released from prison after serving less than nine months of his sentence. He proceeded deliberately to build up the failure of 8 and 9 November into one of the great propaganda legends of the movement.

CHAPTER THREE

THE YEARS OF WAITING

1924–31

Fifty miles west of Munich, in the small town of Landsberg, Hitler served his prison term with some forty other National Socialists. They had an easy and comfortable life, ate well—Hitler became quite fat in prison—had as many visitors as they wished, and spent much of their time out of doors in the garden. On his thirty-fifth birthday, which fell shortly after the trial, the parcels and flowers he received filled several rooms. He had a large correspondence in addition to his visitors, and as many newspapers and books as he wished. Much of the time, however, from July onwards he shut himself up in his room to dictate *Mein Kampf*, first to Emil Maurice and later to Rudolf Hess, who came from Austria voluntarily to share his leader's imprisonment.

Max Amann, who was to publish the book, had hoped for sensational revelations of the November putsch. But Hitler was too canny for that; there were to be no recriminations. His own title for the book was *Four and a Half Years of Struggle against Lies, Stupidity, and Cowardice*, reduced by Amann to *Mein Kampf—My Struggle*. Even then Amann was to be disappointed. For the book contains very little autobiography, but is filled with turgid discussions of Hitler's ideas, written in a verbose style which is both difficult and dull to read.

Dietrich Eckart, Feder, and Rosenberg had all published books or pamphlets, and Hitler was eager to prove that he too, even though he had left school without a certificate, had read and thought deeply, acquiring his own *Weltanschauung*. This accounts for the pretentiousness of the style, the use of long words and constant repetitions, all the tricks of a half-educated man seeking to give weight to his words.

Before his arrest Hitler had managed to send a pencilled note to Alfred Rosenberg with the brief message: 'Dear Rosenberg, from now on you will lead the movement.' This was a surprising choice. Rosenberg was no man of action and as a leader he was ineffective, but this was precisely what attracted Hitler: Rosenberg as his deputy would represent no danger to his own position.

What was to be done now that the Party had been dissolved and Hitler was in prison? Hitler's answer, however camouflaged, was simple: Nothing. He had no wish to see the Party revive its fortunes without him. There were long and sometimes bitter arguments between Hitler and his visitors at Landsberg in 1924 on the issues of combining with other similar movements, including the Völkisch group, an extremist wing of the German Nationalists. Hitler was both suspicious and evasive. He tried by every means to delay decisions until he was released.

A further cause of disagreement was the S.A. Röhm had set to work to weld together again the disbanded forces of the Kampfbund. The Frontbann, as it was now called, grew rapidly, for Röhm was an able organizer with untiring energy; he soon had some thirty thousand men enrolled.

But the greater Röhm's success, the more uneasy Hitler became. His activities threatened Hitler's chances of leaving prison. The Bavarian Government arrested some of the subordinate leaders of the Frontbann, and Hitler's release on parole was delayed.

Hitler was no less worried by the character Röhm was giving to the new organization which had replaced and absorbed the old S.A. The two men had never agreed about the function of the Stormtroops.

For Hitler they were to be instruments of political intimidation and propaganda subordinate to the Party. Röhm demanded that 'the defence organizations should be given appropriate representation in the parliamentary group and that they should not be hindered in their special work.' In December when new elections for the Reichstag were held, Röhm did not find a place on the Nazi list.

By the end of Hitler's year in prison these quarrels and disagreements had reached such a pitch that it appeared possible to write off the former Nazi Party as a serious force in German or Bavarian politics. The Reichstag elections of December 1924 confirmed this. The votes cast for the Nazi-Völkisch bloc fell by more than half, and secured less than 5 per cent of the total Reichstag seats.

Much of the blame for this state of affairs fell on Hitler—with considerable justice. Ludecke writes, 'He was deliberately fostering the schism in order to keep the whip-hand over the party.' And he succeeded. The plans for a united Völkisch Front came to nothing. Ludendorff and Röhm left in disgust, and no powerful Nazi group was created in the Reichstag under the leadership of someone else. The price of this disunity was heavy, but for Hitler it was worth paying. By the time he came out of prison the Party had broken up almost completely—but it had not found an alternative leader, there was no rival to oust.

Hitler's first move on leaving prison on 20 December 1924 was to call on the Minister-President of Bavaria and leader of the strongly Catholic and particularist Bavarian People's Party, Dr. Heinrich Held. The putsch, Hitler admitted, had been a mistake; his one object was to assist the Government in fighting Marxism; he had no use for Ludendorff's and the North Germans' attacks on the Catholic Church, and he had every intention of respecting the authority of the State. Held's attitude was one of scepticism tinged with contempt, but he agreed to raise the ban on the Party and its newspaper. 'The wild beast is checked,' was Held's comment. 'We can afford to loosen the chain.'

On 27 February 1925, Hitler gathered the few who remained faithful for a mass meeting in the Bürgerbräukeller. Max Amann conducted the meeting. Drexler, Strasser, Röhm, and Rosenberg stayed away. Besides Amann, Hitler's only prominent supporters were Streicher and Esser, Gottfried Feder and Frick.

Hitler had not lost his gifts as an orator. When he finished speaking at the end of two hours there was loud cheering from the four thousand who filled the hall.

With the re-founding of the Nazi Party, Hitler set himself two objectives. The first was to establish his own absolute control over the Party by driving out those who were not prepared to accept his leadership without question. The second was to build up the Party and make it a force in German politics within the framework of the constitution.

The process was to prove even slower than Hitler had expected; the times were no longer so favourable as they had been in 1920–3. Hitler's speech on 27 February had laid great stress on the need to concentrate opposition against a single enemy—Marxism and the Jew. But he had added, in an aside which delighted his audience: 'If necessary, by one enemy many can be meant.' In other words, under cover of fighting Marxism and the Jew, the old fight against the State would be resumed. The authorities were alarmed and immediately prohibited him from speaking in public in Bavaria. This prohibition was soon extended to other German states as well. It lasted until May 1927 in Bavaria and September 1928 in Prussia, and was a severe handicap for a leader whose greatest asset was his ability as a speaker.

An even more serious handicap was the improvement in the position of the country, which began while Hitler was in prison. A new reparations agreement—the Dawes Plan—was negotiated, and this was followed in turn by the evacuation of the Ruhr; the Locarno Pact, guaranteeing the inviolability of the Franco-German and Belgian-German frontiers: the withdrawal of allied troops from the first zone of the demilitarized Rhineland, and Germany's entry into the League of Nations by unanimous vote of the League Assembly on 8 September 1926.

In the second presidential elections in April the Nazis supported Field-Marshal von Hindenburg, who had been brought in at the last minute by the Nationalists. Hindenburg won by a narrow margin, but the Nazis had little cause for congratulation. Hindenburg, the greatest figure of the old Army, a devoted Monarchist, a Conservative, and a Nationalist, did more than anything else could have done to reconcile conservative Germans to the Republican régime.

Such success as the Nazis had at this time was due less to Hitler than to Gregor Strasser, who was threatening to take Hitler's place as the effective leader of the Party and was breaking new ground in the north of Germany and the Rhineland, where the Party had hitherto failed to penetrate. A powerfully built man with a strong personality, Strasser was an able speaker and an enthusiast of radical views who laid as much stress on the anti-capitalist points in the Nazi programme as on its nationalism. He was critical of Hitler's attitude and little disposed to submit to his demands for unlimited authority in the Party.

Given Party leadership in North Germany, Strasser, with the help of his brother, Otto, rapidly built up an organization which, while nominally acknowledging Hitler as leader, soon began to develop into a separate party. Gregor Strasser, who was a Reichstag deputy, founded a newspaper, the *Berliner Arbeit-*

szeitung, edited by Otto Strasser, and a fortnightly periodical, *Nationalsozialistische Briefe*, intended for Party officials. As editor of the *Briefe* and Gregor's private secretary, the Strassers secured a young Rhinelander, then still under thirty, a man of some education who had attended a number of universities, and written novels and film scripts which no one would accept. His name was Paul Josef Goebbels.

Hitler had little sympathy with the Strassers' anti-capitalism and their demand for the breaking up of big estates, which embarrassed him in his search for wealthy backers. But while Hitler spent his time in Berchtesgaden, Gregor and Otto Strasser were actively at work extending their influence in the movement.

On 22 November 1925, the Strassers called together a meeting of the North German district leaders in Hanover. Hitler was represented by Gottfried Feder, but it was only by a bare majority that Feder was admitted to the meeting at all, after Goebbels had demanded his ejection.

The split between the Strassers and Hitler crystallized round a question which excited much feeling in Germany in 1925–6, whether the former German royal houses should be expropriated and whether their possessions should be regarded as their own private property or as the public property of the different states. Gregor and Otto Strasser sided with working-class opinion against the princes, while Hitler supported the propertied classes. At this time he was receiving three-quarters of his income from the divorced Duchess of Sachsen-Anhalt, and he denounced the agitation as a Jewish swindle. The Hanover meeting voted to follow the Strasser line. When Feder protested in Hitler's name, Goebbels jumped to his feet: 'In these circumstances I demand that the petty bourgeois Adolf Hitler be expelled from the National Socialist Party.' More important still, the Hanover meeting accepted the Strassers' programme and resolved to substitute it for the Twenty-five Points of the official programme adopted in February 1920. This was open revolt.

Hitler took time to meet the challenge, but when he did move he showed his skill in the way he out-manoeuvred Strasser without splitting the Party. On 14 February 1926 he summoned a conference in his turn, this time in the South German town of Bamberg. Hitler deliberately avoided a Sunday, when the North German leaders would have been free to attend in strength. As a result the Strasser wing of the Party was represented only by Gregor Strasser and Goebbels. In the south Hitler had made the position of District Leader (Gauleiter) a salaried office, a step which left the Gauleiters free to attend solely to Party business and made them much more dependent upon himself. He could thus be sure of a comfortable majority in the meeting at Bamberg.

Strasser was outnumbered from the beginning, and half-way through the meeting Goebbels stood up and declared that, after listening to Hitler, he was convinced that Strasser and he had been wrong, and that the only course was to admit their mistake and come over to Hitler. Having won his point, Hitler did all he could to keep Strasser in the Party. His conciliatory tactics proved successful. The Strasser programme was abandoned, a truce patched up and the unity of the Party preserved. This was not the end of the Strasser episode, but Hitler had handled his most dangerous rival with skill and papered over the breach between himself and the radical wing of the Party.

Hitler had still to face other difficulties in the Party. There was persistent criticism and grumbling at the amount of money the Leader and his friends took out of Party funds for their own expenses, and at the time he spent away from headquarters in Berchtesgaden, or driving around in a large motor-car at the Party's expense. Quarrelling, slander, and intrigue over the most petty and squalid issues seemed to be endemic in the Party.

To keep these quarrels within bounds, Hitler set up a Party court in 1926, the Uschla (Committee for Investigation and Settlement), which became an effective instrument for his tighter control over the Party.

In July 1926, Hitler felt strong enough to hold a mass rally of the Party at Weimar, in Thuringia, one of the few States in which he was still allowed to speak. Five thousand men took part in the march past, with Hitler standing in his car and returning their salute, for the first time, with outstretched arm. Hoffman's photographs made it all look highly impressive, and a hundred thousand copies of the *Völkischer Beobachter* were distributed throughout the country. It was the first of the Reichsparteitage later to be staged, year after year, at Nuremberg.

Goebbels was now whole-heartedly Hitler's man. In November Hitler appointed him as Gauleiter of 'Red' Berlin, an assignment which was to stretch to the full his remarkable powers as an agitator. Thus Hitler not only strengthened the movement in a key position, but provided another check against the independence of the Strassers, who had kept their own press and publishing house in Berlin. Goebbels, whose desertion to Hitler was regarded as rank treachery by the Strasser group, employed every means in his power to reduce their influence and following. In 1927 he founded *Der Angriff* as a rival to the Strassers' paper, and used the S.A. to beat up their most loyal supporters. Appeals to Hitler by Gregor and Otto Strasser produced no effect: he declared he had no control over what Goebbels did.

The old trouble with the S.A. soon reappeared. In November 1926, Hitler reformed the S.A. and found a new commander in Captain Pfeffer von Salomon, but the ex-officers still thought only in military terms.

Both the Berlin and Munich S.A. leadership had to be purged. The Munich S.A. had become notorious for the homosexual habits of Lieutenant Edmund Heines and his friends: it was not for his morals, however, or his record as a murderer, that Hitler threw him out in May 1927, but for lack of discipline and insubordination.

But whatever steps Hitler took, the S.A. continued to follow its own independent course. Pfeffer refused to admit Hitler's right to give orders to his Stormtroops. These men were not interested in politics; what they lived for was precisely this 'playing at soldiers' Hitler condemned. In time Hitler was to find an answer in the black-shirted S.S., a hand-picked corps d'élite very different from the ill-disciplined S.A. mob of camp followers. But it was not until 1929 that Hitler found the right man in Heinrich Himmler, who had been Gregor Strasser's adjutant at Landshut and later his secretary. It took Himmler some years before he could provide Hitler with what he wanted, an instrument of complete reliability with which to exercise his domination over the Party and eventually over the German nation.

Yet, if the Party still fell far short of Hitler's monolithic ideal, 1927 and 1928 saw a continuation of that slow growth in numbers and activity which had begun in 1926. By 1928 the Party organization was divided into two main branches: one directed by Gregor Strasser and devoted to attacking the existing régime, the other directed by Constantin Hierl and concerned with building up in advance the cadres of the new State.

Propaganda was a separate department. In November 1928 Goebbels was made propaganda chief and worked directly under Hitler. At the end of 1927 another familiar figure, Hermann Göring, established himself in Berlin, living by his wits and his social connexions. Hitler, looking for just such contacts in upper-class Berlin, soon renewed his association with

Göring. In May 1928 Göring and Goebbels were both elected to the Reichstag on the short Nazi list of twelve deputies, together with Strasser, Frick, and General von Epp. Hitler himself, ineligible because he was not a German citizen, never stood as a candidate for the Reichstag. He had resigned his Austrian citizenship in 1925 and did not become a naturalized German until 1932 on the eve of his candidature for the German Presidency.

But all Hitler's efforts to obtain power in these years were dwarfed into insignificance by the continued success of the Republican régime. Its political successes were matched by an economic recovery which touched every man and woman in the country. The basis of this recovery was the huge amount of foreign money lent to Germany, and the re-establishment of the currency seemed to have made her a sound financial risk again.

Not only the German Government, but the States, the big cities, even the Churches, as well as industry and business, borrowed at high rates and short notice, spending extravagantly without much thought of how the loans were to be repaid except by borrowing more. In this way Germany made her reparation payments promptly, and at the same time financed the rationalization and re-equipment of her industry, great increases in social services of all kinds and a steady rise in the standard of living of all classes.

Against more food, more money, more jobs, and more security, all Hitler's and Goebbels's skill as agitators made little headway. Hitler's instinct was right. The foundations of this sudden prosperity were exceedingly shaky, and Hitler's prophecies of disaster, although he was wrong in predicting a new inflation, were to be proved right. But, in 1927 and 1928, few in Germany wanted to listen to such gloomy threats.

In September 1928, Hitler called a meeting of the Party leaders in Munich and talked to them frankly.

Much of his speech was taken up with attempting to belittle Stresemann's achievement in foreign policy, but he did not disguise the difficulties which lay ahead. Above all, they had to strengthen the individual Party comrade's confidence in the victory of the movement. The one striking quality of Hitler's leadership in these years is that he never let go, never lost faith in himself and was able to communicate this, to keep the faith of others alive, in the belief that some time a crack would come and the tide at last begin to flow in his favour.

Hitler's first chance came in 1929, and it came in the direction he had foreseen, that of foreign policy.

The occasion was the renewal of negotiations for a final settlement of reparations. At the Hague in August 1929 Stresemann succeeded in linking the two questions of reparations and evacuation, and in persuading the French to agree that the withdrawal of the occupying forces should begin in September, five years ahead of time, and be completed by the end of June 1930.

Before Stresemann died on 3 October 1929, he had overcome the opposition of the French, but the Germans still remained to be convinced. On 9 July 1929, a national committee had been formed to organize a campaign for a plebiscite rejecting the new reparations settlement and the 'lie' of Germany's war-guilt which represented the legal basis of the Allies' claims. The Press and parties of the German Right united in a most violent campaign to defeat the Government and to use the issues of foreign policy and reparations for their ultimate purpose of overthrowing, or at least damaging, the hated Republic. It was by means of this campaign that Hitler first made his appearance on the national stage of German politics.

The leader of the agitation was Alfred Hugenberg, a bigoted German nationalist. An ambitious, domineering and unscrupulous man of sixty-three, Hugenberg had made a fortune out of the inflation and with it bought up a propaganda empire, a whole network of newspapers and news agencies, and a controlling interest in the big U.F.A. film trust. These he used not so much to make money as to push his own views. In 1928 he took over the leadership of the German National Party.

Hugenberg could count on the support of the Stahlhelm, by far the largest of the German ex-servicemen's organizations, under the leadership of Franz Seldte; of the Pan-German League, and of powerful industrial and financial interests, represented by Dr Albert Voegler, General Director of the big United Steel, and later by the President of the Reichsbank, Dr. Hjalmar Schacht. What they lacked was mass support, someone to go out and rouse the mob. Hitler and Hugenberg met, but Hitler was not easily persuaded to come in, partly because of the opposition which he could expect to meet from the radical Strasser group. But the advantages of being able to draw on the big political funds at the disposal of Hugenberg converted him. He put his price high: complete independence in waging the campaign and a large share of the Committee's resources to enable him to do it. For his representative on the Joint Finance Committee Hitler deliberately chose Gregor Strasser: when others in the Party complained, he laughed and told them to wait until he had finished with his allies.

In September 1929, Hugenberg and Hitler published a draft 'Law against the Enslavement of the German People'. After repudiating Germany's responsibility for the war, it demanded the end of all reparations, and the punishment of the Chancellor, the Cabinet, and their representatives for high treason if they agreed to new financial commitments. The bill was sharply defeated, and on 13 March 1930 President Hindenburg put his signature to the Young

Plan laws, which provided that Germany would pay reparations for a further fifty-nine years. The fury of the Hugenberg and Nazi Press and their open attacks on the President revealed the bitterness of their defeat.

But the defeat for Hugenberg and his 'Freedom Law' was no defeat for Hitler. In the preceding six months he had succeeded for the first time in breaking into national politics and showing something of his ability as a propagandist. To millions of Germans who had scarcely ever heard of him before, Hitler had now become a familiar figure, thanks to a publicity campaign entirely paid for by Hugenberg's rival party.

More important still, he had attracted the attention of those who controlled the political funds of heavy industry and big business to his remarkable gifts as an agitator. From now on Hitler could count upon increasing interest and support from at least some of those who had money to invest in nationalist, anti-democratic and anti-working-class politics.

With this money Hitler began to put the Party on a new footing. He took over the Barlow Palace, an old mansion on the Briennerstrasse in Munich, and had it remodelled as the Brown House. A grand staircase led up to a conference chamber, furnished in red leather, and a large corner room in which Hitler received his visitors beneath a portrait of Frederick the Great.

Hitler himself had moved to a large nine-roomed flat on Prinzregentenstrasse, one of Munich's fashionable streets, and he was now seen more frequently in Munich, occasionally in the company of his favourite niece, Geli Raubal, who had a room in the new flat. At this time Hitler enjoyed more private life than at any other, and he was often to refer to it nostalgically. 'In the evening I would put on a dinner jacket or tails to go to the opera. We made excursions by car. . . . My super-charged Mercedes was a joy to all. Afterwards, we would prolong the evening in the company of the actors. . . . From all points of view, those were marvellous days.'

At the Party conference which followed the alliance with Hugenberg Hitler had had to meet a good deal of criticism, voiced by Gregor Strasser, of the dangers of being tarred with the reactionary brush and losing support by too close association with the 'old gang', the old ruling class of pre-war Germany, the industrialists, the Junkers, the former generals, and higher officials who were the backbone of the National Party. His critics had underestimated Hitler's unscrupulousness, that characteristic duplicity, now first exhibited on this scale. With considerable skill he turned his futile campaign with Hugenberg against the Young Plan to great political advantage for himself and the Party, then not only dropped the alliance with Hugenberg and the Nationalists as unexpectedly as he had made it, but proceeded to attack them.

The astonishing recovery Germany had made since the ending of 1923 was abruptly ended in 1930 under the impact of the World Depression. That 1930 was also the year in which Hitler and the Nazi Party for the first time became a major factor in national politics is not fortuitous. Ever since he came out of prison at the end of 1924 Hitler had prophesied disaster, only to see the Republic steadily consolidate itself. Those who had ever heard of Adolf Hitler called him a fool. Now disaster cast its shadow over the land again, and the despised prophet entered into his inheritance.

Hitler neither understood nor was interested in economics, but he was alive to the social and political consequences of events which affected the life of every family in Germany. The most familiar index of these social consequences is the figure for unemployment: more than six millions in 1932 and again in 1933. Translate these figures into terms of men standing hopelessly on the street corners of every industrial town in Germany; of houses without food or warmth; of boys and girls leaving school without any chance of a job, and one may begin to guess something of the incalculable human anxiety and embitterment burned into the minds of millions of ordinary German working men and women.

The social consequences of the depression were not limited to the working class. In many ways it affected the middle class and the lower middle class just as sharply, for they were threatened with the loss not only of their livelihood, but of their respectability. The middle classes had no trade unions or unemployment insurance, and poverty carried a stigma of degradation for them. When the small property holder, shopkeeper, or business man was forced to sell at depreciated values, anti-capitalist feeling against the combines, the trusts, and department stores spread widely amongst a class which had once owned, or still owned, property itself.

Nor was the impact of the slump limited to the towns. In many parts of Germany the peasants and farmers were in an angry and desperate mood, unable to get a fair return for their work, yet hard pressed to pay the interest on mortgages and loans or be turned out of their homes.

Millions of Germans saw the apparently solid framework of their existence cracking and crumbling. In such circumstances men are no longer amenable to the arguments of reason. In such circumstances men entertain fantastic fears, extravagant hatreds, and extravagant hopes. In such circumstances the extravagant demagogy of Hitler began to attract a mass following as it had never done before.

It was already clear that the economic crisis would produce a political crisis as well—a crisis of the régime. The greatest weakness of the Weimar Republic from the beginning had been its failure to provide a stable party basis for government. The party leaders, absorbed in manoeuvring and bargaining for advantages, were not displeased with this situation. Weak governments suited them to this extent that it made those in power more accessible to party pressure and blackmail. But the moment the country was faced with a major crisis, differences on the share of sacrifice each class was to bear—whether unemployment pay and wages were to be cut, taxes raised, a capital levy exacted, tariffs increased, and help given to landowners and farmers—were allowed to become so bitter that the methods of parliamentary government, which in Germany meant the construction of a coalition by a process of political bargaining, failed to provide the strong government which the country so obviously needed.

Such a situation was much to the advantage of the Nazis, who had been unremitting in their attacks on the parliamentary republic and democratic methods of government. They had already launched a propaganda campaign to win support among the first class to feel the depression, the farmers. Hitler was impressed by Walther Darré, an agricultural expert who had recently written a book on the peasantry as the 'Life Source of the Nordic Race.' He appointed Darré as the Party's agricultural adviser with the commission to draw up a peasant programme, which was published over Hitler's name on 6 March 1930. It was marked not only by practical proposals to give economic aid to the farming population, but also by its insistence upon the peasantry as the most valuable class in the community.

In the case of agriculture it was simple to play for the support of both the big landowners and the peasants, since these had a common economic interest in the demand for protection and higher prices, and a common grievance in their neglect by parties which were too preoccupied with the urban population of Germany. But when it came to industry, business, and trade it was not so easy, for here there was an open clash of interests and bitter antagonism between the workers and the employers, no less than between the small trader and the big companies and department stores. Hitler needed the support of both big business interests because they controlled the funds and the masses because they had the votes. But in origin the National Socialists had been a radical anti-capitalist party, and this side of the Nazi programme was not only taken seriously by many loyal Party members but was of increasing importance in a period of economic depression.

How seriously Hitler took the socialist character of National Socialism was to remain one of the main causes of disagreement and division within the Nazi Party up to the summer of 1934; this was well illustrated in 1930 by the final breach between Hitler and Otto Strasser.

When Gregor Strasser moved to Munich, his brother Otto remained in Berlin, and through his paper, the *Arbeitsblatt* (still the official Nazi journal in the north), and his publishing house maintained an independent radical line which irritated and embarrassed Hitler. In April 1930, the trade unions in Saxony declared a strike, and Otto Strasser came out in full support of their action. The industrialists made perfectly plain to Hitler that unless the Party at once repudiated Strasser's stand there would be no more subsidies. Hitler enforced an order that no member of the Party was to take part in the strike, but he was unable to silence Strasser's papers. On 21 May Hitler suddenly appeared in Berlin and invited Otto Strasser to meet him for a discussion at his hotel.

According to Strasser's account, Hitler's tactics were a characteristic mixture of bribery, appeals, and threats. Hitler offered to take over the Strasser publishing house and make Otto his Press Chief for the entire Reich; if he refused he would be driven out of the Party. Strasser accused Hitler of wanting to 'strangle the social revolution,' and Hitler retorted that Strasser's Socialism was nothing but Marxism: 'There are no revolutions but racial revolutions.' Strasser had demanded the nationalization of industry and asked Hitler what he would do with the great German armament firm of Krupps if he came to power. Hitler replied: 'Of course I should leave it alone. Do you think I should be so mad as to destroy Germany's economy?'

At the end of June Hitler wrote to Goebbels instructing him to drive Otto Strasser and his supporters from the Party. Goebbels obliged with alacrity. Otto Strasser stuck to his Socialist principles, published his talks with Hitler, broke with his brother Gregor (who stayed with Hitler), and set up a party later known as the Black Front. The dispute over the socialist objectives of National Socialism was not yet settled, but Hitler had not lost by making clear his own attitude. In September 1930 the Nazi success at the National elections astonished the world. It was Hitler, not Strasser, who captured the mass vote, while the Black Front dwindled into insignificance and its founder sought refuge over the frontier.

The government was dissolved in July 1930, and in the following election campaign the Nazis used every trick of propaganda to attract attention and win votes. In the big towns there was a marked increase in public disorder in which the S.A. took a prominent part. Slogans painted on walls, posters, demonstrations, rallies, mass meetings, crude and unrestrained demagogy, anything that would help to create an impression of energy, determination and success was pressed into use. Hitler's appeal in the towns was especially to the middle class hit by the depression.

At the same time the Nazis devoted much time and attention to the rural voter. Many who were voting for the first time responded eagerly to attacks on the 'System' which left them without jobs, and to the display of energy, the demand for discipline, sacrifice, action and not talk, which was the theme of Nazi propaganda.

In 1930 the mood of a large section of the German nation was one of resentment. Hitler, with an almost inexhaustible fund of resentment in his own character to draw from, offered them a series of objects on which to lavish all the blame for their misfortunes. It was the Allies, especially the French, who were to blame; the Republic, with its corrupt and self-seeking politicians; the money barons, the bosses of big business, the speculators and the monopolists; the Reds and the Marxists, who fostered class hatred and kept the nation divided; above all, the Jews, who fattened and grew rich on the degradation and weakness of the German people. The old parties and politicians offered no redress; they were themselves contaminated with the evils of the system they supported. Germany must look to new men, to a new movement, to make her strong and feared, to restore to her people the dignity, security and prosperity which were their birthright.

To audiences weighed down with anxiety and a sense of helplessness Hitler cried: If the economic experts say this or that is impossible, to hell with economics. What counts is will, and if our will is hard and ruthless enough we can do anything. The Germans are the greatest people on earth. It is not your fault that you were defeated in the war and have suffered so much since. It is because you were betrayed in 1918 and have been exploited ever since by those who are envious of you and hate you; because you have been too honest and too patient. Let Germany awake and renew her strength, let her remember her greatness and recover her old position in the world, and for a start let's clear out the old gang in Berlin.

This is a fair summary of the sort of speech Hitler and his lieutenants made in hundreds of meetings in the summer of 1930. It showed a psychological perception of the mood of a large section of the German people which was wholly lacking from the campaigns of the other parties.

In the middle of September thirty million Germans went to the polls. The results surprised even Hitler. Nazi seats in the Reichstag leaped from 12 to 107. From ninth the Nazis had become the second Party in the State.

Overnight, Hitler had become a politician of European importance. The foreign correspondents flocked to interview him. *The Times* printed his assurances of goodwill at length, while in the *Daily Mail* Lord Rothermere welcomed his success as a reinforcement of the defences against Bolshevism.

Now that the Nazis had won this great electoral success the question arose as to what use they were going to make of it. In a speech Hitler made at Munich ten days after the election he left it unclear whether he meant to use the Nazi faction in the Reichstag to discredit democratic institutions and bring government to a standstill, following this with a seizure of power by force; or whether he intended to come to power legally as a result of success in the elections and postpone any revolutionary action until after he had secured control of the machinery of the State.

Almost certainly Hitler meant to have his revolution—but after, not before, he came to power. He was too impressed by the power of the State to risk defeat in the streets, as he had, against his better judgement, in November 1923. The revolutionary romanticism of the barricades was out of date. Power had to be obtained legally.

For various reasons Hitler was unwilling to say this too openly. He had to consider its effect on his own party; many were attracted to it by the promise of violence and hoped for a March on Berlin to seize power. This very threat in itself was a useful lever in persuading those who controlled access to power—the Army and the President's advisers—to cooperate with him. To repudiate revolution altogether was to throw away his best chance of coming to power legally. Finally, if the tactics of legality failed, he might be faced with the alternatives of political decline or making a putsch in earnest.

Two particular problems were bound up with the question of legality in the National Socialist movement up to 1934: the relations of the Nazi Party and the Army, and the role to be played by the S.A. The two questions are in fact only different sides of the same penny, but it will be easier to deal with them separately.

Since Röhm's resignation, relations between the Nazis and the Army had been bad. In an effort to control the S.A., Hitler had forbidden them in 1927 to have any connexion with the Army, and the Ministry of Defence had retorted by forbidding the Army to accept National Socialists as recruits or to employ them in arsenals and supply depots.

Yet Hitler was very much aware that the Army's support, or at least its neutrality, was the essential key to his success. In March 1929, he delivered a speech at Munich on the subject of National Socialism and the Armed Forces which was both a challenge to the Army and a bid for its favour. Hitler began by attacking the idea which General von Seeckt had made the guiding principle of the new Army—that the Army must stand apart from politics. This, Hitler declared, was simply to put the Army at the service of the Republican régime, which had stabbed the old Army in the back in 1918 and betrayed Germany to her enemies.

Hitler's speech was published verbatim in a special

Army issue of the *Völkischer Beobachter,* and Hitler followed it up by articles in a new Nazi monthly, the *Deutscher Wehrgeist* (The German Military Spirit), in which he argued that by its attitude of hostility towards nationalist movements like the Nazis the Army was betraying its own traditions and cutting the ground away from under its own feet. Hitler's arguments, which showed again his uncanny skill in penetrating the minds of those he sought to influence, were not without effect, especially among the younger officers, who saw little prospect of promotion in an army limited by the Treaty to a hundred thousand men, and who were attracted by Hitler's promises that he would at once expand and restore the Army to its old position in the State if he came to power.

Three such young officers were charged with spreading Nazi propaganda and were tried before the Supreme Court at Leipzig on 23 September. Since Hitler was now the leader of the second most powerful Party in the country, the Army leaders were extremely interested to discover what his attitude towards the Army would be. As a witness, Hitler did not miss his opportunity to make his effect on the Army. He went out of his way to reassure them about the S.A. Stormtroops' exclusively political function, and insisted, 'I have always held the view that every attempt to disintegrate the Army was madness.' He promised a 'great German People's Army' and the formation of a new government by constitutional means. Hitler's statement at the Leipzig trial provided the basis for his subsequent negotiations with the Army leaders and their eventual agreement to his assumption of power.

Hitler's talk of legality, however, was only a tactic to persuade the generals and the other guardians of the State to hand over power without forcing him to seize it, for everything about his movement proclaimed its brazen contempt for law. Hitler had therefore to take care that in his preoccupation with tactics he did not so far compromise the revolutionary character of his movement as to rob it of its attractive power.

The danger point was the S.A., which was to become the expression of the Party's revolutionary purpose. One of the favourite S.A. slogans was: 'Possession of the streets is the key to power in the State,' and from the beginning of 1930 the political struggle in the Reichstag and at elections was supplemented —in part replaced—by the street fights of the Party armies.

In the course of one of these gang feuds in February 1930, a young Berlin S.A. leader, Horst Wessel, was shot by the Communists. Goebbels skilfully built him up into the prototype of the martyred Nazi idealist, and Wessel's verses provided the S.A. with their marching song, the famous *Horst Wessel Lied.* In the first six months of 1930 the authorities issued a number of prohibitions against outdoor meetings, parades, and uniforms to check the growth of public disorder. But these measures proved ineffective; forbidden to wear their brown shirts, the Nazis paraded in white. Night after night they and the Communists marched in formation, broke up rival political meetings, beat up opponents, and raided each other's 'territory'. As the unemployment rose, the number of recruits mounted: the S.A. offered a meal and a uniform, companionship and excitement.

The key to the Nazi campaign was incessant activity, a sustained effort of propaganda and agitation all the year round. In this the S.A. had an essential part to play. But it was propaganda that Hitler had in mind; the S.A. were to be the shock troops of a revolution that was never to be made. Hitler's problem was to keep the spirit of the S.A. alive without allowing it to find an outlet in revolutionary action; to use them as a threat of civil war, yet never to let them get so far out of hand as to compromise his plan of coming to power without a head-on collision with the forces of the State, above all with the Army.

At the end of 1930 Hitler had considerable cause for satisfaction. Party membership was rising towards the four hundred thousand figure. In the Reichstag the Nazi—every man in brown uniform—had already shown their strength and their contempt for Parliament by creating such disorder that the sittings had to be frequently suspended. In the streets the S.A., by calculated hooliganism, had forced the Government to ban the anti-militarist film *All Quiet on the Western Front*.

Hitler was in no danger of underestimating the opposition to his leadership which still existed in the Party. Success alone would silence criticism, but now success no longer seemed impossible. He had reached the threshold of power.

At the beginning of January 1931, Ernst Röhm, already reinstated as a Party member, was requested by Hitler to take over new duties as Chief of Staff of the S.A. He immediately set to work to make the S.A. by far the most efficient of the Party armies. The organization was closely modelled on that of the Army, with its own headquarters and General Staff quite separate from the organization of the Party, and its own training college for S.A. and S.S. leaders opened at Munich in June 1931. When Röhm took over, the S.A. numbered roughly a hundred thousand men; a year later Hitler could claim three hundred thousand.

The direction of the Party in the years 1931 and 1932 was for all practical purposes in the hands of six men—Hitler himself, Röhm, Gregor Strasser, Göring, Goebbels, and Frick. Röhm was important as an organizer and for his contacts with the Army. Göring, with his wide range of acquaintances and his good-humoured charm, became Hitler's chief political 'contact-man' in the capital, with a general commission to negotiate with other parties and groups. The leader of the Nazi Party in the Reichstag was Dr. Wilhelm Frick. An early and convinced National Socialist, he was useful to Hitler as a good administrator and a man who knew thoroughly the machinery and the mentality of the German civil service. Both Gregor Strasser and Goebbels were able speakers. Strasser possessed the personality to be a leader in his own right if he bestirred himself; Goebbels, undersized, lame and much disliked for his malicious tongue, was useful for his abounding energy and fertility of ideas.

So highly organized a machine must have cost large sums of money to run. A good deal of money, of course, came from the Party itself—from membership dues, from the sale of Party newspapers and literature, from the admission charges and collections at the big meetings. There is no doubt that the Party made heavy demands on its members—even the unemployed S.A. men had to hand over their unemployment-benefit money in return for their food and shelter. Almost certainly the proportion of revenue which was raised by the Party itself has been underestimated. But there were also subsidies from interested supporters.

The main support for the Nazis came from a powerful group of coal and steel producers in the Rhineland and Westphalia, including Emil Kirdorf, the biggest figure in the Ruhr coal industry, and Fritz Thyssen and Albert Voegler of the United Steel Works. But however much Hitler received from industrialists and bankers, he was not a political puppet created by the capitalists. They were to discover, like the conservative politicians and the generals, that, contrary to the popular belief, bankers and business men are too innocent for politics when the game is played by a man like Hitler.

In speaking of the Nazi movement as a 'party' there is a danger of mistaking its true character. For the Nazi Party was no more a party, in the normal democratic sense of that word, than the Communist Party is today; it was an organized conspiracy against the State. The Party's programme was important to win support, and, for psychological reasons, it had to be kept unalterable and never allowed to become a subject for discussion. But the leaders' real object was to get their hands on the State. They were the gutter élite, avid for power, position, and wealth; the sole object of the Party was to secure power by one means or another.

No State, if it was resolved to remain master in its own house, could tolerate the threat such an organization implied. If the people in authority in Germany at this time had been really determined to smash the Nazi movement they would have found the means. Why did they lack the will and the determination? To this there are several answers.

First, Hitler's tactics of legality enabled him to win the maximum advantage from the democratic constitution of the Weimar Republic. He was shrewd enough to realize that he would be the loser, not the gainer, in any attempt to resort to force, whereas so long as he kept within the letter of the law he could fetter the authorities with their own slow-moving legal processes.

Second, the government in Germany in 1931 and 1932 had enough difficulties already. Throughout the winter of 1930–1 the economic crisis, far from lifting, bore down more heavily on the German people. Hitler found no difficulty in laying the blame for all the economic distress of the country on the Government's policy, particularly as Germany was still saddled with reparation payments.

Finally, a stinging rebuff in foreign policy occurred when France, supported by Italy and Czechoslovakia, took the German Foreign Minister's proposal to form an Austro-German Customs Union as a move towards the political and territorial union of Austria with Germany which was expressly forbidden by the Treaties of Versailles and St Germain. The measures taken by the French helped to precipitate the German financial crisis of the summer and forced the German Foreign Minister to announce that the project was being abandoned. The result was a sharp humiliation on the Government of Chancellor Heinrich Brüning and the inflaming of national resentment in Germany.

Faced with such difficulties, both in domestic and foreign policy, any government was likely to hesitate before adding to its problems by the uproar which the suppression of the Nazi Party, the second largest in the Reichstag, would inevitably have entailed, so long as Hitler was clever enough to avoid any flagrant act of illegality. The refusal of the German parties to sink their differences and jointly assume responsibility for the unpopular measures which had to be taken drove Brüning into a dangerous dependence on support of the President and the Army. The attitude of both towards the Nazis was equivocal. Here was the third reason for the reluctance to take action against the Nazis.

From the beginning of 1930, General Wilhelm Groener, the Minister of Defence, a man of integrity and experience, had been uneasily conscious that a good many members of the Officer Corps were becoming sympathetic to the Nazis. The nationalist appeal of Nazi propaganda and its promise of a powerful Germany with an expanded Army were beginning to have their effect. The Army could still be relied on to support Brüning if Hitler attempted to make a putsch, but it was not at all certain that the Government would be able to count on the Army if it was a question of suppressing the Nazi Party without the pretext of revolt.

In October 1931 the President, Field-Marshal von Hindenburg, was eighty-four. Such political judgement as he had ever had was failing. What he cared about most of all was the German Army in which he had spent his life. Faithfully reflecting opinion in the Army, Hindenburg too was opposed to the use of force against the Nazis. He would only agree to act if there was some unequivocal act of rebellion on their part.

Even more important was the opinion of Major-General Kurt von Schleicher, who had made himself virtually the authoritative voice of the Army in politics. Schleicher, a General Staff Officer—able, charming, and ambitious—was far more interested in politics and intrigue than in war. He was the head of a new department in his Ministry, the Ministeramt, which handled all matters common to both the Army and Navy and acted as liaison between the armed services and other ministries. Schleicher used this key position to make himself one of the most powerful political figures in Germany. Through an old friendship with Hindenburg's son, he had an entrée to the old man, who listened to and was impressed by what he said.

Schleicher's object was to secure a strong government which, in place of futile coalitions, would master the economic and political crisis and prevent the need for the Army to intervene to put down revolution. In particular he feared an attack by Poland, if the German Army should be occupied in dealing with simultaneous Nazi and Communist risings.

Schleicher, therefore, shared fully—and was partly responsible for—the reluctance of Groener and Hindenburg to take any initiative against the Nazis. But he went further: impressed by the Nazi success at the elections and by their nationalist programme, he began to play with the idea of winning Hitler's support for Brüning and converting the Nazi movement with its mass following into a prop of the existing government, instead of a battering ram directed against it.

He began by removing the ban on the Army's employment of National Socialists in arsenals and supply depots and the prohibition of Nazi enlistment in the Army. In return Hitler reaffirmed his adherence to the policy of legality by forbidding the S.A. to take part in street-fighting. During the succeeding months Schleicher had several talks with Röhm, who was eager as always to work with the Army. By the latter half of 1931 he was ready to try to secure Hitler's agreement to Hindenburg's re-election as a first step to drawing the Nazis into support of the Government and taming their revolutionary ardour.

Nothing could have suited Hitler better. He had built up a remarkable organization, the strength of which grew steadily, but the question remained how he was to change the success he had won into the hard coin of political power. One way of adding to the Nazi vote was to combine with Hugenberg's German National Party. On 9 July 1931, Hitler and Hugenberg met in Berlin and issued a statement to the effect they would henceforward cooperate for the overthrow of the existing 'System'. But alliance with the Nationalists, with their strongly upper-class character, was bound to lead to much discontent in the radical wing of the Party. Although Hitler continued to make intermittent use of the Nationalist alliance, it was only when no other course presented itself.

If Hitler was to carry his policy of legality to success it could only be done in one way, a possibility created by the peculiar system under which Germany was now governed. Because Chancellors were unable to find a stable parliamentary majority or to win an election, they were forced to use the President's emergency powers, thus transferring political power from the nation to the little group of men round the President. If Hitler could persuade these men to take him into partnership and make him Chancellor, with the right to use the President's emergency powers—a presidential, as opposed to a parliamentary, government—then he could dispense with the clear electoral majority which still eluded him and with the risky experiment of a putsch.

Neither Schleicher nor the President was at all satisfied with the existing situation. They did not believe that the President's emergency powers could be made into a permanent basis for governing the country. They were looking for a government which, while prepared to take resolute action to deal with the crisis, would also be able to win mass support in the country, and, if possible, secure a majority in the Reichstag. Brüning had failed to win such a majority at the elections. Schleicher, therefore, began to look elsewhere for the mass support which he felt to be necessary for the presidential government.

Hitler had two assets: the Nazi success at the elections was a promise of the support he would be able to provide if he was brought in, and the organized violence of the S.A. was a threat of the revolution he might make if he were left out. Hitler's game, therefore, from 1931 to 1933 was to use the revolution he was unwilling to make and the mass support he was unable to turn into a majority, the first as a threat, the second as a promise, to persuade the President and his advisers to take him into partnership and give him power.

This is the key to the complicated and tortuous political moves of the period between the autumn of 1931 and 30 January 1933, when the game succeeded and Hindenburg appointed Adolf Hitler as Chancellor. During this period there were successive negotiations between the Nazi leaders and the little group of men who bore the responsibility for the experiment of presidential government. Hitler did not at the time see this as the only means by which he could come to power legally. He continued to speculate on the possibility of a coalition with the Nationalists—even at one time with the Centre—or, better still, on the chances of winning an outright majority at the next elections. Each time the negotiations broke down he turned again to these alternatives, but always with an eye toward resuming negotiations.

Years ago, in Vienna, Hitler had admired the tactics of Karl Lueger and had summed them up in two sentences in *Mein Kampf*: 'In his political activity, Lueger attached the main importance to winning over those classes whose threatened existence tended to stimulate rather than paralyse their will to fight. At the same time he took care to avail himself of all the instruments of authority at his disposal, and to bring powerful existing institutions over to his side, in order to gain from these well-tried sources of power the greatest possible advantage for his own movement.' Hitler was well on the way to 'winning over those classes whose existence was threatened'; now he faced the task of 'bringing the powerful existing institutions over to his side', above all the Army and the President. The years of waiting were at an end.

CHAPTER FOUR

THE MONTHS OF OPPORTUNITY

October 1931–30 January 1933

The first contacts between Hitler and the men who disposed of power in Germany were scarcely auspicious. At the beginning of the autumn of 1931 Schleicher had a meeting with Hitler, arranged with Röhm's help, and subsequently persuaded the Chancellor, Brüning, and President Hindenburg to see him.

Brüning asked for Hitler's support until the reparations question was settled and Hindenburg reelected as President. After this had been accomplished he was willing to retire and allow someone else more acceptable to the parties of the Right to take his place. Instead of giving a direct answer, Hitler launched into a monologue, the main point of which was that when he came to power he would not only get rid of Germany's debts but would re-arm and, with England and Italy as his allies, force France to her knees. He failed to impress the Chancellor, and the meeting ended inconclusively.

In his interview with President Hindenburg on 10 October, the first occasion on which the two men had met, Hitler was nervous and ill-at-ease. His niece, Geli Raubal, with whom he was in love, had committed suicide three weeks before, and he had wired to Göring, who was at the bedside of his dying wife in Sweden, to return and accompany him. Nazi accounts of the meeting are singularly reticent, but Hitler obviously made the mistake of talking too much and trying to impress the old man with his demagogic arts; instead he bored him. Hindenburg is said to have grumbled to Schleicher afterwards that Hitler was an odd fellow who would never make a Chancellor, but, at most, a Minister of Posts.

Altogether it was a bad week for Hitler. The day after his interview with the President he took part in a great demonstration of the Right-wing 'National' opposition at Harzburg. Hugenberg, representing the Nationalists; Franz Seldte and Theodor Düsterberg, the leaders of the Stahlhelm; Dr. Schacht and General von Seeckt; Graf Kalkreuth, the president of the Junkers' Land League, and half a score of figures from the Ruhr and Rhineland industries, all joined in passing a solemn resolution uniting the parties of the Right. They demanded the immediate resignation of Brüning's Government, followed by new elections. Hitler agreed to take part in the Rally with great reluctance, and Frick felt obliged to defend the decision to the Nazi contingent with a speech in which he said openly that they were only using the Nationalists as a convenient ladder to office. Hitler felt oppressed by his old lack of self-confidence in face of all these frock-coats, top-hats, Army uniforms, and formal titles. This was the *Reaktion* on parade, and the great radical Tribune was out of place. To add to his irritation, the Stahlhelm arrived in much greater numbers than the S.A., and Hugenberg and Seldte stole the limelight. Hitler declined to take part in the official procession, read his speech in a perfunctory fashion, and left before the Stahlhelm marched past. The united front of the National Opposition had virtually collapsed before it was established.

On 13 October, Brüning presented to the Reichstag a reconstituted government in which General Groener, the Minister of Defence—at Schleicher's suggestion—took over the Ministry of the Interior, and the Chancellor himself became Foreign Minister. In face of the Nationalists' and Nazis' demands for his resignation, Brüning appeared to be taking on a new lease of political life, with renewed proofs of the support of the Army and the President.

Hitler expressed his frustration and fury at the course of events in an open letter to the Chancellor (published on 14 October) in which he attacked the policy of the Government as a disastrous betrayal of German interests.

Having delivered his broadside, Hitler went off on 17 October to Brunswick, where more than a hundred thousand S.A. and S.S. men tramped past the saluting base for six hours, and the thundering cheers mollified his wounded vanity. Thirty-eight special trains and five thousand lorries brought the Brown Shirts pouring into Brunswick. Hitler presented twenty-four new standards, and at night a great torchlight parade lighted up the countryside. This was a show the like of which neither Hugenberg and the Stahlhelm nor the Government could put on: while they continued to talk of the need for popular support, Hitler already had it.

Events continued to flow in Hitler's favour. In December 1931, the figure of registered unemployment passed the five-million mark. It was a grim winter in Germany. Brüning's emergency measures were poor comfort to a people suffering from the primitive misery of hunger, cold, lack of work, and lack of hope, and Brüning, with his aloof and reserved manner, was not the man to put across a programme of sacrifice and austerity.

The Nazis gained steadily in strength. Their membership rose to more than 800,000 at the end of 1931. Following their success in the Oldenburg provincial elections in May and at Hamburg in September, they swept the board at the Hessian elections in November. The threat and the promise were gaining in weight.

These facts were not lost on General von Schleicher, who continued his talks with Hitler in November and December. Schleicher was more and more impressed with the need to bring Hitler into the game and make use of him.

Hitler meanwhile kept up the attack on Brüning as the embodiment of all the evils of the 'System' by which Germany had been governed since 1918. An open letter he wrote to Brüning in December is interesting for a frank statement by Hitler of what he meant by legality: 'You refuse, as a "statesman", to admit that if we come to power legally we could then break through legality. Herr Chancellor, the fundamental thesis of democracy runs: "All power issues from the People." The constitution lays down the way by which a conception, an idea, and therefore an organization, must gain from the people the legitimation for the realization of its aims. But in the last resort it is the People itself which determines its Constitution.

'Herr Chancellor, if the German nation once empowers the National Socialist Movement to introduce a Constitution other than that which we have today, then you cannot stop it. . . . When a Constitution proves itself to be useless for its life, the nation does not die—the Constitution is altered.'

Here was a plain enough warning of what Hitler meant to do when he got power, yet Schleicher, Papen, and the rest were so sure of their own ability to manage this ignorant agitator that they only smiled and took no notice.

Brüning had fewer illusions, but all his plans depended upon being able to hold out until economic conditions improved, or he could secure some success in foreign policy. His ability to do this depended in turn upon the re-election of Hindenburg as President. Brüning believed that he could rely on Hindenburg to support him and continue to sign the decrees he laid before him. The old man, failing in health, reluctantly agreed only when the Chancellor promised to try to secure an agreement with the Party leaders in the Reichstag to prolong the presidential term of office without re-election. In any case, a bitter electoral contest for the Presidency at such a time was something to be avoided. And so Brüning, too, agreed to further negotiations with Hitler in order to win him over to his plan.

Hitler was in Munich, in the offices of the *Völkischer Beobachter* when the summons came. He is reported to have crashed his fist down on the telegram in exultation: 'Now I have them in my pocket. They have recognized me as a partner in their negotiations.'

Brüning's proposal was substantially the same as in the previous autumn: Hitler was asked to agree to a prolongation of Hindenburg's presidency for a year or two, until the country had begun its economic recovery and the issues of reparations and the German claim to equality of rights in armaments had been settled. In return, Brüning renewed his offer to resign as soon as he had settled the question of reparations.

Hitler asked for time to consider his reply. Hugenberg, who was also consulted by the Chancellor, as leader of the Nationalists, was strongly opposed to prolonging Hindenburg's term of office, arguing that it could only strengthen Brüning's position. Goebbels, who took the same view, wrote in his diary: 'Brüning only wants to stabilize his own position indefinitely. . . . The contest for power, the game of chess, has begun. . . . It will be a fast game, played with intelligence and skill. The main point is that we hold fast, and waive all compromise.' Gregor Strasser's view was that Hindenburg would be unbeatable in any election the Nazis might force on the Government, and that it was in the Party's interests to accept a temporary truce. But Röhm as well as Goebbels argued that it would be a fatal mistake for the Party to appear to avoid a chance to go to the nation, especially after the recent successes in the provincial elections. Long and anxious debates followed among the Nazi leaders. In the end Röhm's point of view was accepted.

Hugenberg's reply to Brüning's proposal was a blank refusal. Hitler also rejected it, but tried to drive a wedge between Chancellor and President by writing direct to the President over Brüning's head, warning him that the Chancellor's plan was an infringement of the Constitution; adding, however, that he himself was willing to support Hindenburg's re-election if the President would repudiate Brüning's proposal. Hitler offered to make Hindenburg the joint presidential candidate of the Nazis and the Nationalists, if the old man would agree to dismiss Brüning, form a Right-wing 'National' government, and hold new elections for the Reichstag and the Prussian Diet. The newly elected Reichstag, in which Hitler was confident of a majority for the Nazi and Nationalist parties, would then proceed to prolong his term of office.

When Hindenburg refused, Hitler launched a violent attack on Brüning. Brüning in turn accused Hitler of playing party politics at the expense of Germany's chances of improving her international position, and Hitler retorted that nothing could be more beneficial to German foreign policy than the overthrow of the 'System' by which Germany had been governed since 1918.

After this exchange any hopes of avoiding an election for the presidency were at an end. For a second time the attempt to do a deal with Hitler had failed. Brüning threw all his energy into the campaign. Schleicher was equally set on securing the President's re-election, since the position and powers of the Presidency were the basis of his plans. For this reason he was willing to support Brüning, but after that, General von Schleicher considered, a lot of things might happen. The President himself, nettled by the refusal of the Right-wing parties to support the prolongation of his office, finally agreed to offer himself for re-election. On the Government side of the fence, therefore, the breakdown of the negotiations had been followed by at least a temporary consolidation of forces in Brüning's favour.

This was far from being the case in the Nazi camp. Now that his attempt to split Hindenburg and Brüning had failed, Hitler had to face an awkward decision. Was he to risk an open contest with the old President? Hindenburg, or rather the Hindenburg legend, was a formidable opponent. Failure might destroy the growing belief in Nazi invincibility: on the other hand, dare they risk evading the contest?

For a month Hitler hesitated, and Goebbels's diary is eloquent on the indecision and anxiety of the Nazi leaders. Not until 22 February was Goebbels allowed to announce Hitler's candidature to a packed Nazi meeting at the big Berlin Sportpalast. Goebbels writes that when he made the announcement, 'a storm of deafening applause rages for nearly ten minutes. Wild ovations for the Leader. The audience rises with shouts of joy. They nearly raise the roof. . . . People laugh and cry at the same time.'

Hitler, after hesitating for a month, now staked everything on winning, and flung himself into the campaign with a whole-hearted conviction of success. Even before Hitler finally broke off the negotiations with Brüning, Goebbels was already at work preparing for the election campaign. One of his greatest anxieties had been the financing of the campaign. On 5 January he wrote despairingly: 'Money is wanting everywhere. . . .' A month later he was much more cheerful: 'Money affairs improve daily. The financing of the electoral campaign is practically assured.' One of the reasons for this sudden change of tone in Goebbels's references to finance was a visit Hitler had paid to Düsseldorf, the capital of the German steel industry, on 27 January.

At a meeting arranged by Fritz Thyssen, Hitler spoke to the Industry Club. It was the first time that many of the West German industrialists present had met Hitler, and their reception of him was cool and reserved. Yet Hitler, far from being nervous, spoke for two and a half hours without pause, and made one of the best speeches of his life. In it is to be found every one of the stock ideas out of which he built his propaganda, brilliantly dressed up for the audience of businessmen he was addressing.

Hitler began by attacking Brüning's view that the dominant consideration in German politics at this time ought to be the country's foreign relations. The determining factor in national life was the inner worth of a people and its spirit. In Germany, however, this inner worth had been undermined by setting up the false values of democracy and the supremacy of mere numbers in opposition to the creative principle of individual personality.

Hitler chose his illustrations with skill. Private property, he pointed out, could only be justified on the ground that men's achievements in the economic field were unequal. 'But it is absurd to build up economic life on the conceptions of achievement, of the value of personality and on the authority of personality, while in the political sphere you deny this authority and thrust in its place the law of the greatest number—democracy.' Not only was it inconsistent, it was dangerous, for the philosophy of egalitarianism would in time be extended from politics to economics, as it already had been in Bolshevik Russia: 'In the economic sphere Communism is analogous to democracy in the political sphere.'

Hitler dwelt at length on the threat of Communism, for it was something more, he said, than 'a mob storming about in some of our streets in Germany, it is a conception of the world which is in the act of subjecting to itself the entire Asiatic continent'. Unless it were halted it would 'gradually shatter the whole world . . . and transform it as completely as did Christianity'. Already, thanks to the economic crisis, Communism had gained a foothold in Germany. Unemployment was driving millions of Germans to look on Communism as the 'logical theoretical counterpart of their actual economic situation'. This was the heart of the German problem—not the result of foreign conditions, 'but of our internal aberration, our internal division, our internal collapse'. And this state of affairs was not to be cured by the economic expedients embodied in emergency decrees, but by the exercise of political power. It was not economics but politics that formed the prime factor in national life: 'There can be no economic life unless behind this economic life there stands the determined political will of the nation absolutely ready to strike—and to strike hard.'

The same, Hitler went on, was true of foreign policy. 'The Treaty of Versailles in itself is only the consequence of our own slow inner confusion and aberration of mind.'

It was no good appealing for national unity and sacrifice for the State 'when only fifty per cent of a people are ready to fight for the national colours, while fifty per cent have hoisted another flag which stands for a State which is to be found only outside the bounds of their own State.

'Unless Germany can master this internal division in *Weltanschauungen* no measures of the legislature can stop the decline of the German nation.'

Recognizing this fact, the Nazi movement had set out to create a new outlook which would re-unite and re-vitalize the German people. 'Here is an organization which is filled with an indomitable, aggressive spirit. . . . Today we stand at the turning-point of Germany's destiny. . . . Either we shall succeed in working out a body-politic hard as iron from this conglomeration of parties, associations, unions, and *Weltanschauungen*, from this pride of rank and madness of class, or else, lacking this internal consolidation, Germany will fall in final ruin. . . .

'Remember that it means sacrifice when today many hundreds of thousands of S.A. and S.S. men . . . have to buy their uniforms, their shirts, their badges, yes, and even pay their own fares. But there is already in all this the force of an ideal—a great ideal! And if the whole German nation today had the same faith in its vocation as these hundred thousands, if the whole nation possessed this idealism, Germany would stand in the eyes of the world otherwise than she stands now!'

When Hitler sat down the audience, whose reserve had long since thawed, rose and cheered him wildly. Large contributions from the resources of heavy industry soon flowed into the Nazi treasury. As the Army officers saw in Hitler the man who promised to restore Germany's military power, so the industrialists came to see in him the man who would defend their interests against the threat of Communism and the claims of the trade unions, giving a free hand to private enterprise and economic exploitation in the name of the principle of 'creative individuality'.

The election campaign for the Presidency was notable for the bitterness with which it was fought, for the extraordinary confusion of parties, and for the character of the Nazi campaign, a masterpiece of organized agitation which attempted to take Germany by storm. Every constituency down to the most remote village was canvassed, and the walls of the towns were plastered with screaming Nazi posters; films of Hitler and Goebbels were made and shown everywhere (an innovation in 1932); gramophone records were produced and sent through the post. But, true to Hitler's belief in the superiority of the spoken word, the main Nazi effort went into organizing a chain of mass meetings at which the principal Nazi orators, Hitler, Goebbels, Gregor Strasser, worked their audiences up to hysterical enthusiasm by mob oratory of the most unrestrained kind.

The result was baffling. When the polls were closed on the evening of 13 March Hitler had nearly one-third of the total votes in Germany. But all the Nazi efforts left them more than seven million votes behind Hindenburg. This was outright defeat, and Goebbels was in despair.

By a quirk of chance, however, Hindenburg's vote was slightly short of the absolute majority required. A second election had therefore to be held. Before morning on 14 March special editions of the *Völkischer Beobachter* were on the streets carrying Hitler's new election manifesto: 'The first election campaign is over, the second has begun today. I shall lead it.'

It was an uphill fight, with Hitler driving a tired and dispirited Party, but the ingenious mind of Goebbels hit on a novel electioneering device. The leader should cover Germany by plane—'Hitler over Germany'. On 3 April the flight began with four mass meetings in Saxony, at which Hitler addressed a quarter of a million people. He visited twenty different towns in a week from East Prussia to Westphalia, from the Baltic to Bavaria. On 8 April, when a violent storm raged over Western Germany and all other air traffic was grounded, the leader flew to Düsseldorf and kept his engagement, with the whole Nazi Press blaring away that here at last was the man with the courage Germany needed.

Defeat was certain, but by his exacting performance Hitler pushed up his vote again on 10 April by more than two millions. The President was safely home with a comfortable 53 per cent yet by tenacity and boldness Hitler had avoided disaster, capturing votes not only from the Nationalists, but also from the Communists, whose vote fell by over a million. The day after the election Goebbels wrote in his diary: 'The campaign for the Prussian State elections is prepared. We go on without a breathing space.'

Once again, however, the awkward question presented itself: how was electoral success, which, however remarkable, still fell far short of a clear majority, to be turned to political advantage? On 11 March Goebbels noted: 'Talked over instructions with the S.A. and S.S. commanders. Deep uneasiness is rife everywhere. The notion of an uprising haunts the air.' On the other side, Gregor Strasser, who had opposed fighting the presidential campaign from the beginning, now renewed his argument that the chances of success for the policy of legality were being thrown away by Hitler's 'all-or-nothing' attitude and his refusal to make a deal, except on his own exaggerated terms. What was the point of Hitler's virtuoso performance as an agitator, Strasser asked, if it led the Party, not to power, but into a political cul-de-sac?

For the moment Hitler had no answer to either side. It was the Government which took the initiative and used its advantage to move at last against the S.A.

At the end of November 1931 the State authorities of Hesse had secured certain documents drawn up by the legal adviser to the Nazi Party in Hesse, Dr. Werner Best, after secret discussions among a small group of local Nazi leaders. These papers contained a draft of the proclamation to be issued by the S.A. in the event of a Communist rising, and suggestions for emergency decrees to be issued by a provisional Nazi government after the Communists had been defeated, including the immediate execution of those who resisted the Nazi authorities, who refused to cooperate or who were found in possession of arms. Amongst the measures proposed was the abolition of the right to private property, of the obligation to pay debts of interest on savings, and of all private incomes. The S.A. was to be given the right to administer the property of the State and of all private citizens; all work was to be compulsory, without reward, and people were to be fed by a system of food cards and public kitchens. Provision was added for the erection of courts-martial under Nazi presidents.

The discovery of these plans caused a sensation, and seriously embarrassed Hitler, who declared (probably with justice) that he had known nothing of them and, had he known, would have disavowed them. Despite pressure from the Prussian State Government, however, the Reich Government declined to take action against the Nazis.

Evidence of Nazi plans for a seizure of power continued to accumulate. However much Hitler underlined his insistence upon legal methods, the character of the S.A. organization was such that the idea of a putsch was bound to come naturally to men whose politics were conducted in an atmosphere of violence and semi-legality. On the day of the first presidential election Prussian police, raiding Nazi headquarters, found copies of Röhm's orders and marked maps which confirmed the report that the S.A. had been prepared to carry out a *coup d'état* if Hitler secured a majority. Near the Polish frontier other orders were captured instructing the local S.A. in Pomerania not to take part in the defence of Germany in the event of a surprise Polish attack.

As a result of these discoveries the State governments, led by Prussia and Bavaria, presented Groener (the Reich Minister of the Interior) with an ultimatum. Either the Reich Government must act against the S.A. or, they hinted, they would take independent action themselves. Groener felt obliged to act. On 10 April, the day of the second election, a meeting presided over by the Chancellor confirmed Groener's view, and on the 14th a decree was promulgated dissolving the S.A., the S.S., and all their affiliated organizations.

Röhm for a moment thought of resistance; after all, the S.A. now numbered four hundred thousand men. But Hitler was insistent: the S.A. must obey.

Overnight the Brown Shirts disappeared from the streets. But the S.A. organization was left intact to appear as ordinary Party members. Brüning and Groener would get their answer, Hitler declared, at the Prussian elections.

Prussia was by far the largest of the German states, embracing nearly two-thirds of the whole territory of the Reich, with a population of forty out of a total of sixty-five millions. It had been the stronghold of German democracy, and the Prussian Ministry of the Interior had been more active than any other official agency in trying to check Nazi excesses. To capture a majority in Prussia, therefore, would be a political victory for the Nazis second only in importance to securing a majority in the Reichstag.

On 24 April at the Prussian elections and other State elections, some four-fifths of Germany went to the polls. In spite of the Nazi propaganda machine, Hitler's second series of highly publicized flights over Germany, and the support of the Nationalists, the Nazis fell short of the majority for which they had hoped. Three times the trumpet had sounded and still the walls refused to fall.

At this moment there appeared a *deus ex machina* in the shape of General von Schleicher, prepared to discuss once again the admission of the Nazis by the back door.

General Schleicher had resumed his relations with Röhm before the presidential elections. He appears at this time to have been playing with the idea of detaching the S.A. from Hitler, and bringing them under the jurisdiction of the State as the militia Röhm had always wanted to make them. Unknown to Hitler, it had already been agreed between Röhm and Schleicher that, in the event of a war-emergency, the S.A. would come under the command of the Army. Schleicher, however, was still attracted by the alternative idea of bringing Hitler himself into the

Government camp. In either case, the prohibition of the S.A. was bound to embarrass his plans.

Schleicher went behind Groener's back and persuaded Hindenburg to write an irritable letter complaining about the activities of the Social Democratic Reichsbanner, with the implication that the prohibition of the S.A. had been one-sided. The letter had been made public almost before Groener received it. A malicious whispering campaign against him now began, and on 10 May Göring delivered a violent attack on him in the Reichstag. When Groener, a sick man, attempted to reply, he met a storm of abuse from the Nazi benches. Scarcely had he sat down, exhausted by the effort, when he was blandly informed by Schleicher, the man he regarded almost as his own son, that the Army no longer had confidence in him, and that it would be best for him to resign. Brüning loyally defended Groener, but there were such scenes of uproar in the Reichstag that the Chamber had to be cleared by the police. The next day Groener resigned. The Nazis were jubilant.

Groener's fall, treacherously engineered by Schleicher, was a grave blow to German democracy. One of the greatest weaknesses of the Weimar Republic was the equivocal attitude of the Army towards the republican régime. Groener was the only man amongst the Army's leaders who had served it with wholehearted loyalty, and there was no one to replace him.

But Groener's departure was only a beginning. Schleicher had now made up his mind that the chief obstacle to the success of his plan for the deal with the Nazis was Brüning, the butt of Nazi attacks on the 'System'. With the same cynical disloyalty with which he had stabbed Groener in the back, Schleicher now set about unseating Brüning.

Brüning was not in a strong position to defend himself. He had failed to secure a stable majority in the Reichstag, and had so far failed to restore prosperity to Germany. His great hope—the cancellation of reparations and the recognition of Germany's right to equality in armaments—had been frustrated by the postponement of the Reparations Conference at Lausanne until June 1932 and the opposition of the French. Ironically, his one great success, the re-election of the President, weakened rather than strengthened his position. For, with that safely accomplished, Brüning no longer appeared indispensable, and, under the careful coaching of Schleicher and other candid friends, the old man had come to feel resentment against the Chancellor as the man whose obstinacy had forced him to endure an election campaign, and to stand as the candidate of the Left against his own friends on the Right.

Moreover, Brüning, because of his economic and social policies, had made enemies who enjoyed great influence with the President, including industrialists and the powerful Junker class. Finally, Schleicher, claiming to speak with the legendary authority of the Army, announced that the Army no longer had confidence in the Chancellor. A stronger man was needed to deal with the situation, and he already had a suitable candidate ready in Franz von Papen. He added the all-important assurance that the Nazis had agreed to support the new Government.

Ostensibly Hitler played no part in the manoeuvres which led to Brüning's dismissal. Schleicher offered him the overthrow of the Brüning Cabinet, the removal of the ban on the S.A. and S.S., and new elections for the Reichstag in return for tacit support, the 'neutrality' of the Nazis towards the new presidential cabinet which Papen was to form. Such a promise cost Hitler nothing. Time would show who was to do the double-crossing, Schleicher or the Nazis. Meanwhile Hitler's agreement provided Schleicher with a winning argument for Hindenburg. Papen would be able to secure what Brüning had failed to get, Hitler's support, without taking him into the Cabinet.

Groener's fall on 13 May raised the hopes of the

Nazi leaders high. On the 18th Goebbels wrote in his diary: 'For Brüning alone winter seems to have arrived. He is being secretly undermined and is already completely isolated. He is anxiously looking for collaborators—"My kingdom for a Cabinet Minister!" . . . Our mice are busily at work gnawing through the last supports of Brüning's position.' 'Rat' would perhaps have been a better word to describe the part played by General von Schleicher.

At the end of May Hindenburg requested the Chancellor's resignation, and Brüning resigned. That fatal reliance on the President which he had been forced to accept as the only way out of the political deadlock had produced a situation in which governments could be made and unmade by the simple grant or withdrawal of the President's confidence. Who bore the responsibility for allowing such a situation to arise will long be a matter of controversy, but the result was plain enough: it was the end of democratic government in Germany. The key to power over a nation of sixty-five million people was now openly admitted to lie in the hands of an aged soldier of eighty-five and the little group of men who determined his views.

On 29 May Hitler and Göring saw the President. Hindenburg informed them briefly that he intended to appoint Papen as Chancellor and understood that Hitler had agreed to support him. Was this correct? Hitler answered: 'Yes.'

The new Chancellor, Franz von Papen, a man in his fifties, came from a Catholic family of the Westphalian nobility. He had belonged to the right cavalry regiment (he was a celebrated gentleman-rider) and now to the right clubs, the Herrenklub and the Union. He had great charm, a wide acquaintance in the social world, connexions with both German and French industry, and considerable political ambitions. So far these ambitions had not been taken seriously by anyone else. Schleicher was attracted to the improbable choice of Papen as Chancellor by the belief that he would prove a pliant instrument in his hands, but he seriously underestimated Papen's ambition, tenacity, and unscrupulousness. The choice startled everyone and pleased few, with the important exception of the President, who was delighted with the company of a Chancellor who knew how to charm and flatter so well that he soon established relations with him such as no other minister had ever had.

If Schleicher believed that Papen would be able to rally a coalition of the Centre and the Right he was soon disillusioned. Only with great difficulty was it possible to collect a Cabinet of men willing to serve under Papen. Of its ten members, none of whom was a political figure of the first rank, seven belonged to the nobility with known Right-wing views. From the beginning there was not the least chance of Papen avoiding an overwhelming defeat if he met parliament; the power of the 'Cabinet of Barons' was openly and unashamedly based upon the support of the President and the Army.

Of the four parties in Germany which commanded mass support, the Communists and the Social Democrats were bound to oppose Papen's government; his own party, the Centre, had excommunicated him, and only the Nazis remained as a possible ally, bought at the price of two concessions: the dissolution of the Reichstag and the lifting of the ban on the S.A. The question was whether this temporary arrangement could be turned into a permanent coalition.

Both sides were willing to consider such a proposal —Hitler because this was the only way in which he could come to power if he failed to win an outright majority; and the group around the President, Papen and Schleicher, because this offered the only prospect of recruiting popular support for their rule and the best chance, they believed, of taking the wind out of the Nazi sails. The elements of a deal were present all the time; the question was, on whose terms—Hitler's or Papen's? When Papen could not get Nazi support on his terms, he left them to cool their heels, calculating that the strain on the Party of continued frustration would force Hitler to reduce his demands. Hitler, on his side, tried to stick it out without capitulating. This is the underlying pattern of events in the latter half of 1932. Superimposed on it is a second pattern created by the fact that both sides became divided on the right tactics to pursue; on one side this is represented by a split between Papen and Schleicher, on the other side by the quarrel between Hitler and Gregor Strasser.

Papen dissolved the Reichstag on 4 June, and fixed the new elections for the last day of July. Even this brief delay aroused Nazi suspicions; and when the lifting of the S.A. ban was postponed until the middle of the month, relations between Hitler and the new Government became strained. On 5 June Goebbels wrote in his diary: 'We must disassociate ourselves at the earliest possible moment from the temporary bourgeois Cabinet.' When Hitler saw Papen on the 9th, he made no pretence of his attitude. 'I regard your Cabinet,' he told the Chancellor, 'only as a temporary solution and will continue my efforts to make my party the strongest in the country. The Chancellorship will then devolve on me.' There was considerable grumbling in the Party at a 'compromise with Reaction'. Unless the Nazis were to be tarred with the same brush, they had to assert their independence.

When the ban on the S.A. was lifted, Ernst Thälmann, the Communist leader, described it as an open provocation to murder. This proved to be literally true, for, in the weeks which followed, murder and violence became everyday occurrences in the streets of the big German cities. The fiercest fighting was between the Nazis and the Communists: of eighty-six people killed in July 1932, thirty were Communists and thirty-eight Nazis.

On the flimsy pretext that the Prussian Government could not be relied on to deal firmly with the Communists, Papen used the President's emergency powers on 20 July to depose the Prussian Ministers, appointing himself as Reich Commissioner for Prussia. By this action Papen hoped partly to conciliate the Nazis, partly to steal some of the Nazi thunder against 'Marxism'. To carry out his plan Papen had stretched the constitutional powers of the President to the limit, but the two largest working-class organizations in Germany, the Social Democratic Party and the trade unions, had not put up even a token resistance in face of Papen's *coup d'état,* a significant pointer to the opposition (or lack of it) which Hitler might expect to meet if he came to power.

The removal of the Prussian Government was a heavy blow to those who still remained loyal to the Weimar Republic. The republican parties were shown to be on the defensive and lacking the conviction to offer more than a passive resistance. However much Papen and Schleicher might claim the credit of this show of energy for the new government, in fact any blow which discredited democratic and constitutional government must bring advantage to the Nazis and the Communists, the two extremist parties. The impression that events favoured the triumph of one or other form of extremism was strengthened, and helped both parties to win votes at the coming elections.

The elections were held on the last day of July 1932. Goebbels had been making his preparations since the beginning of May, and the fourth election compaign in five months found the Nazi organization at the top of its form. The argument that things must change, and the promise that, if the Nazis came to power, they would, proved a powerful attraction in a country driven to the limit of endurance by two years of economic depression and mass unemployment, made worse by the inability of the Government to relieve the nation's ills. It was the spirit of revolt engendered by these conditions to which Nazism gave expression, unhampered by the doctrinaire teaching and class exclusiveness of Communism.

The whole familiar apparatus of Nazi ballyhoo was brought into play—placards, Press, sensational charges and counter-charges, mass meetings, demonstrations, S.A. parades. As a simple feat of physical endurance, the speaking programme of men like Hitler and Goebbels was remarkable. Again Hitler took to the skies, and in the third 'Flight over Germany' visited and spoke in close on fifty towns in the second half of July. Delayed by bad weather, Hitler reached one of his meetings, near Stralsund, at half past two in the morning. A crowd of thousands waited patiently for him in drenching rain. When he finished speaking they saluted the dawn with the mass-singing of *Deutschland über Alles.* This was more than clever electioneering. The Nazi campaign could not have succeeded as it did by the ingenuity of its methods alone, if it had not at the same time corresponded and appealed to the mood of a considerable proportion of the German people.

When the results were announced on the night of 31 July the Nazis had outstripped all their competitors, and with close on fourteen million votes and 230 seats in the Reichstag had more than doubled the support they had won at the elections of September 1930. They were now by far the largest party in Germany, their nearest rivals, the Social Democrats, polling just under eight million votes, the Communists five and a quarter million, and the Centre four and a half. Although the Nazi vote (37.3 per cent) still fell short of the clear majority for which they had hoped, Hitler felt himself to be in a very strong position to make a deal with Papen and Schleicher. The combined strength of the two extremist parties, the Nazi and the Communist, added up to more than 50 per cent of the Reichstag, sufficient to make government with parliament impossible, unless the Nazis could be brought to support the Government. With a voting strength of 13,700,000 electors, a party membership of over a million and a private army of 400,000 S.A. and S.S., Hitler was the most powerful political leader in Germany, knocking on the doors of the Chancellery at the head of the most powerful political party Germany had ever seen.

Inflamed by the election campaign, and believing that the long-awaited day was within sight, the S.A. threatened to get out of hand. In the first nine days of August a score of incidents was reported every day, culminating on 9 August in the murder at Potempa, a village in Silesia, of a Communist, Konrad Pietrzuch, who was brutally kicked to death by five Nazis in front of his mother. The same day Papen's Government announced the death penalty for clashes which led to people being killed. The Nazis at once protested indignantly.

Aware of the highly charged feeling in the Party, Hitler took time before he moved. He was in a mood for 'all-or-nothing'. On 5 August he saw General von Schleicher and put his demands before him: the Chancellorship for himself, and other Nazis at the head of the Prussian State Government, the Reich, and Prussian Ministries of the Interior (which controlled the police). With these were to go the Ministry of Justice and a new Ministry of Popular Enlightenment and Propaganda, which was reserved for Goebbels. An Enabling Bill, giving Hitler full power to govern by decree, would be presented to the Reichstag; if the Chamber refused to pass it, it would be dissolved. Hitler came away in high hopes that the

General would use all his influence to secure the Chancellorship for him. He was so pleased that he suggested to Schleicher a tablet should be affixed to the walls of the house to commemorate their historic meeting. He then returned to Berchtesgaden to await events.

On 9 August, Strasser and Frick joined him there with disquieting news. The violent behaviour of the S.A. and some of the wilder election and post-election statements were making people ask if the Nazis were fit to have power. Business and industrial circles were becoming worried lest a Hitler Chancellorship should lead to radical economic experiments on the lines Gottfried Feder and Gregor Strasser had often threatened. Still no word came from Berlin.

On 11 August Hitler decided to bring matters to a head. Sending messengers ahead to arrange for him to see the Chancellor and the President, he motored to Berlin, which he reached late in the evening of the 12th, and drove to Goebbels's house. Röhm had already visited Papen and Schleicher and had asked bluntly who was to be Chancellor. Had Hitler misunderstood Schleicher? The answer Röhm had been given was none too satisfactory. After Goebbels told him the news, Hitler paced up and down for a long time, uneasily calculating his chances. The decisive meeting with Papen and Schleicher was fixed for the next day at noon.

Papen was less impressed by Hitler's success than might have been expected. Hitler had failed to win the majority he hoped for, and Papen could argue that the results of the elections and the divisions in the Reichstag were such as to justify the continuation of a presidential cabinet, independent of the incoherent Party groupings. Indeed, Papen saw no reason at all why he should resign in Hitler's favour. He enjoyed the favour of the President as no one ever had before, and the President certainly had no wish to exchange the urbane and charming Papen for a man whom he disliked. Papen, like most other political observers, was convinced that the Nazis had reached their peak and from now on would begin to lose votes. If he was still prepared to do a deal with Hitler it must be on his, and not Hitler's, terms.

Schleicher's attitude too had changed. When Hitler met the General and Papen together on the 13th, the most they were prepared to offer him was the Vice-Chancellorship, together with the Prussian Ministry of the Interior for one of his lieutenants. Hitler's claim to power as the leader of the largest party in the Reichstag was politely set aside. The President, Papen told him, insisted on maintaining a presidential cabinet in power and this could not be headed by a Party leader like Hitler. Hitler rejected Papen's offer out of hand, lost his temper and began to shout. He must have the whole power, nothing less. He talked wildly of mowing down the Marxists, of a St Bartholomew's Night, and of three days' freedom of the streets for the S.A. Papen and Schleicher were shocked by the raging uncontrolled figure who now confronted them. They were scarcely reassured by his declaration that he wanted only as much power as Mussolini had claimed in 1922. While Hitler meant by this a coalition government, including non-Fascists, such as Mussolini had originally formed, they understood him to be claiming a dictatorship in which he would govern alone without them—and, as the history of Hitler's Chancellorship in 1933 was later to show, they were fundamentally right.

Hitler left in a rage of disappointment, and drove back to Goebbels's flat. When a telephone call came from the President's Palace at three o'clock, the caller was told that there was no point in Hitler coming, as a decision had already been arrived at. But the President insisted. Nothing, it was said, would be finally decided till he had seen Hitler—and Hitler, angry and shaken, went.

The President received him standing up and leaning on his stick. His manner was cold. Hitler's argument that he sought power by legal means, but to obtain his ends must be given full control over government policy, made no impression on the old man. According to Otto Meissner, head of the Presidential Chancery, who was present, the President retorted that in the present tense situation he could not take the risk of transferring power to a new Party which did not command a majority and which was intolerant, noisy, and undisciplined. 'Hindenburg added that he was ready to accept Hitler and his movement in a coalition government, the precise composition of which could be a subject of negotiation, but that he could not take the responsibility of giving exclusive power to Hitler alone. . . . Hitler, however, was adamant in his refusal to put himself in the position of bargaining with the leaders of the other parties and of facing a coalition government.'

Before the interview was over Hindenburg took the chance to remind Hitler of the promise, which he had now broken, to support Papen's Government. In the words of the communiqué, 'he gravely exhorted Herr Hitler to conduct the opposition on the part of the N.S. Party in a chivalrous manner, and to bear in mind his responsibility to the Fatherland and to the German people.' For once, the Nazi propaganda machine was caught off its guard, and the Government's damaging version of the meeting was on the streets before the Nazis realized what was happening. It spoke of Hitler's 'demand for entire and complete control of the State'; described the President's refusal to hand over power to 'a movement which had the intention of using it in a one-sided manner'; referred explicitly to Hitler's disregard of the promises of support he had given before the election, and repeated Hindenburg's warning to him on the way to conduct opposition. Hitler's humiliation was complete.

If ever Hitler needed confidence in his own judgement, it was now. A false move could have destroyed his chances of success, and it was easy to make such a move. The policy of legality appeared discredited and bankrupt. Hitler had won such electoral support as no other party had had in Germany since the First World War, he had kept strictly to the letter of the Constitution and knocked on the door of the Chancellery, only to have the door publicly slammed in his face. The way in which his demands had been refused touched Hitler on a raw spot; once again he had been treated as not quite good enough, an uneducated, rough sort of fellow whom one could scarcely make Chancellor. His old hatred and contempt for the bourgeoisie and their respectable politicians flared up. He was angry and resentful, feeling he had walked into a trap and was being laughed at by the superior people who had made a fool of him. In such a mood there was a great temptation to show them he was not bluffing, to give the S.A. their head, and let the smug bourgeois politicians see whether he was just a 'revolutionary of the big mouth', as Goebbels had once called Strasser.

There was strong pressure from the Party in the same direction. The S.A. had always disliked the policy of legality, and had only been constrained to submit to it with difficulty. Now that legality had led to an open set-back and humiliation they were even more restive. The difficulties with which Hitler was confronted are vividly illustrated by the case of the Potempa murderers. The five S.A. members responsible for the murder of the Communist miner, Pietrzuch, were sentenced to death on 22 August. The S.A. were furious: this was to place the nationally minded Nazis and the anti-national Communists on the same footing. Hitler had therefore to choose between offending public opinion and travestying his own policy of legality if he came out on the side of the murderers, or risking a serious loss of confidence on the part of the S.A. if he failed to intervene on their behalf. His answer was to send a telegram to the five murderers: 'My comrades: in the face of this most monstrous and bloody sentence I feel myself bound to you in limitless loyalty. From this moment, your liberation is a question of our honour. To fight against a government which could allow this is our duty.' He followed this with a violent manifesto in which he attacked Papen for deliberately setting on foot a persecution of the 'nationally minded' elements in Germany.

Although the Nazi Press and Nazi speeches show an increasing radicalism from August up to the second Reichstag elections in November, and although Hitler came out in uncompromising opposition to Papen's Government, he still refused to depart from his tactics of legality, or to let himself be provoked into the risk of attempting a seizure of power by force.

Shortly after the Potempa incident Hermann Rauschning, one of the leaders of the Danzig Senate, visited Hitler, whom he found moody and preoccupied. His silence was interspersed with excited and violent comments, many of them on the character of the next war. Much of it was prophetic: he laid great stress upon the psychological and subversive preparations for war—if these were carried out with care, peace would be signed before the war had begun. 'The place of artillery preparation for frontal attack will in future be taken by revolutionary propaganda, to break down the enemy psychologically before the armies begin to function at all. . . . How to achieve the moral break-down of the enemy before the war has started—that is the problem that interests me. . . . We shall provoke a revolution in France as certainly as we shall *not* have one in Germany. The

French will hail me as their deliverer. The little man of the middle class will acclaim us as the bearers of a just social order and eternal peace. None of these people any longer want war or greatness.' Rauschning could get little out of Hitler about the current political situation. Only when they came to discuss Danzig did Hitler show any interest in the actual position in Germany. His first question was whether Danzig had an extradition agreement with Germany, and it was soon clear that his mind was occupied with the possibility of having to go underground, if the Government should move against the Party and ban it. In that case Danzig, with its independent status under the League of Nations, might well offer a useful asylum.

Desultory contacts with the Government continued through the rest of the summer and into the autumn, but they led nowhere. In August and September the Nazis made an approach to the Centre Party: together they could command a majority in the Reichstag. One practical result of these talks was the election of Göring to the presidency of the Reichstag by the combined votes of the Nazis, the Centre, and the Nationalists on 30 August.

The climax of the weeks of intrigue and manoeuvring came on 12 September, when the Reichstag met for the first full session since the elections at the end of July. Foreseeing trouble, Papen procured a decree for the Chamber's dissolution from the President in advance. With this up his sleeve, he felt in complete command of the situation. The actual course of events, however, took both sides by surprise. When the session opened, before a crowded audience in the diplomatic and public galleries, the Communist deputy Ernst Torgler moved a vote of censure on the Government as an amendment to the Order of the Day. It had been agreed amongst the other parties that one of the Nationalist deputies should formally oppose it. When the moment came, however, the Nationalists made no move, and amid a puzzled and embarrassed silence Frick rose to his feet to ask for half an hour's delay. At a hurried meeting, Göring, Hitler, Strasser, and Frick decided to out-smart the Chancellor, vote with the Communists, and defeat the Government.

Immediately the deputies had taken their seats; again Göring, as President, announced that a vote would be taken at once on the Communist motion of no-confidence. Papen, rising in protest, requested the floor. But Göring, studiously affecting not to see the Chancellor, looked in the other direction, and the voting began. White with anger, Papen produced the traditional red portfolio which contained the decree of dissolution, thrust it on Göring's table, then ostentatiously marched out of the Chamber. Still Göring had no eyes for anything but the voting. The Communist vote of no-confidence was carried by 513 votes to 32, and Göring promptly declared the Government overthrown. As for the scrap of paper laid on his desk, which he now found time to read, it was, he declared, obviously worthless since it had been countersigned by a Chancellor who had now been deposed.

Whether—as the Nazis affected to believe—the elaborate farce in the Reichstag had really damaged Papen or not, for the moment the Chancellor had the advantage. For Papen insisted that, as the decree of dissolution had already been signed and placed on the table before the vote took place, the result of the motion was invalid. The Reichstag was dissolved, after sitting for less than a day, and the Nazis faced the fifth major electoral contest of the year.

Privately they were only too well aware that Papen was right and that they must count on a reduced vote. Hitler refused to consider a compromise, and accepted Papen's challenge, but there was no disguising the fact that this would be the toughest fight of all.

One of the worst difficulties was lack of money. Four elections since March had eaten deep into the Party's resources, and the invaluable contributions from outside had lately begun to dwindle. Hitler's refusal to come to terms, his arrogant claim for the whole power, his condonation of violence at Potempa, the swing towards Radicalism in the campaign against the 'Government of Reaction'—all these factors, combined, no doubt, with strong hints from Papen to industrial and business circles not to ease the blockade, had placed the Party in a tight spot.

In these circumstances it was only Hitler's determination and leadership that kept the Party going. His confidence in himself never wavered. When the Gauleiters assembled at Munich early in October he used all his arts to put new life and energy into them. 'He is great and surpasses us all,' Goebbels wrote enthusiastically. 'He raises the Party's spirits out of the blackest depression. With him as leader the movement must succeed.'

The Nazi leaders, however, were under no illusions about the election results. The fifth election of the year found a mood of stubborn apathy growing among the German people, a feeling of indifference and disbelief, against which propaganda and agitation beat in vain. It was precisely on this that Papen had calculated, and his calculation was not far wrong. For the first time since 1930 the Nazis lost votes. Their seats in the Reichstag were reduced to 196 out of 584, although they were still the largest party in the Chamber.

Papen was delighted with the results, which he regarded as a moral victory for his government and a heavier defeat for Hitler than the figures actually showed. If Hitler wanted power, thought Papen, he had better come to terms before his electoral assets dwindled still further.

At first it looked as if Hitler would be forced to accept Papen's terms. However, it was now Papen who overplayed his hand, with unexpected results. Determined, in spite of the electoral set-back, not to walk into another trap, Hitler sat tight and refused to be drawn by Papen's indirect approaches.

On 13 November Papen wrote officially to Hitler suggesting that they should bury their differences and renew negotiations for a concentration of all the nationally minded parties. Hitler's reply, an open rebuff, ruled out the possibility of further negotiations between himself and Papen at this stage. Indeed, he had already issued a manifesto immediately after the elections in which he had charged Papen with the responsibility for the increase in the Communist vote. By his reactionary policy, Hitler declared, Papen was driving the masses to Bolshevism. There could be no compromise with such a régime.

Papen was perfectly prepared to plunge the country into still another election in order to force the Nazis to their knees, but he unexpectedly encountered opposition from Schleicher. Not only was Schleicher irritated by Papen's increasing independence and the close relationship he had established with the President, but he began to see in Papen's personal quarrel with Hitler an obstacle to securing that concentration of the 'national' forces which was, in Schleicher's view, the only reason for ever having made Papen Chancellor. He was more than ever alarmed at the prospect of a civil war in which both the Communists and the Nazis might be on the other side of the barricade. It did not take long for him to reach the conclusion that Papen was becoming more of a hindrance than an asset to his objective of a deal with the Nazis.

Schleicher found support for his views in the Cabinet, and Papen was urged to resign, in order to allow the President to consult the Party leaders and try to find a way out of the deadlock. With considerable shrewdness Papen swallowed his anger and agreed; he was confident that negotiations with Hitler and the other Party leaders would end in failure, and he would return to office with his hand strengthened. He would then be able to insist on whatever course he saw fit to recommend. Hindenburg, obviously irritated by the whole affair, saw no reason at all why he should part with Papen, and had become increasingly suspicious of Schleicher, which augured well for the success of Papen's calculations. None the less, on 17 November, Papen tendered the resignation of his Cabinet, and the President, on his advice, requested Hitler to call on him.

Events followed the course Papen had foreseen. On 18 November Hitler arrived in Berlin and spent some hours in discussion with Goebbels, Frick, and Strasser; Göring was hastily summoned from Rome, where he had been engaged in talks with Mussolini. The next day Hitler drove to the Palace. 'You have declared,' the President said, 'that you will only place your movement at the disposal of a government of which you, the leader of the Party, are the head. If I consider your proposal, I must demand that such a Cabinet should have a majority in the Reichstag. Accordingly, I ask you, as the leader of the largest party, to ascertain, if and on what conditions, you could obtain a secure workable majority in the Reichstag on a definite programme.'

On the face of it this was a fair offer, but it was so designed as to make it impossible for Hitler to succeed. For Hitler could not secure a majority in the Reichstag. In any case, what Hitler wanted was to be made, not a parliamentary Chancellor, but a presidential Chancellor, with the same sweeping powers as the President had given to Papen. Hindenburg sternly refused: if Germany had to be governed by the emergency powers of a presidential Chancellor, then there was no point in replacing Papen; the only argument in favour of his resignation was that Hitler would be able to provide something which Papen had failed to secure, a parliamentary majority. Hitler could only retort that the negotiations had been foredoomed to fail in view of Hindenburg's resolve to keep Papen, whatever the cost. Once again the policy of legality had led to public humiliation; once again the Leader returned from the President's palace empty-handed and out-manoeuvred.

Discussions between the President and other Party leaders produced no better result. But at this point Papen's calculations began to go wrong. For Schleicher, too, had not been idle, and through Gregor Strasser he was now sounding out the possibility of the Nazis joining a Cabinet in which, not Papen, but Schleicher himself would take the Chancellorship. The offer was communicated to Hitler in Munich. According to the Nazi version, Hitler declined to be drawn in by Schleicher's move. He called a conference of his chief lieutenants at Weimar on 1 December. Strasser came out strongly in favour of joining a Schleicher Cabinet and found some support from Frick. Göring and Goebbels, however, were opposed to such a course, and Hitler accepted their point of view.

Meanwhile, on the evening of 1 December, Schleicher and Papen saw Hindenburg together. Papen proposed that he should resume office, prorogue the Reichstag indefinitely, and prepare a reform of the constitution to provide for a new electoral law and the establishment of a second Chamber. Until that could be carried out he would proclaim a state of emergency, govern by decree, and use force to smash any opposition. Schleicher's objections were threefold: such a course was unconstitutional; it involved a danger of civil war, since the vast majority of the nation had declared themselves emphatically opposed to Papen in two elections; and it was unnecessary. He announced that he was convinced he himself could obtain a parliamentary majority in the Reichstag.

Papen emerged triumphant when Hindenburg entrusted him with the task of forming a new government, but the next day Schleicher announced that the Army no longer had confidence in Papen and was not prepared to take the risk of civil war which Papen's policy would entail. He provided detailed evidence in support of his argument. Once again the Army had shown itself to be the supreme arbiter in German politics. Hindenburg withdrew his commission to Papen.

Papen had only two consolations, but they were to prove substantial. At last Schleicher, the man who had used his influence behind the scenes to unseat Groener, Brüning, and now Papen, was forced to come out into the open and assume personal responsibility for the success or failure of his plans. On 2 December General von Schleicher became the last Chancellor of pre-Hitler Germany, and—Papen's second consolation—he took office at a time when his credit with the President, on which he had drawn so lavishly in the past year, was destroyed. The old man, who had tolerated Schleicher's earlier intrigues, neither forgot nor forgave the methods by which Schleicher turned out Papen. Let Kurt von Schleicher succeed if he could; but if he failed, and turned to the President for support, he need expect no more mercy than he had shown his victims.

Schleicher had now to make good his claim that he could succeed where Papen had failed, and produce that national front, including the Nazis, which had been his consistent aim for two years. For all his love of intrigue and lack of scruple, Schleicher was an intelligent man. He had never fallen into the error of supposing that 'strong' government by itself was a remedy for the crisis, nor did he underestimate the force which lay behind such extremist movements as the Nazis and the Communists. His aim, stated again and again in these years, was to harness one of these movements, the Nazis, to the service of the State.

Schleicher's closest contact in the Nazi Party at this time was Gregor Strasser. If Hitler represented the will to power in the Party, and Röhm its preference for violence, Gregor Strasser represented its idealism. To Strasser National Socialism was a real political movement, and he took its programme seriously. He was the leader of the Nazi Left-wing which, to the annoyance of Hitler's industrialist friends, still dreamed of a German Socialism and still won votes for the Party by its anti-capitalist radicalism. But Strasser, if he was much more to the Left than the other Party leaders, was also the head of the Party Organization, more in touch with feeling throughout the local branches than anyone else, and more impressed than any of the other leaders by the setbacks of the autumn. He became convinced that the only course to save the Party from going to pieces was to make a compromise and get into power at once, even as part of a coalition. He saw the Party's chance to influence government policy and carry out at least a part of its programme being sacrificed to Hitler's ambition and his refusal to accept anything less than 'the whole power'.

The day after Schleicher became Chancellor he sent for Gregor Strasser and made an offer to the Nazis. Having failed to get Hitler to discuss a deal, Schleicher suggested that Strasser himself should enter his Cabinet as Vice-Chancellor and Minister-President of the Prussian State Government. If he accepted, Strasser could take over Schleicher's plans for dealing with unemployment and help to establish cooperation with the trade unions. The offer to Strasser was a clever move on Schleicher's part. Not only was it attractive to Strasser as a way out of the Party's difficulties, but it would almost certainly split the Party leadership. If Hitler stood out, Strasser might agree to come into the Cabinet on his own responsibility, and carry his following out of the Party.

On 5 December, at a conference of the Party leaders, Strasser again found support from Frick, the leader of the Nazi group in the Reichstag. Göring and Goebbels, however, were hotly opposed and carried Hitler with them. Hitler laid down terms for discussion with Schleicher and deliberately excluded Strasser from negotiating with the Chancellor. On 7 December Hitler and Strasser had a further conversation, in the course of which Hitler bitterly accused Strasser of bad faith, of trying to go behind his back and oust him from the leadership of the Party. Strasser angrily retorted that he had been entirely loyal. He went back to his room and wrote Hitler a long letter in which he resigned from his position in the Party. He reviewed the whole course of their relationship since 1925, attacked the irresponsibility and inconsistency of Hitler's tactics, and prophesied disaster if he persisted in them.

The threat to his own authority in the Party touched Hitler more closely than the loss of votes or the failure of negotiations had ever done, and there is no doubt that he was shaken by Strasser's revolt. Goebbels wrote in his diary: 'We are all rather downcast, in view of the danger of the whole Party falling to pieces. . . . For hours the Leader paces up and down the room in the hotel. Suddenly he stops and says: "If the Party once falls to pieces, I shall shoot myself without more ado!" '

But Strasser had always lacked the toughness to challenge Hitler outright. Now, instead of rallying the latent opposition to Hitler in the Party, he cursed the whole business and took his family off for a holiday in Italy.

Strasser's disappearance gave Hitler time to recover his confidence and quell any signs of mutiny. A declaration condemning Strasser in the sharpest terms was submitted to a full meeting of the Party leaders and Gauleiters on 9 December. Hitler used all his skill to appeal to the loyalty of his old comrades and brought tears to their eyes. With a sob in his voice he declared that he would never have believed Strasser guilty of such treachery. At the end of this emotional *tour de force* 'the Gauleiters and Deputies,' Goebbels records, 'burst into a spontaneous ovation for the leader. All shake hands with him, promising to carry on until the very end and not to renounce the great Idea, come what may. . . . The feeling that the whole Party is standing by him with a loyalty never hitherto displayed has raised his spirits and invigorated him.'

Schleicher continued his talks with the other Party leaders, including representatives of the trade unions. The failure to bring in the Nazis at this stage did not unduly depress him. On 15 December he expounded his program for tackling inflation, agricultural difficulties, and unemployment. But Schleicher failed to overcome the distrust and hostility of the Social Democrats and the trade unions, or even of the Centre, which, remembering his part in the overthrow of Brüning, was not converted to his support by his advocacy of a policy not unlike Brüning's own. At the same time, the industrialists disliked his conciliatory attitude towards labour; the farmers were furious at his reduction of agricultural protection; the East Elbian landowners denounced his plans for land settlement as 'agrarian Bolshevism'.

Secure in his belief that his enemies were unable to combine against him, Schleicher made the great mistake of underestimating the forces opposed to him. So far as the Nazis were concerned there were good grounds for believing them to be a declining force. Their most immediate problem was shortage of funds. The Nazi organization was highly expensive to run. The Party was filled with thousands of officials who kept their places on the Party pay-roll often without clearly defined functions, and the S.A., the hard core of which consisted of unemployed men who lived in S.A. messes and barracks, must have cost immense sums.

More serious was the sense of defeatism and demoralization in the Party. At the end of 1932, two and a half years after the first great election campaign, Goebbels wrote in his diary: 'This year has brought us eternal ill-luck. . . . The past was sad, and the future looks dark and gloomy; all chances and hopes have quite disappeared.'

Suddenly, at the turn of the year, Hitler's luck changed, and a chance offered itself. The varied antagonisms which Schleicher had aroused found a common broker in the unexpected figure of Franz von Papen, and on 4 January Papen and Hitler met quietly in the house of a Cologne banker, Kurt von Schröder. The meeting was arranged through Wilhelm Keppler, one of the Nazi 'contact-men' with the world of business and industry.

First, misunderstandings had to be removed. Papen slipped out of the responsibility for Hitler's humiliation by putting all the blame on Schleicher for Hindenburg's refusal to consider Hitler as Chancellor. But what Papen had really come to talk about was the prospect of replacing Schleicher's Government: he suggested the establishment of a Nationalist and Nazi coalition in which he and Hitler would be joint Chancellors. Hitler said he would have to be head of the Government, but Papen's supporters could be ministers if they agreed to go along with such changes as the elimination of the Social Democrats, Communists, and Jews from leading positions in Germany. Schröder reports that 'Papen and Hitler reached agreement in principle so that many of the points which had brought them in conflict could be eliminated and they could find a way to get together.'

Next day, to the embarrassment of both the participants, the meeting was headline news in the Berlin papers, and awkward explanations had to be given.

Papen denied that the meeting was in any way directed against Schleicher.

It is certainly wrong to suppose that the Hitler-Papen Government, which was to replace Schleicher, was agreed upon at Cologne; much hard bargaining lay ahead, and Schleicher's position had still to be more thoroughly undermined. But the first contact had been made; the two men had found common ground in their dislike of Schleicher and their desire to be revenged on him, each had sounded out the other's willingness for a deal. Hitler, moreover, received the valuable information that Schleicher had not been given the power to dissolve the Reichstag by the President. With Papen's blessing and Schröder's help, arrangements were now made to pay the Nazis' debts. Schröder was also one of a group of industrialists and bankers who, in November 1932, sent a joint letter to Hindenburg urging him to give Hitler the powers to form a presidential cabinet.

The Nazis could do little to help forward the intrigue against Schleicher—that had to be left to Papen—but it was important to remove the impression of their declining strength. For this purpose Hitler decided to concentrate all the Party's resources on winning the elections in the tiny state of Lippe. When the Nazis were rewarded by an electoral victory, so loud was the noise made by the Nazi propaganda band that, even against their own better judgement, the group round the President were impressed.

The Nazis then proceeded to follow their success at Lippe by staging a mass demonstration in front of the Communist headquarters in Berlin. 'We shall stake everything on one throw to win back the streets of Berlin,' Goebbels wrote. The Government, after some hesitation, banned the Communists' counter-demonstration, and on 22 January, with a full escort of armed police, ten thousand S.A. men paraded on the Bülowplatz and listened to a ranting speech by Hitler. Goebbels exulted, 'The Communists have suffered a great defeat. . . . This day is a proud and heroic victory for the S.A. and the Party.'

By 20 January it was clear that Schleicher's attempt to construct a broad front representing all but the extremist parties had failed. One after another all the German Party leaders turned down Schleicher's approaches. The Nationalists had been alienated by the Chancellor's schemes for land colonization and by the threat to publish a secret Reichstag report on the scandals of the *Osthilfe*, the 'loans' which successive governments had made available to distressed landowners in the eastern provinces. They finally broke with Schleicher on 21 January and turned to the Nazis. Hitler had already seen Hugenberg, the Nationalist leader, on the 17th, and the final stage of negotiations for a Nazi-Nationalist Coalition opened on the evening of the 22nd in Joachim von Ribbentrop's house at Dahlem.

At the meeting Papen, Meissner, and the President's son, Oskar von Hindenburg, faced Hitler, Göring, the Nazis' principal negotiator, and Frick. One important gain Hitler made that night was to win over Oskar von Hindenburg, with whom he had a private conversation of an hour. It is believed that Hitler secured his support by a mixture of bribes and blackmail, possibly threatening to start proceedings to impeach the President and to disclose Oskar's part in the *Osthilfe* scandals and tax evasion on the presidential estate.

On the 23rd, Schleicher went to see the President. His hopes of splitting the Nazi Party had been frustrated; he admitted that he could not find a parliamentary majority and he asked for power to dissolve the Reichstag and govern by emergency decree. Hindenburg refused. Ironically, Schleicher had reached the same position as Papen at the beginning of December, when he had forced Papen out because the latter wanted to fight Hitler, and had himself urged the need to form a government which would have the support of the National Socialists. The positions were exactly reversed, for it was now Papen who was able to offer the President the alternative which Schleicher had advocated in December, the formation of a government with a parliamentary majority in which the Nazi leader would himself take a responsible position. With the knowledge that this alternative was now being prepared behind Schleicher's back, the President again refused his request on 28 January for power to dissolve the Reichstag, and left the Chancellor with no option but to resign. At noon the same day, Hindenburg officially entrusted Papen with the negotiations to provide a new government.

It was still uncertain whether it would be possible to bring Hitler and Hugenberg into the same coalition. Eager at any cost to prevent a Papen Chancellorship, and still convinced that the only practical course was to bring Hitler into the Government, Schleicher sent the Commander-in-Chief of the Army, General von Hammerstein, to warn Hitler that they might still both be left out in the cold by Papen. In that case Schleicher put forward the suggestion of a Hitler-Schleicher coalition to rule with the united support of the Army and the Nazis. Hitler, however, who was still hoping to hear that agreement had been reached for a full coalition between Papen, Hugenberg, and himself, returned a non-committal reply.

Much more alarming to Hitler was the possibility that the Army, under the leadership of Schleicher and Hammerstein, might at the last moment prevent the formation of the proposed coalition.

The keys to the attitude of the Army were held by the President and by General Werner von Blomberg. Hindenburg had agreed to the formation of a Ministry in which Hitler was to be Chancellor and had nominated Blomberg to serve as Minister of Defence under Hitler. If Blomberg, who had been in touch with Hitler, accepted the President's commission, Hitler could be virtually sure of the Army. Blomberg had recently been serving as chief military adviser to the German delegation at the Disarmament Conference and had been hurriedly recalled without Schleicher's or Hammerstein's knowledge. Fortunately for Hitler, Blomberg accepted his new commission from the President, and the threat of a last-minute repudiation by the Army was thereby avoided.

It is possible that fear of what Schleicher might do helped Papen and Hugenberg to make up their minds and hastily compose their remaining differences with the Nazis. At any rate, on the morning of Monday the 30th, after a sleepless night during which he sat up with Göring and Goebbels to be ready for any eventuality, Hitler received the long-awaited summons to the President. The deal which Schleicher had made the object of his policy, and for which Strasser had worked, was accomplished at last, with Schleicher and Strasser left out.

During the morning a silent crowd filled the street between the Kaiserhof and the Chancellery. Shortly after noon a roar went up from the crowd: the Leader was coming. He ran down the steps to his car and in a couple of minutes was back in the Kaiserhof. As he entered the room his lieutenants crowded to greet him. The improbable had happened: Adolf Hitler, the petty official's son from Austria, the down-and-out of the Home for Men, the *Meldegänger*, of the List Regiment, had become Chancellor of the German Reich.

BOOK II

CHANCELLOR
1933–9

CHAPTER FIVE

REVOLUTION AFTER POWER
30 January 1933–August 1934

Nazi propaganda later built up a legend which represented Hitler's coming to power as the upsurge of a great national revival. The truth is more prosaic. Despite the mass support he had won, Hitler came to office in 1933 as the result, not of any irresistible revolutionary or national movement sweeping him into power, nor even of a popular victory at the polls, but as part of a shoddy political deal with the 'Old Gang' whom he had been attacking for months past. Hitler did not seize power; he was jobbed into office by a backstairs intrigue.

Far from being inevitable, Hitler's success owed much to luck and even more to the bad judgement of his political opponents. As Hitler freely admitted afterwards, the Party's fortunes were at their lowest ebb when the unexpected intervention of Papen offered them a chance they could scarcely have foreseen.

Before he came to power Hitler never succeeded in winning more than 37 per cent of the votes in a free election. Had the remaining 63 per cent of the German people been united in their opposition he could never have hoped to become Chancellor by legal means; he would have been forced to choose between taking the risks of a seizure of power by force or the continued frustration of his ambitions. He was saved from this awkward dilemma by two factors: the divisions and ineffectiveness of those who opposed him, and the willingness of the German Right to accept him as a partner in government.

The inability of the German parties to combine in support of the Republic had bedevilled German politics since 1930, when Brüning had found it no longer possible to secure a stable majority in the Reichstag or at the elections. The Communists openly announced that they would prefer to see the Nazis in power rather than lift a finger to save the Republic. Despite the violence of the clashes on the streets, the Communist leaders followed a policy approved by Moscow which gave priority to the elimination of the Social Democrats as the rival working-class party.

The Social Democrats, though more alive to the Nazi threat, had long since become a conservative trade-union party without a single leader capable of organizing a successful opposition to the Nazis. Though loyal to the Republic, since 1930 they had been on the defensive, had been badly shaken by the Depression and were hamstrung by the Communists' attacks.

The Catholic Centre maintained its voting strength to the end, but it was notoriously a Party whose first concern was to make an accommodation with any government in power in order to secure the protection of its particular interests. In 1932–3 the Centre Party was so far from recognizing the danger of a Nazi dictatorship that it continued negotiations for a coalition with the Nazis and voted for the Enabling Law which conferred overriding powers on Hitler after he had become Chancellor.

In the 1930s there was no strong middle-class Liberal Party in Germany. The middle-class parties which might have played such a role—the People's

Party and the Democrats—had suffered a more severe loss of votes to the Nazis than any other German parties, and this is sufficient comment on the opposition they were likely to offer.

But the heaviest responsibility of all rests on the German Right, who not only failed to combine with the other parties in defence of the Republic but made Hitler their partner in a coalition government. The old ruling class of Imperial Germany had never reconciled itself to the loss of the war or to the overthrow of the monarchy in 1918. They were remarkably well treated by the Republican régime which followed. Many of them were left in positions of power and influence; their wealth and estates remained untouched by expropriation or nationalization; the Army leaders were allowed to maintain their independent position; the industrialists and business men made big profits out of a weak and complaisant government, while the help given to the Junkers' estates was one of the financial scandals of the century. All this won neither their gratitude nor their loyalty. Whatever may be said of individuals, as a class they remained irreconcilable, contemptuous of and hostile to the régime they continued to exploit. What the German Right wanted was to regain its old position in Germany as the ruling class; to destroy the hated Republic and restore the monarchy; to put the working classes 'in their places'; to rebuild the military power of Germany; to reverse the decision of 1918 and to restore Germany—their Germany—to a dominant position in Europe. Blinded by interest and prejudice, the Right forsook the role of a true conservatism, abandoned its own traditions and made the gross mistake of supposing that in Hitler they had found a man who would enable them to achieve their ends. A large section of the German middle class, powerfully attracted by Hitler's nationalism, and many of the German Officer Corps followed their lead.

The formation of a Nazi-Right coalition was based

on the belief that Hitler, once he had been brought into the government, could be held in check and tamed. At first sight the terms to which Hitler had agreed appeared to confirm this belief.

He was not even a presidential chancellor; Hindenburg had been persuaded to accept Hitler on the grounds that this time he would be able to provide a parliamentary majority. No sooner was the Cabinet formed than Hitler started negotiations to bring the Centre Party into the coalition. When these negotiations did not lead to agreement Hitler insisted that new elections must be held in order to provide a parliamentary basis for the coalition in the form of an electoral majority.

Papen saw nothing but cause for self-congratulation on his own astuteness. He had levelled scores with General von Schleicher, yet at the same time realized Schleicher's dream, the harnessing of the Nazis to the support of the State—and this, not on Hitler's, but on his own terms. For Hitler, Papen assured his friends, was tied hand and foot by the conditions he had accepted. True, Hitler had the Chancellorship, but the real power, in Papen's view, rested with the Vice-Chancellor, himself.

It was the Vice-Chancellor, not the Chancellor, who enjoyed the special confidence of the President; it was the Vice-Chancellor who held the key post of Minister-President of Prussia, with control of the Prussian administration and police; and the Vice-Chancellor who had the right, newly established, to be present on all occasions when the Chancellor made his report to the President. Furthermore, only three of the eleven Cabinet posts were held by Nazis, and apart from the Chancellorship both were second-rate positions.

With these arguments Papen overcame Hindenburg's reluctance to make Hitler Chancellor. In this way they would obtain that mass support which the 'Cabinet of Barons' had so notoriously lacked. Hitler was to play his old role of barker for a circus-show in which he was now to have a place as partner and his name at the top of the bill, but in which the real decisions would be taken by those who outnumbered him in the Cabinet. This was *Realpolitik* as practised by Papen, a man who—as he prided himself—knew how to distinguish between the reality and the shows of power.

Rarely has disillusionment been so complete or so swift to follow. Those who, like Papen, believed they had seen through Hitler were to find they had badly underestimated both the leader and the movement. For Hitler's originality lay in his realization that effective revolutions, in modern conditions, are carried out with, and not against, the power of the State: the correct order of events was first to secure access to that power and then begin his revolution. Hitler never abandoned the cloak of legality; instead he turned the law inside out and made illegality legal.

In six months, Hitler and his supporters were to demonstrate a cynicism and lack of scruple—qualities on which his partners particularly prided themselves—which left Papen and Hugenberg gasping for breath. At the end of those six months they were to discover, like the young lady of Riga, the dangers of going for a ride on a tiger.

At five o'clock on the afternoon on Monday 30 January, Hitler presided over his first Cabinet meeting. The Cabinet was still committed to seeking a parliamentary majority by securing the support of the Centre Party, and Göring duly reported on the progress of his talks with the leader of the Centre, Monsignor Kaas. If these failed, then, Hitler suggested, it would be necessary to dissolve the Reichstag and hold new elections. Hugenberg saw the danger of letting Hitler conduct an election campaign with the power of the State at his command, but he reluctantly agreed to the plans after Hitler solemnly promised him that the composition of the coalition government would not be altered, whatever the results of the elections.

The next day, when Hitler saw Monsignor Kaas, he took good care that the negotiations with the Centre should fail. On the advice of Papen, Hindenburg agreed once more to sign a decree dissolving the Reichstag 'since the formation of a working majority has proved impossible'. The Centre Party protested to the President that this was not true, but the decree had been signed, the date for the new elections fixed and the first and most difficult of the obstacles to Hitler's success removed. Papen and Hugenberg had allowed themselves to be gently guided into the trap. Goebbels wrote confidently in his diary, 'The struggle is a light one now, since we are able to employ all the means of the State. Radio and Press are at our disposal. We shall achieve a masterpiece of propaganda. Even money is not lacking this time.'

In order to leave no doubts of the expectations they had, Göring summoned about twenty-five of Germany's leading industrialists to his palace on the evening of 20 February. After Hitler spoke, Göring bluntly appealed to them: 'The sacrifice asked for is easier to bear if it is realized that the elections will certainly be the last for the next ten years, probably even for the next hundred years.' It was agreed to raise an election fund of three million Reichsmarks to be divided between the partners in the coalition, but there was little doubt that the Nazis would get the lion's share.

Throughout the election campaign Hitler refused to outline any programme for his Government. At Munich he said: 'Programmes are of no avail, it is the human purpose which is decisive.' The Nazi campaign was directed against the record of the fourteen years of party government in Germany, which had 'piled mistake upon mistake, illusion upon illusion.' What had the Nazis to put in its place? He was no democratic politician, Hitler virtuously replied, to trick the people into voting for him by a few empty promises. 'I ask of you, German people, that after you have given the others fourteen years you should give us four.'

Hitler did not rely on the spoken word alone. Although the other parties were still allowed to function, their meetings were broken up, their speakers assaulted and beaten, their posters torn down and their papers continually suppressed. Official figures admitted fifty-one people killed during the election campaign and several hundreds injured. This time the Nazis did not mean to be robbed of power by any scruples about fair play or free speech.

Papen believed he had tied Hitler down by restricting the number of Cabinet posts held by the Nazis to a bare minimum, but the real key to power in the State—control of the Prussian police force and of the Prussian State Administration—lay with Göring, Prussian Minister of the Interior. By the curious system of dual government which existed in Germany, this office carried out the work of administering two-thirds of Germany. In the critical period of 1933–4, no man after Hitler played so important a role in the Nazi revolution as Göring. His energy and ruthlessness, together with his control of Prussia, were indispensable to Hitler's success. The belief that Göring would be restrained by Papen as Minister-President of Prussia proved ill-founded. Göring issued orders and enforced his will, as if he were already in possession of absolute power.

The moment Göring entered office he began a drastic purge of the Prussian State service. He paid particular attention to sweeping out the senior police officers, replacing them with S.A. or S.S. leaders. In February, after urging the police to show no mercy to the activities of 'organizations hostile to the State'—that is to say, the Communists—Göring continued: 'Police officers who make use of fire-arms in the execution of their duties will, without regard to the consequences of such use, benefit by my protection; those who . . . fail in their duty will be punished in accordance with the regulations.' In other words, when in doubt shoot.

On 22 February Göring published an order establishing an auxiliary police force. Fifty thousand men were called up, among them twenty-five thousand from the S.A. and fifteen thousand from the S.S. All they had to do was to put a white arm-band over their brown shirts or black shirts: they then represented the authority of the State. It was the equivalent of handing over police powers to the razor and cosh gangs. This was 'legality' in practice.

The day after Hitler became Chancellor, Goebbels noted in his diary: 'In a conference with the Leader we arrange measures for combating the Red terror.' On 24 February the police raided Communist H.Q. in Berlin. An official communiqué reported the discovery of plans for a Communist revolution. The publication of the captured documents was promised in the immediate future. They never appeared, but on the night of 27 February the Reichstag building mysteriously went up in flames.

While the fire was still spreading, the police arrested a young Dutch Communist, Marianus van der Lubbe, who was found in the deserted building.

Göring at once declared that van der Lubbe was only a pawn in a major Communist plot to launch a campaign of terrorism. The arrest of Communist leaders, including the Bulgarian Georgi Dimitroff, followed at once, and the Reichstag Fire Trial was held in Leipzig with all the publicity the Nazis could contrive. The publicity, however, badly misfired. Not only did Dimitroff defend himself with skill, but the prosecution failed completely to prove any connexion between van der Lubbe and the other defendants. The trial ended in a fiasco with the acquittal and release of the Communist leaders, leaving the unhappy van der Lubbe to be hurriedly executed.

It is still a matter of controversy whether the burning of the Reichstag was, in fact, planned and carried out by the Nazis themselves, as has long been believed, or whether van der Lubbe acted on his own. At any rate, there is no doubt who profited by the fire. The next day Hitler promulgated a decree signed by the President 'for the protection of the People and the State.' The decree was described 'as a defensive measure against Communist acts of violence.' It suspended the guarantees of individual liberty under the Weimar Constitution, authorized the Reich Government if necessary to take over full powers in any federal State, increased the penalty for the crimes of high treason, poisoning, arson, and sabotage to one of death, and instituted the death penalty, or hard labour for life, in the case of conspiracy to assassinate members of the Government, or grave breaches of the peace.

Armed with these all-embracing powers, Hitler and Göring were in a position to take any action they pleased against their opponents. They cleverly postponed the formal proscription of the Communist Party until after the elections, so that the working-class vote should continue to be divided between the rival parties of the Communists and the Social Democrats. But acts of terrorism against the Left-wing parties were now intensified.

In the last week of the election campaign, the Nazi propaganda machine redoubled the force of its attack on the 'Marxists', producing the most hair-raising accounts of Communist preparations for insurrection and a 'blood-bath', for which the Reichstag Fire and the arrest of van der Lubbe were used to provide substantiation. Hitler stormed the country in a last hurricane campaign, declaring his determination to stamp out Marxism and the parties of the Left without mercy. For the first time the radio carried his words into every corner of the country.

The campaign reached its climax on Saturday 4 March, the 'Day of the Awakening Nation', when Hitler spoke in Königsberg, the ancient coronation town and capital of the separated province of East Prussia. Hitler declared: 'One must be able to say once again: German People, hold your heads high and proudly once more! You are no longer enslaved and in bondage, but you are free again and can justly say: We are all proud that through God's powerful aid we have once more become true Germans.'

As Hitler finished speaking bonfires blazed out on the hill-tops, all along the 'threatened frontier' of the east. It was the culmination of a month in which the tramping columns of S.A. troops, the torchlight parades, the monster demonstrations, cheering crowds, blaring loudspeakers, and mob-oratory, the streets hung with swastika flags, the open display of brutality and violence, with the police standing by in silence—all had been used to build up the impression of an irresistible force which would sweep away every obstacle in its path.

In face of all this it is a remarkable fact that still the German people refused to give Hitler the majority he sought. With the help of his Nationalist allies, Hitler had a bare majority in the new Reichstag, but it did not escape the attention of the Nazi leaders that with the proscription of the Communist deputies they would have a clear parliamentary majority themselves, without the need of the Nationalist votes. The chances of Papen, Hugenberg, and the Nationalists acting as an effective brake on their partners in the coalition appeared slight.

Hitler's dictatorship rested on the constitutional foundation of a single law: the so-called Enabling Law, *Gesetz zur Behebung der Not von Volk und Reich* (Law for Removing the Distress of People and

Reich). As it represented an alteration of the Constitution, a majority of two-thirds of the Reichstag was necessary to pass it, and Hitler's first preoccupation after the elections was to secure this. One step was simple: the eighty-one Communist deputies could be left out of account, those who had not been arrested so far would certainly be arrested if they put in an appearance in the Reichstag. Negotiations with the Centre were resumed and, in the meantime, Hitler showed himself in his most conciliatory mood towards his Nationalist partners. Both the discussions in the Cabinet and the negotiations with the Centre revealed the same uneasiness at the prospect of the powers the Government was claiming. But the Nazis held the whip-hand with the decree of 28 February, and Hitler was prepared to promise anything at this stage to get his bill through, with the appearances of legality preserved intact.

The Enabling Bill which was laid before the House on 23 March 1933 gave the Government the power for four years to enact laws without the cooperation of the Reichstag. It specifically stated that this power should include the right to deviate from the Constitution and to conclude treaties with foreign states. It also provided that laws to be enacted by the Government should be drafted by the Chancellor, and should come into effect on the day after publication.

As the deputies arrived, they had to pass through a solid rank of black-shirted S.S. men encircling the building; inside, the corridors and walls were lined with brown-shirted S.A. troops.

Hitler's opening speech was restrained. He spoke of the disciplined and bloodless fashion in which the revolution had been carried out, and of the spirit of national unity which had replaced the party and class divisions of the Republic. The Government, he declared, would make use of the powers of the Enabling Bill only when they were 'essential for carrying out vitally necessary measures.' There was no threat to the Reichstag, the President, the States or the Churches. 'The Government,' he concluded, 'offers to the parties of the Reichstag the opportunity for friendly cooperation. But it is equally prepared to go ahead in face of their refusal and of the hostilities which will result from that refusal. It is for you, gentlemen of the Reichstag, to decide between war and peace.'

The leader of the Social Democrats, Otto Wels, spoke next. There was silence as he walked to the tribune, but from outside came the baying of the Stormtroopers chanting: 'We want the Bill—or fire and murder.' It needed courage to stand up before this packed assembly—most of the Communists and about a dozen of the Social Democrat deputies had already been thrown into prison—and to tell Hitler and the Nazis to their faces that the Social Democratic Party would vote against the Bill. Wels spoke with moderation; to be defenceless, he added, was not to be without honour. But the very suggestion of opposition had been enough to rouse Hitler to a fury. Brushing aside Papen's attempt to restrain him, he mounted the tribune a second time and gave the Reichstag, the Cabinet, and the Diplomatic Corps a taste of his real temper, savage, mocking, and brutal. 'I do not want your votes,' he spat at the Social Democrats. 'Germany will be free, but not through you. Do not mistake us for bourgeois. The star of Germany is in the ascendant, yours is about to disappear, your death-knell has sounded.'

The rest of the speeches were an anti-climax. Monsignor Kaas, still clinging to his belief in Hitler's promises, rose to announce that the Centre Party would vote for the Bill, a fitting close to the shabby policy of compromise with the Nazis which the Centre had followed since the summer of 1932. Then came the vote, and excitement mounted. When Göring declared the figures—for the Bill, 441; against, 94—

the Nazis leaped to their feet and with arms outstretched in salute sang the Horst Wessel song.

Outside in the square the huge crowd roared its approval. The Nazis had every reason to be delighted: with the passage of the Enabling Act, Hitler secured his independence, not only from the Reichstag but also from the President. He had full power to set aside the Constitution. The street gangs had seized control of the resources of a great modern State, the gutter had come to power.

In March 1933, however, Hitler was still not the dictator of Germany. The process of *Gleichschaltung*—'coordination'—by which the whole of the organized life of the nation was to be brought under the single control of the Nazi Party, had still to be carried out. Hitler and Frick had not waited for the passage of the Enabling Act to take steps to bring the governments of the States firmly under their control. On the evening of 9 March von Epp, with full authority from Berlin, carried out a *coup d'état* in Munich. The Held Government was turned out, and Nazis appointed to all the principal posts.

Similar action was taken in the other States. In April, Hitler nominated Reich Governors (Reichsstatthälter) in every State, and gave them the power to appoint and remove State Governments, to dissolve the Diets, to prepare and publish State laws, and to appoint and dismiss State officials. All eighteen of the new Reich Governors were Nazis, usually the local Gauleiters. In Prussia the new law afforded an opportunity to turn out Papen. Hitler now appointed himself Reichsstatthälter for Prussia and promptly delegated his powers to Göring as Prussian Minister-President.

On the first anniversary of Hitler's accession to power, 30 January 1934, a Law for the Reconstruction of the Reich rounded off this work of subordinating the federal States to the authority of the central Government. The State Diets were abolished; the sovereign powers of the States transferred to the Reich; and the Reichsstatthälter and State Governments placed under the Reich Government. This was the culmination of a year of *Gleichschaltung*, in which all representative self-government from the level of the States downwards through the whole system of local government had been stamped out.

The process of *Gleichschaltung* did not stop with the institutions of government. If Hitler meant to destroy Marxism in Germany he had obviously to break the independent power of the huge German trade-union movement, the foundation on which the Social Democratic Party rested. The Nazis cleverly camouflaged their intentions by declaring May Day a national holiday, and holding an immense workers' rally in Berlin which was addressed by Hitler. On the morning of the next day the trade-union offices all over the country were occupied by S.A. and S.S. troopers. Many union officials were arrested, beaten, and thrown into concentration camps. All the unions were then merged into a new German Labour Front, headed by Robert Ley.

Hitler gave assurances to the workers when he addressed the First Congress of German Workers on 10 May. But the intention behind his talk of honouring labour and abolishing the class war were not long concealed. Before the month was out a new law ended collective bargaining and appointed Labour Trustees, under the Government's orders, to settle conditions of work.

The Social Democrats attempted to carry on loyally for a time, but their efforts proved futile. On 10 May Göring ordered the occupation of the Party's buildings and newspaper offices, and the confiscation of their funds, and on 22 June banned the Social Democratic Party as an enemy of the people and State.

The remaining parties represented a more delicate problem, but this did not long delay their disappearance. Not even Hitler's partners in the coalition, the Nationalists, were spared.

On 14 July the Official Gazette contained the brief announcement, signed by Hitler, Frick and Gürtner:

'The German Government has enacted the following law, which is herewith promulgated:

'Article I: The National Socialist German Workers' Party constitutes the only political Party in Germany.

'Article II: Whoever undertakes to maintain the organizational structure of another political Party or to form a new political Party will be punished with penal servitude up to three years or with imprisonment up to three years, if the action is not subject to a greater penalty according to other regulations.'

With the suppression of the parties, the basis of the coalition which had brought Hitler into power disappeared. As so often later in his foreign policy, Hitler resorted to his favourite tactic of surprise, of doing just the things no one believed he would dare to do, with a bland contempt for convention or tradition. In a few weeks he had banned the Communist and Social Democratic Parties, dissolved the Catholic Centre and the Right-wing Nationalists, and taken over the Stahlhelm and the trade unions, six of the most powerful organizations in Germany—and, contrary to all expectations, nothing had happened. With equal success he had ridden rough-shod over the rights of the federal States. The methods of gangsterism applied to politics, the crude and uninhibited use of force in the first, not in the last, resort, produced startling results.

Any opposition in the Cabinet crumpled up before the wave of violence which was eliminating all the political landmarks in the German scene. Papen, shorn of his power as Reich Commissioner in Prussia, was a shrunken figure. Hitler no longer paid attention to the rule that the Vice-Chancellor must always be present when he saw the President; indeed, he rarely bothered to see the President at all, now that he had the power to issue decrees himself.

Thus by the summer of 1933 Hitler was complete master of a Government which was independent alike of Reichstag, President, and political allies. All Papen's calculations of January, his assurance that once the Nazi Party was harnessed to the State it would be tamed, had proved worthless. For Hitler had grasped a truth which eluded Papen, the political dilettante, that the key to power no longer lay in the parliamentary and presidential intrigues by means of which he had got his foot inside the door—and by means of which Papen still hoped to bind him—but, outside, in the masses of the German people. Papen, deceived by Hitler's tactics of legality, had never grasped that the revolutionary character of the Nazi movement would only be revealed after Hitler had come to power, and was now astonished and intimidated by the forces he had released.

For it is a mistake to suppose, as Papen did, that because Hitler came to power by the backstairs there was no genuine revolutionary force in the Nazi Party. The S.A. regarded Hitler's Chancellorship as the signal for that settling of accounts which they had been promised for so long. With the long-drawn-out economic depression and the accompanying political uncertainty and bitterness, the revolutionary impulse of the S.A. was bound to strike echoes in a large section of the German people. This wave of revolutionary excitement which passed across Germany in 1933 took several forms.

Its first and most obvious expression was violence. The violence of the period between the Reichstag Fire and the end of the year was on a different scale from anything that had happened before. The Government itself deliberately employed violence and intimidation as a method of governing, using such agencies as the Gestapo (the Prussian Secret State Police established by Göring), and the concentration camps opened at Oranienburg, Dachau, and other places. At the same time, the open contempt for justice and order shown by the State encouraged normally suppressed impulses of cruelty, envy, and revenge. The traditional sanctions of the police and the courts were withdrawn, and common crime from robbery to murder brazenly disguised as 'politics'. The only measure taken by the Government was to issue amnesties for 'penal acts committed in the national revolution'.

Violence, if it repelled, also attracted many, especially among the younger generation. It was indeed a characteristic part of revolutionary idealism. For 1933, like other revolutionary years, produced great hopes, a sense of new possibilities, the end of frustration, the beginning of action, a feeling of exhilaration and anticipation after years of hopelessness. Hitler succeeded in releasing pent-up energies in the nation, and in re-creating a belief in the future of the German people. It is wrong to lay stress only on the element of coercion, and to ignore the degree to which Hitler commanded a genuine popular support in Germany. To suppose that the huge votes which he secured in his plebiscites were solely, or even principally, due to the Gestapo and the concentration camps is to miss what Hitler knew so well, the immense attraction to the masses of force plus success.

Side by side with this went the familiar and seamy accompaniment of all revolutionary upheavals, the rush to clamber on the band wagon and the scramble for jobs and advantages. The purge of the civil service, the closing of the professions to Jews in 1933, the creation of new posts in government, in industry and business, whetted the appetites of the unsuccessful, the ambitious, and the envious. Most of the men who now held power in Germany and the thousands of Nazis who had become mayors of cities, deputies, government officials, and heads of departments belonged to one of these classes. The six million unemployed in Germany, who had not disappeared overnight when Hitler came to power, represented a revolutionary pressure that was not easily to be damned.

It was by harnessing these forces of discontent and revolt that Hitler had created the Nazi movement, and as late as the middle of June 1933 he was still prepared to tell a gathering of Nazi leaders in Berlin: 'The German Revolution will not be complete until the whole German people has been fashioned anew, until it has been organized anew, and has been reconstructed.'

But there was a point beyond which this process could not go without seriously endangering the efficiency of the State and of the German economy. The two dangers to which Hitler had to pay particular attention were the disruption of the economic organization of the country, and attempts to interfere with the inviolability of the Army.

Men like Gottfried Feder believed that the time had come to put into practice the economic clauses of the Party's original programme, with its sweeping proposals for nationalization, profit-sharing, and the abolition of unearned incomes.

Although Hitler was indifferent to economic questions, he saw that radical economic experiments at such a time would throw the German economy into a state of confusion, and would prejudice, if not destroy, the chances of cooperation with industry and business to end the Depression and bring down the unemployment figures.

Hitler summoned the Gauleiters to Berlin on 13 July and told them: 'Political power we had to conquer rapidly and with one blow; in the economic sphere other principles of development must determine our action. Here progress must be made step by step without any radical breaking up of existing conditions which would endanger the foundations of our own life. . . .'

July 1933 marked a turning point in the development of the revolution. Hitler now spoke of a new phase which he described as 'educating the millions who do not yet in their hearts belong to us'.

Hitler's own wish to bring the revolution to an end, for the time being at least, and to consolidate its gains, is plain enough. However, he was far from convincing all his followers of the necessity of his new policy. Röhm, the S.A. Chief of Staff, spoke in the name of the hundreds of thousands of embittered Nazis who had been left out in the cold, and wanted no end to the revolution until they too had been provided for. At the beginning of August Göring, in line with the change of policy, announced the dismissal of the S.A. and S.S. auxiliary police. On 6 August, before a parade of eighty thousand S.A. men, Röhm gave his answer: 'Anyone who thinks that the tasks of the S.A. have been accomplished will have to get used to the idea that we are here and intend to stay here, come what may.'

From the summer of 1933 to the summer of 1934 this quarrel over the Second Revolution was to form the dominant issue in German politics. Demands to renew and extend the Revolution grew louder and more menacing. Röhm, Goebbels, and many of the S.A. leaders made open attacks on *Reaktion*, that comprehensive word which covered everyone the S.A. disliked, from capitalists and Junkers, Conservative politicians and stiff-necked generals, to the respectable bourgeois citizen with a job and the civil service bureaucrats. The S.A. looked back nostalgically to the spring of the previous year, when the gates to the Promised Land had been flung open, and Germany had appeared to be theirs to loot and lord it over as they pleased. Now, they grumbled, the Nazis had gone respectable, and many who had secured a Party card only the day before were allowed to continue with their jobs, while deserving *Alte Kämpfer* were left out on the streets. In characteristically elegant language the S.A. began to talk of clearing out the pigsty, and driving a few of the greedy swine away from the troughs.

While the S.A., which was a genuine mass movement with strong radical and anti-capitalist leanings, became restive, Hitler was as strongly opposed as ever to Röhm's inveterate desire to turn the S.A. into soldiers and to remodel the Army. There were particularly strong reasons why he wished to avoid alienating the Army leaders at this time. The willingness of the Army to see Hitler become Chancellor and the benevolent neutrality of the Army while he arrogated more and more power to himself were decisive factors in the establishment of the Nazi régime. The key figure in guaranteeing the friendly attitude of the Army was General von Blomberg, but Hitler also made a powerful appeal to the Army by two promises: German military strength would be restored by rearmament, and the Army would not be called upon to intervene in a civil war.

Hitler's relations with Blomberg became closer as he began to take the first steps in rebuilding the military power of Germany. Looking ahead, he recognized the importance of having the Army again on his side, if he was to succeed Hindenburg, and he was anxious that nothing should disturb the confidence of the Army leaders in the new régime.

Röhm took a different view. By the end of 1933 the S.A. numbered between two and three million men, and Röhm stood at the head of Laforce more than ten or twenty times the size of the regular Reichswehr. The S.A. leaders, ambitious and hungry for power, saw in their organization the revolutionary army which should provide the military power of the New Germany. They were contemptuous of the rigid military hierarchy of the professional Army and were avid for the prestige, power, and pickings they would acquire by supplanting the generals. Their motives were as crude as their manners, but undeniably a number of them were tough, possessed ability, and commanded powerful forces. The Army generals, however, were adamant in their refusal to accept the S.A. on an equal footing.

Hitler went out of his way to reassure the generals that he remained loyal to them, but the problem of the S.A. remained. If it was not to be incorporated into the Army, as Röhm wanted, what was to become of it? The S.A. was the embarrassing legacy of the years of struggle. In it were collected the 'old fighters' who had been useful enough for street brawling, but for whom the Party had no further use when it came to power and took over the State; the disillusioned radicals, resentful at Hitler's compromise with existing institutions; the ambitious, who had failed to get the jobs they wanted, and the unsuccessful, who had no jobs at all.

Thus relations between the S.A. and the Army became a test case involving the whole question of the so-called Second Revolution—the point at which the revolution was to be halted—and the classic problem of all revolutionary leaders once they have come to power, the liquidation of the Party's disreputable past.

Hitler first attempted to solve this problem by conciliation and compromise. A law promulgated on 1 December made Röhm and Hess members of the Reich Cabinet. Röhm's appointment repaired an omission which had long been a grievance with the S.A. At the beginning of the New Year Hitler addressed a letter to Röhm of unusual friendliness, thanking him for his 'imperishable services' and assuring him of his friendship. More attention was now paid to the needs and grievances of the 'old fighters', and members of the Party or S.A. who had suffered sickness or injury in the political struggle for the national movement were granted State pensions.

Röhm, however, was not to be silenced by such sops. In February he proposed in the Cabinet that the S.A. should be used as the basis for the expansion of the Army, and that a single Minister should be appointed to take charge of the Armed Forces of the State. The obvious candidate for such a post was Röhm himself. The Army High Command presented a unanimous opposition to such a proposal and appealed to the President, as the guardian of the Army's traditions, to put a stop to Röhm's attempted interference.

Hitler declined to take Röhm's side in the dispute, and the plan was allowed to drop for the moment. Although temporarily checked, Röhm kept up his pressure on the Army, and his relations with General von Blomberg, the Minister of Defence, grew strained.

At the end of March Hitler and von Blomberg were secretly informed that President Hindenburg could not be expected to live very much longer. Within a matter of months, perhaps of weeks, the question of the succession would have to be settled. It had long been the hope of conservative circles that Hindenburg's death would be followed by a restoration of the monarchy, and this was the President's own wish. Hitler, however, was now opposed both to a restoration and to the perpetuation of the existing situation. So long as the independent position of the President existed alongside his own, so long as the President was Commander-in-Chief of the Armed Forces, and so long as the oath of allegiance was taken to the President and not to himself, Hitler's power was something less than absolute. He was determined that, when Hindenburg died, he and no one else should succeed to the President's position. It was to Adolf Hitler that the Armed Forces should take the new oath of allegiance. The first and most important step was to make sure of the Army, whose leaders claimed to represent the permanent interests of the nation independently of the rise and fall of governments and parties.

In the second week of April an opportunity presented itself. On 11 April Hitler left Kiel on the cruiser *Deutschland* to take part in naval manoeuvres. He was accompanied by General von Blomberg, Colonel General Freiherr von Fritsch, the Commander-in-Chief of the German Army, and Admiral Erich Raeder, the Commander-in-Chief of the German Navy. It is believed to have been during the course of this short voyage that Hitler came to terms with the generals: the succession for himself, in return for the suppression of Röhm's plans and the continued inviolability of the Army's position as the sole armed force in the State. A conference of senior Army officers under Fritsch's chairmanship endorsed Blomberg's decision in favour of Hitler after—but only after—the terms of the *Deutschland* Pact had been communicated to them.

The news of Hitler's offer to cut down the number of the S.A., which leaked out and was published in Prague, sharpened the conflict between Röhm and the Army. Röhm had powerful enemies inside the Party as well as in the Army. Göring, who had been made a general by Hindenburg, once in power gravitated naturally towards the side of privilege and authority, and was on the worst of terms with the Chief of Staff of the S.A. He began to collect a powerful police force 'for special service', which he kept ready under his own hand at the Lichterfelde Cadet School near Berlin. On 1 April 1934, Heinrich Himmler, already head of the Bavarian police and Reichsführer of the black-shirted S.S., was unexpectedly appointed by Göring as head of the Prussian Gestapo. Göring found in Himmler an ally against a common enemy, for the first obstacle Himmler sought to remove from his path was Ernst Röhm. Himmler and his S.S. were still a part of the S.A. and subordinate to Röhm's command, although the rivalry between the S.A. and S.S. was bitter, and Röhm's relationship with Himmler could hardly have been less cordial. Röhm's only friends in the Party leadership were Goebbels and—paradoxically enough—the man who had him murdered, Hitler.

The situation with which Hitler had to deal was produced by the intersection of the problems of the Second Revolution, of the S.A. and the Army, and of the succession to President von Hindenburg. It was the problem of the succession which introduced a note of urgency by making Hitler's own position vulnerable, for if he was to secure their support for his succession to the Presidency, the Army leadership was determined to exact in return the removal of the S.A. threat to take over the Army and renew the revolution.

Hitler chose to stand by his agreement with the Army and repudiate the Revolution, but he disguised his decision as action forced on him not by pressure from the Right, but by disloyalty and conspiracy on the Left.

On 4 June Hitler sent for Röhm and had a conversation with him which lasted for five hours. According to Hitler's later account, he warned Röhm against any attempt to start a Second Revolution. At the same time as he assured Röhm that he had no intention of dissolving the S.A., Hitler reproached him with the scandal created by his own behaviour and that of his closest associates in the S.A. leadership, who had become notorious for their disorder, luxury and sexual perversion. A day or two later Hitler ordered the S.A. to go on leave for the month of July, returning to duty on 1 August. During their leave they were forbidden to wear their uniforms or to take part in any demonstrations or exercises. Röhm announced on 7 June that he himself was about to take a period of sick leave. Hitler agreed to attend a conference of S.A. leaders to discuss the future of the movement at Wiessee, near Munich, on 30 June. It was a rendezvous which Hitler did not fail to keep.

On 13 July Hitler gave his version of what happened between 8 June and 30 June. According to his speech, Röhm had renewed his old relations with General von Schleicher, and the two men had agreed on a concrete programme: the present régime in Germany could not be supported; the Army and all national associations must be united in a single band under Röhm; Papen must be removed, and Schleicher himself become Vice-Chancellor.

Since Röhm was not sure that Hitler would agree, he made preparations to carry out his plan by a *coup*, the main role in which was to be played by the S.A. Staff Guards. To complete the conspiracy, Hitler continued, Schleicher and his former assistant in the Defence Ministry, General von Bredow, got in touch with 'a foreign Power' (later identified as France), and Gregor Strasser, who had retired into private life after Hitler's Chancellorship, was brought into the plot.

After his talk with Hitler on 4 June, Röhm—still according to Hitler's version—planned to take Hitler captive, hoping to use his authority to call out the S.A. and paralyse the other forces in the State. The action taken at the end of June was directed, Hitler claimed, to forestalling Röhm's putsch which was about to be staged in a matter of hours.

Part of this story can with some certainty be rejected as untrue from the beginning. On the very day he was supposed to be storming the Chancellery in Berlin, Röhm was seized in bed at the hotel in

Wiessee where he was taking a cure and awaiting Hitler's arrival for the conference they had arranged. Most of the other S.A. leaders were either on their way to Wiessee or had actually arrived. Karl Ernst, the S.A. leader in Berlin (whom Hitler represented as one of the most important figures in the plot), was taken prisoner at Bremen, where he was about to leave by boat for a honeymoon in Madeira. The whole story of an imminent *coup d'état* was a lie, either invented later by Hitler as a pretext for his own action, or possibly made use of at the time by Göring and Himmler to deceive Hitler and force him to move against Röhm. There is no evidence to support the claim that the conspirators of June 1934 were Röhm and the S.A. The plot was hatched by Göring and Himmler, the enemies of Röhm; the treachery and disloyalty were not on Röhm's side, but on theirs and Hitler's; and if ever men died convinced—not without reason—that they had been 'framed', it was the men who were shot on 30 June 1934.

Throughout June 1934 there was an ominous tension in Berlin, heightened by rumours and much speculation. At the end of May both Brüning and Schleicher were warned that, in the event of a purge, their lives were in danger. Brüning took the advice seriously and left for Switzerland; Schleicher went no farther than the Starnbergersee, and returned in time to be shot.

On 14 June Hitler flew to Venice for the first of many celebrated conversations with Mussolini. Mussolini, at the height of his reputation and resplendent with uniform and dagger, patronized the worried Hitler, who appeared in a raincoat and a soft hat. Mussolini was not only pressing on the subject of Austria, where Nazi intrigues were to lead to trouble before the summer was out, but frank in his comments on the internal situation in Germany. He advised Hitler to put the Left wing of the Party under restraint, and Hitler returned from Venice depressed and irritable.

No part is more difficult to trace in this confused story than that played by Gregor Strasser—if indeed he played any part at all other than that of victim. Hitler had apparently renewed touch with Strasser earlier in the year. According to Otto Strasser, Hitler offered Gregor the Ministry of National Economy before Hitler left for Venice, and about the same time, Goebbels had been seeing Röhm secretly.

These attempts to keep in touch with Strasser, the one-time leader of the Left wing of the Party, and with Röhm, the leader of the S.A., in which radicalism was endemic, were evidently related to a conflict still going on in Hitler's mind. It is easy to believe that he was eager to avoid dealing a heavy blow to the Party, by delaying action in the hope that Hindenburg might die suddenly, or that in some other way the crisis could be solved without irrevocable decisions.

At this stage Hitler was given a sharp reminder of the realities of the situation from an unexpected quarter. Papen had dropped into the background since the spring of 1933, but he remained Vice-Chancellor and still enjoyed the special confidence of the old President. The divisions within the Party offered him a chance of re-asserting his influence, and for the last time he made use of his credit with the President to stage a public protest against the recent course, and, even more, against the prospective course, of events in Germany. The protest was made in the course of a speech at the University of Marburg on 17 June and crystallized the anxieties and uncertainties of the whole nation. The Vice-Chancellor talked about the Second Revolution, warning, 'Whoever toys irresponsibly with such ideas should not forget that a second wave of revolution might be followed by a third, and that he who threatens to employ the guillotine may be its first victim. . . . At some time the movement must come to a stop and a solid social structure arise. . . .'

Papen also discussed the shortcomings of Nazi propaganda, concluding, 'It is time to join together in fraternal friendship and respect for all our fellow countrymen, to avoid disturbing the labours of serious men and to silence fanatics.'

Hitler immediately made scathing references to 'the pygmy who imagines he can stop with a few phrases the gigantic renewal of a people's life'. But Papen's protest was not so easily brushed aside. Goebbels took immediate steps to ban its publication, but copies were smuggled out of Germany and published abroad, creating a sensation which did not fail to penetrate to Germany. When Papen appeared in public at Hamburg on 24 June he was loudly cheered. It was evident that he had spoken for a great part of the nation.

On 20 June Papen went to see Hitler and demand the removal of the ban on publishing his speech. In a stormy interview Papen threatened his own and the resignation of the other conservative ministers in the Cabinet. Hitler saw quite clearly that he was face to face with a major crisis. When he flew to Neudeck on 21 June to see the ailing President, he was met by General von Blomberg, with an uncompromising message: either the Government must bring about a relaxation of the state of tension or the President would declare martial law and hand over power to the Army. Hitler was allowed to see the President only for a few minutes, but the interview, brief though it was, sufficed to confirm Blomberg's message. The Army was claiming the fulfilment of its bargain, and by now Hitler must have realized that more was at stake than the succession to the Presidency: the future of the whole régime was involved.

It is impossible to penetrate Hitler's state of mind

in the last week of June. Obviously he must have been aware of the preparations which were now rapidly put in hand and have agreed to them at least tacitly, yet to the very last day he seems to have hesitated to take the final step. At this stage it was not Hitler but Göring and Himmler who gave the orders and prepared to eliminate their rivals in the Party leadership. In the background the Army made its own arrangements. On 25 June the Commander-in-Chief, General von Fritsch, placed the Army in a state of alert, ordering all leave to be cancelled and the troops to be confined to barracks. On 28 June the German Officers' League expelled Röhm, and on 29 June the *Völkischer Beobachter* carried a signed article by General von Blomberg, which left no doubt that the coming purge had the Army's blessing.

While Hitler was conveniently out of town on the 28th, Göring and Himmler ordered their police commandos and S.S. to hold themselves in readiness.

Far away from the tension and rumours of Berlin, on the shores of the Tegernsee, Röhm continued to enjoy his sick leave with his usual circle of young men. So little was he aware of what was being planned that he had left his Staff Guards in Munich. His carelessness and confidence are astonishing. Yet, even in Berlin, the local S.A. leader, Karl Ernst, who was uneasily aware of something in the wind and alerted the Berlin S.A. on the afternoon of 29 June, was so far misled as to believe the danger was a putsch by the Right directed against Hitler. Ernst never understood what had happened, even after his arrest, and died shouting: '*Heil Hitler.*'

On the 29th Hitler, still keeping away from Berlin, made a tour of labour camps in Westphalia, and in the afternoon stopped at Godesberg on the Rhine, where he brought himself to take the final decision. Goebbels, who in the past few days had hurriedly dropped his contacts with Röhm, brought the news that the Berlin S.A., although due to go on leave the next day, had been suddenly ordered to report to their posts. At two o'clock on the morning of the 30th Hitler flew to Munich. Before leaving he had telegraphed to Röhm to expect him at Wiessee the next day.

The purge had already begun when Hitler landed in Munich at four o'clock on Saturday morning. On the evening of the 29th Major Walther Buch and a group of men including dim figures from Hitler's old days in Munich arrested the local S.A. leaders on the pretext that they were about to carry out a *coup d'état.* At the Ministry of the Interior, where S.A. Obergruppenführer August Schneidhuber and his deputy were held under guard, Hitler, who had now worked himself up into a fury, tore off their insignia with his own hand and cursed them for their treachery.

In the early morning of the 30th a column of cars tore down the road from Munich to Wiessee where Röhm and Heines were still in their beds at the Hanselbauer Hotel. Edmund Heines, the convicted murderer who had been reinstated in the S.A. as Obergruppenführer for Silesia, was found sleeping with one of Röhm's young men; he is said to have been dragged out and shot on the road.

Back in Munich, seven to eight hundred men of Sepp Dietrich's S.S. Leibstandarte Adolf Hitler had been brought in from their barracks—the Army providing the transport—and ordered to provide a shooting squad at the Stadelheim Prison. It was there that Röhm was now shot by order of the man whom he had launched on his political career and who seven months before had written to thank him for his imperishable services.

In Berlin the executions, directed by Göring and Himmler, began on the night of 29–30 June and continued throughout the Saturday and Sunday. The chief place of execution was the Lichterfelde Cadet School, and once again the principal victims were the leaders of the S.A. But in Berlin the net was cast more widely. When the bell rang at General von Schleicher's villa and the general went to the door, he was shot down where he stood and his wife with him. His friend, General von Bredow, was shot on his doorstep the same evening. Gregor Strasser, arrested at noon on the Saturday, was executed in the Prinz Albrechtstrasse Prison. Papen's office was wrecked, he himself was kept under house arrest for four days, two of his advisers, Bose and Edgar Jung, were shot, and two others arrested.

Late on Saturday, Hitler returned from Munich. Göring, Himmler, Frick, and a group of police officers stood watching for the plane. As it dived out of the sky and rolled across the field a guard of honour presented arms. The first to step out was Hitler. Without saying a word, he shook hands with the group on the airfield; the silence was broken only by the repeated click of heels. He walked slowly past the guard of honour, and not until he had started to walk towards his car did he begin to talk to Göring and Himmler.

The executions went on all day Sunday—while Hitler gave a tea-party in the Chancellery garden. Only on Monday morning did the shooting cease, when the German people, shaken and shocked, returned to work, and Hindenburg addressed his thanks to the Chancellor for his 'determined action and gallant personal intervention, which have nipped treason in the bud'.

How many were killed has never been settled. Hitler in his speech to the Reichstag admitted fifty-eight executed and another nineteen who had lost their lives. The *White Book* later published in Paris gave a total of four hundred and one.

The largest group of victims belonged to the S.A. Another group was formed by Schleicher and his wife, Bredow, Strasser and Papen's two assistants.

Many of those murdered had little, if any, connexion with Röhm or the S.A., and fell victims to private quarrels. Kahr, who had played a big role in 1923, but had since retired, was found in a swamp near Dachau; his body was hacked to pieces. Father Bernhard Stempfle, who had once revised the proofs of *Mein Kampf*, was discovered in the woods outside Munich; he had been shot 'while trying to escape'. In Hirschberg, Silesia, a group of Jews was murdered, for no other apparent reason than to amuse the local S.S. A music critic named Willi Schmidt was shot mistakenly because he had the same name as someone else on the list.

In an effort to prevent too much becoming known, Goebbels forbade German newspapers to carry obituary notices of those who had been executed or 'had committed suicide'. The ban only led to exaggerated rumours and to the intensification of the feeling of horror and fear. Not until 13 July did Hitler appear before the Reichstag and reveal a part of the story.

Hitler began with a lengthy recital of the achievements of National Socialism, in defence of his policy as Chancellor. When he came to describe the events leading up to 30 June he threw the whole blame on Röhm, who had forced him to act against his own wishes. Hitler gave great prominence to the charges of corruption, favouritism, and homosexuality against Röhm's group, and went out of his way to represent them as betraying the ordinary, decent S.A. man. He spoke of those who had become 'revolutionaries who favoured revolution for its own sake and desired to see revolution established as a permanent condition.' But, Hitler replied, 'for us the Revolution is no permanent condition.' Finally he repeated the promise, which to the Army leaders was the covenant in which they placed their faith: 'In the State there is only one bearer of arms, and that is the Army; there is only one bearer of the political will, and that is the National Socialist Party.'

The Officer Corps, intended only on preserving the privileged position of the Army, and indifferent to what happened in Germany so long as Nazification stopped short of the military institutions of the country, could see no further than the ends of their own noses. The menace of the S.A. was broken for good on the week-end of 30 June, but already a new and far more dangerous challenge to the autonomy of the Army was taking shape. As a reward for their service in the Röhm purge, Himmler's S.S. were now given their independence of the S.A., and placed directly under Hitler's orders with Himmler as Reichsführer S.S. At last Hitler had got what he had always wanted, an absolutely dependable and unquestioning instrument of political action. No group of men was to suffer so sharp a reversal of their calculations as the Army officers, who, in the summer of 1934, ostentatiously held aloof from what happened in Germany and expressed an arrogant satisfaction at the Chancellor's quickness in seeing where the real power in Germany lay.

For anyone less blind than the generals, the way in which Hitler dealt with the threat of a second revolution must have brought consternation rather than satisfaction. Never had Hitler made so patent his total indifference to any respect for law or humanity, and his determination to preserve his power at any cost. Never had he illustrated so clearly the revolutionary character of his régime as in disowning the Revolution.

When President von Hindenburg died on the morning of 2 August, all had been arranged. Within an hour came the announcement that the office of President would henceforward be merged with that of the Chancellor, and that Hitler would become the Head of the State—as well as Supreme Commander-in-Chief of the Armed Forces of the Reich. Among the signatures at the foot of the law announcing these changes were those of Papen, Blomberg, and Schacht: the representatives of Conservatism acquiesced in their own defeat.

The same day the officers and men of the German Army took the oath of allegiance to their new Commander-in-Chief. The form of the oath was significant. The Army was called on to swear allegiance not to the Constitution, or to the Fatherland, but to Hitler personally.

Between March 1933 and August 1934 the balance of power in Germany had shifted decisively in Hitler's favour. In that year and a half he had mastered the machine of State, suppressed the opposition, dispensed with his allies, asserted his authority over the Party and S.A., and secured for himself the prerogatives of the Head of the State and Commander-in-Chief of the Armed Forces. The Nazi revolution was complete: Hitler had become the dictator of Germany.

On 19 August the German people was invited to express by a plebiscite its approval of Hitler's assumption of Hindenburg's office as Führer and Reich Chancellor, the official title by which Hitler was now to be known. The Political Testament of President Hindenburg, much discussed but so far not discovered, was now conveniently produced. According to Papen, Hindenburg had decided to omit any reference to the restoration of the monarchy from his testament but to embody a strong recommendation of such a course in a separate letter to Hitler. Both documents were delivered to Hitler, but the letter was never seen again and no more was heard of a restoration. To remove any doubt, Colonel Oskar von Hindenburg was put up to broadcast on the eve of the plebiscite. 'My father,' he told the German people, 'had himself seen in Adolf Hitler his own direct successor as Head of the German State, and I am acting according to my father's intention when I call on all German men and women to vote for the handing over of my father's office to the Führer and Reich Chancellor.'

On the day of the plebiscite 95.7 per cent of the forty-five and a half million voters went to the polls, and more than thirty-eight million voted 'Yes', 89.93 per cent of the votes cast. Four and a quarter millions had the courage to vote 'No'; another eight hundred and seventy thousand spoiled their papers.

It was an impressive majority, and when the Party Rally was held at Nuremberg in September Hitler was in benign mood. In his proclamation he spoke a good deal about the Nazi revolution which had now, he announced, achieved its object and come to an end. 'Just as the world cannot live on wars, so peoples cannot live on revolutions. . . . Revolutions,' he added, 'have always been rare in Germany. The Age of Nerves of the nineteenth century has found its close with us. In the next thousand years there will be no other revolution in Germany.'

It was an ambitious epitaph.

CHAPTER SIX

THE COUNTERFEIT PEACE

1933–7

Hitler hated the routine work of government, and, once he had stabilized his power, he showed comparatively little interest in what was done by his departmental Ministers except to lay down general lines of policy. In the Third Reich each of the Party bosses, Göring, Goebbels, Himmler, and Ley, created a private empire for himself, while the Gauleiters on a lower level enjoyed the control of their own local pashaliks. Hitler deliberately allowed this to happen; the rivalries which resulted only increased his power as supreme arbiter. Not until his own position, or special interests, were affected did he rouse himself to intervene actively.

Certain subjects, even in internal affairs, always interested Hitler—building plans, and anti-Semitic legislation, for instance—but he rapidly became absorbed in the two fields of foreign policy and preparation for war. He was not interested in administration, or carrying out a programme of reform—he was interested in power. The Party had been the instrument by which he acquired power in Germany; the State was now to be the instrument by which he meant to acquire power in Europe.

The aggressive—or, to use the favourite Nazi word, dynamic—foreign policy which Germany began to follow under Hitler's leadership gave expression to the long-smouldering rebellion of the German people against the defeat of 1918 and the humiliation of the Peace Settlement. Through the sense of national unity which it fostered, it served to strengthen the political foundations of the régime in popular support. Through the revived industrial activity which it stimulated by the rearmament programme, it helped to overcome the economic crisis in which the Republic had foundered. The revolutionary impulse in Nazism was diverted into challenging the existing order outside Germany's frontiers and the creation of a European New Order, in which the big jobs and the privileges would go to the *Herrenvolk*. Above all, such a foreign policy was the logical projection of that unappeased will to power, both in Hitler himself and in the Nazi Party, which was now eager to extend its mastery further.

In the 1920s Hitler wrote in *Mein Kampf*: 'What a use could be made of the Treaty of Versailles! . . . How each one of the points of that Treaty could be branded in the minds and hearts of the German people until sixty million men and women find their souls aflame with a feeling of rage and shame; and a torrent of fire bursts forth as from a furnace, and a will of steel is forged from it, with the common cry: "*Wir wollen wieder Waffen!*—We will have arms again!"' There was little doubt that Hitler's first objective in foreign policy would be to annul the Treaty, and in January 1941 he himself said, with considerable justification: 'My programme was to abolish the Treaty of Versailles. It is nonsense for the rest of the world to pretend today that I did not reveal this programme until 1933, or 1935, or 1937. Instead of listening to the foolish chatter of émigrés, these gentlemen would have been wiser to read what I had written and rewritten thousands of times. No human being has declared or recorded what he wanted more often than I.'

In practice, now that reparations had been ended, this could only mean Germany's right to rearm on terms of full equality with other nations, and the recovery of at least part of the territories lost in

1918–19: the Saar, Alsace-Lorraine, the German colonies, above all Danzig, and the lands incorporated in the new state of Poland.

But this was only a part of Hitler's programme in foreign policy. As Hitler had said quite plainly in *Mein Kampf*, his aim was to extend the frontiers of Germany to include those people of German race and speech who, even in 1914, had lived outside the Reich, the Germans of Austria, and the Sudeten Germans of Czechoslovakia, who, before 1914, had formed part, not of the German Empire, but of the Hapsburg Monarchy.

Hitler was an Austrian, a fact of the greatest importance in understanding his foreign policy. For, in the 1860s, when Bismarck carried out the unification of Germany and founded the German Empire, he deliberately excluded from it the Germans of the Hapsburg Monarchy. After the collapse of the Monarchy these Germans became citizens either of the Austrian Republic or of Czechoslovakia. Amongst these Germans of the old Hapsburg Monarchy there had sprung up before the war an extreme Pan-German nationalism which sought to re-establish a union of all Germans in a single Greater Germany, and which was now violently opposed to the claims of the Czechs and the other former subject peoples of the Monarchy to nationhood and equality with the Germans.

Born on the frontier between Germany and Austria, Hitler felt called upon to reunite the two German states which had been left divided by Bismarck's solution of the German problem. His hatred for the Czechs was the product of his early life in an empire where the Germans felt themselves on the defensive against the rising tide of Slav nationalism, most strongly represented in Hitler's experience by the Czech working men whom he met in Vienna. Here, too, is to be found one of the roots of the distinction Hitler made between the *Volk*, all those of German race and speech, and the State, which need not be co-extensive with the first, or might—as in the case of the old Hapsburg Monarchy and Czechoslovakia—include peoples of different races.

Even this does not exhaust the meaning of Hitler's belief that Germany's 1914 frontiers were inadequate. In the Nazi Party programme, adopted as early as 1920, is to be found the culmination of Hitler's foreign policy in the demand for *Lebensraum,* living room for the future of the *Volk.* Hitler advocated a continental policy of territorial expansion eastwards, seeking *Lebensraum* for Germany in Eastern Europe, in the rich plains of Poland, the Ukraine, and Russia. Such a policy would mean the resumption of the ancient struggle against the Slavs, and its logical consequence was war with Russia. Hitler faced and accepted this as early as the 1920s, and when the German armies invaded the Soviet Union in 1941 it was in execution of a policy whose outlines are already to be found in *Mein Kampf.*

Hitler looked to the conquest of Eastern Europe and Russia for the opportunity to build his New Order, the empire of the *Herrenvolk* based upon the slave labour of the inferior races. Such plans involved the movement of populations, the deliberate depression of whole races to a lower standard of life and civilization, the denial of any chance of education or medical facilities, even, in the case of the Jews, systematic extermination.

In these schemes for redrawing the map of the world and remodelling the distribution of power upon biological principles the authentic flavour of Nazi geopolitics is to be discovered. Hitler's over-inflamed imagination set no bounds to the expansion of Nazi power. As Papen remarked after the war: 'It was on the limitless character of Nazi aims that we ran aground.' In the early 1930s these appeared no more than the fantasies with which Hitler beguiled the early morning hours round the fire in the Berghof;

by the early 1940s, however, the fantastic was on the verge of being translated into reality.

In 1933–4, the prospects of accomplishing even the annexation of Austria, still less of overrunning Russia, appeared remote. Germany was politically isolated. Economically, she was only beginning to recover from the worst slump in her history. Her army, limited to the hundred thousand men permitted by the Treaty, was easily outnumbered by that of France alone. A move in any direction appeared certain to run into the network of alliances with which France sought to strengthen her security. So impressed were the German diplomats and the German generals with the strength of the obstacles in Germany's way that up to the Battle of France in 1940, their advice was always on the side of caution.

Hitler, on the other hand, became more and more sure of himself. He was convinced that he had a far keener appreciation of political—or military—factors than the High Command or the Foreign Office, and he dazzled them by the brilliant success of the bold tactics he adopted. Hitler took office as Chancellor without any previous experience of government. He had no knowledge of any country outside Germany and Austria, and spoke no foreign language. His sole experience of politics had been as a Party leader and agitator. He knew nothing and cared less for official views and traditions; he was suspicious of anyone who might try to instruct him. In the short run, these were assets. He refused to be impressed by the strength of the opposition his schemes were likely to meet, or to be restricted to the conventional methods of diplomacy. He displayed a skill in propaganda and a mastery of deceit, a finesse in exploring the weaknesses of his opponents and a crudeness in exploiting the strength of his own position which he had learned in the struggle for power in Germany and which he now applied to international relations with even more remarkable results.

No man was more of an opportunist, than Hitler, and no man had more luck. But Hitler knew how to turn events to his advantage. He knew what he wanted and he held the initiative. His principal opponents, Great Britain and France, knew only what they did not want—war—and were always on the defensive. The fact that Hitler was ready to risk war, and started preparing for it from the day he came to power, gave him a still greater advantage.

The first and indispensable step was to rearm. Until he had the backing of military power for his diplomacy, Hitler's foreign policy was bound to be restricted in its scope. Until rearmament reached a certain stage Germany was highly vulnerable to any preventive action which France or the other Powers might take under the provisions of the Treaty of Versailles. The overriding objective of German foreign policy, therefore, for the first years of Hitler's régime was to secure the time and freedom to rebuild military power.

Hitler's speeches from this period are masterpieces of careful propaganda. Well aware that there were many abroad—especially in Great Britain—who had long felt uneasy about the shortcomings of the Peace Settlement, he hinged all arguments upon the unequal treatment of Germany after the war. This invoked sympathy for Germany, allowed Hitler to appear as the representative of reason and justice, protesting against the unreasonableness and injustice of Germany's former opponents, and enabled him to turn round and use with great effect against the supporters of the League of Nations all the slogans of Wilsonian idealism, from self-determination to a peace founded upon justice.

In October 1933, when it became clear that the French—uneasily conscious of the inferiority of their manpower and industrial resources to those of Germany—were not prepared to disarm, Hitler pushed his argument a stage further. On 14 October he announced that Germany was driven, by the denial of equal rights, to withdraw from the Disarmament Conference and the League of Nations. Germany had tried to cooperate, but had suffered a bitter disillusionment and humiliation. In sorrow, rather than in anger, he had decided to take this step, which was demanded by the self-respect of the German people.

The withdrawal from the League was not without risks, in view of Germany's military inferiority, and a secret directive was issued to the Armed Forces, in case the League should apply sanctions. It was the first of Hitler's gambles in foreign policy—and it succeeded. Events wholly justified his diagnosis of the state of mind of his opponents—their embarrassment in face of a case which they felt was not without justice; the divided public opinion of Great Britain and France; the eagerness to be reassured and to patch up a compromise, all those elements on which Hitler was to play with such skill time and again. He then issued a proclamation in which he declared force to be useless in removing international differences, affirmed the German people's hopes in disarmament and renewed his offer to conclude pacts of non-aggression at any time.

Hitler's cleverest stroke was to announce, on the same day as the withdrawal from the League, that he would submit his decision at once to a plebiscite. This was to invoke the sanctions of democracy against the democratic nations. All the long-pent-up resentment of the German people against the loss of the war and the Treaty of Versailles was expressed in the vote: 95 per cent approved of Hitler's policy.

Hitler had now manoeuvred himself into the strongest possible position in which to begin German rearmament. When the other Great Powers sought to renew negotiations, Hitler replied that disarmament was clearly out of the question. All that could be hoped for was a convention for the limitation of armaments, and Germany's terms for cooperation would be the recognition of her right to raise an army of three hundred thousand men. Rearmament had in fact already begun, while Great Britain and France had placed themselves in the disadvantageous position, from which they were never to recover until the war, of asking the German dictator what concessions he would accept to reduce his price.

While these negotiations continued, Hitler strengthened his hand in an unexpected direction. No feature of the Treaty of Versailles stirred more bitter feelings in Germany than the loss of territory to the new State of Poland. Relations between Poland and Germany continued to be strained, and nowhere was the rise to power of the Nazis viewed with more alarm than in Warsaw.

It caused a diplomatic sensation, therefore, when, on 26 January 1934, Hitler announced that the first country with which Nazi Germany had concluded a Non-Aggression Pact was Poland. The Pact was never popular in Germany, but it was an astute move for Hitler to make. Uultimately, there was no place for an independent Poland in Hitler's Europe, but Hitler could not move against Poland for years to come. He turned this situation to his advantage, and made an ostentatious parade of his enforced virtue. He was thus able to substantiate his claim to peaceful intentions, and the Pact with Poland was constantly used in Hitler's 'peace' speeches from 1934 to 1936.

But the importance of the Pact was greater than its value as propaganda. Poland was one of the bastions of the French security system in Eastern Europe, but the Poles were becoming restive at the casual way in which they felt they were treated by France. The Polish Government was beginning to turn towards an independent neutrality, in which it was hoped that Poland would be able to balance between her two great neighbours, Germany and Russia. The Nazi offer of a Ten-Year Pact fitted admirably into this new policy, and Hitler was thereby able to make the first breach in the French alliance system and the first display of those tactics of 'one-by-one' with which he was to achieve so much.

This was a good beginning, but there were reminders during 1934 of the dangers of the situation, notably in the case of Austria. The forty thousand Austrian Nazis, who formed a part of the German Party under Hitler's leadership, lived and worked for the day when the *Anschluss* should take place. Local leaders kept up a violent propaganda campaign, backed by imtimidation and acts of terrorism. Anyone in Austria had only to tune in to the Munich radio station to get confirmation of the support the local Nazis were receiving from Germany. An attempt by the Austrian Nazis to capture power by a rising appeared imminent.

German relations with Austria, however, were not simply a family affair, as Hitler tried to insist. France, the ally of Czechoslovakia, and Italy, the patron of Dollfuss's fascist régime in Austria and of Hungary, were bound to be disturbed by the prospect of an *Anschluss* and a consequent Nazi advance to the threshold of the Balkans. On 17 February the governments of France, Great Britain, and Italy published a joint declaration to the effect that they took 'a common view of the necessity of maintaining Austria's independence and integrity in accordance with the relevant treaties'.

On 25 July, while Madame Dollfuss and her family were staying with Mussolini, the Austrian Nazis broke into the Vienna Chancellery and shot Dollfuss, while others occupied the radio station and announced the appointment of Anton Rintelen as Chancellor.

Although the German Legation in Vienna had been heavily implicated in the plot, it is unlikely that Hitler knew what was planned. This was no time for foreign adventures, so soon after the events of 30 June and with the succession to Hindenburg still in the balance. Dollfuss died of his wounds, but the putsch failed. The rebels in Vienna were quickly overpowered, and the leaders, followed by several thousand Austrian Nazis, only escaped by getting across the German frontier. Mussolini, furious at what he regarded as Hitler's bad faith, ordered Italian divisions to the Austrian frontier and sent the Austrian Government an immediate telegram promising Italian support in the defence of their country's independence.

The Nazis had over-reached themselves, and Hitler had promptly to repudiate all connexion with the conspiracy. The initial announcement of the official German News Agency, couched in enthusiastic terms, was hurriedly suppressed; the murderers of Dollfuss were surrendered to the Austrian Government; Hitler appointed Papen to go to Vienna as Minister-Extraordinary in order to repair the damage. The choice of Papen, a Catholic, a Conservative, and Vice-Chancellor in Hitler's Cabinet, was intended to conciliate the Austrians; at the same time it was a convenient way of getting rid of the man who had been lucky to escape with his life on the week-end of 30 June. These hasty measures tided over the crisis and preserved appearances. But it had been made plain enough to Hitler that the opposition to his schemes would have to be divided before it could be overcome.

For the rest of 1934 the unanimity of the other Powers in face of further German adventures was strengthened, rather than weakened, but the year ended not without some cause for congratulation on Hitler's part. On 9 October Louis Barthou, the energetic French Foreign Minister, who stood for a policy of firmness in face of Nazi demands, was assassinated at Marseille. His successor was Pierre Laval, a master of *combinazioni* and shady political deals. Despite appearances, Hitler held to his belief that behind the façade of unity the Powers lacked the will to oppose him or to combine together for long. In 1927 Hitler said to Otto Strasser: 'There is no solidarity in Europe; there is only submission.' It was the essential premise on which all his plans depended; the next year, 1935, was to show how just was his diagnosis.

From the summer of 1934 the principal object of the Western Powers' diplomacy was to persuade Germany to sign a pact of mutual assistance covering Eastern Europe. Just as the Locarno Pact included France, Germany, Belgium, Great Britain, and Italy, each undertaking to come to the immediate aid of France and Belgium, or Germany, if either side were attacked by the other, so this Eastern Locarno would include Russia, Germany, Poland, Czechoslovakia, and the other states of Eastern Europe and would involve the same obligation of automatic assistance in the case of an attack.

Hitler had no intention of entering into any such scheme: it was not aggression that he feared, but checks upon his freedom of action. His opposition was powerfully assisted by that of Poland, whose policy of balancing between Moscow and Berlin fatally overestimated her own strength, and fatally underestimated the danger from Germany.

Hitler courted the Poles assiduously, constantly urging on them the common interest Poland and Germany had in opposing Russia. Göring was used by Hitler in the role of a candid friend of the Poles. In his talks with Polish generals and with Marshal Pilsudski in January 1935, he almost suggested a joint attack on Russia, implying that the Ukraine would become a Polish sphere of influence and North-western Russia would be Germany's. The Poles were wary of such seductive propositions, but during 1935 relations between the two governments became steadily closer. In May Göring visited Cracow for Pilsudski's funeral, and Hitler had a long conversation with the Polish Ambassador. The attention Hitler paid to Polish-German relations was to repay him handsomely.

Meanwhile, the British and French Governments renewed their attempts to reach a settlement with Germany, proposing that the Locarno Pact be strengthened by the conclusion of an agreement to cover unprovoked aggression from the air and be supplemented by two similar pacts of mutual assistance, one dealing with Eastern Europe, the other with Central Europe.

Hitler faced a difficult decision. German rearmament had reached a stage where further concealment would prove a hindrance. It seemed clear from their proposals that the Western Powers would be prepared to waive their objections to German rearmament in return for Germany's accession to their proposals for strengthening and extending collective security. Yet Hitler had to avoid tying his hands. He also had to provide some dramatic stroke of foreign policy to gratify the mood of nationalist expectation in Germany. A bold unilateral repudiation of the disarmament clauses of the Treaty of Versailles would suit him very much better than negotiations with the Western Powers, in which he would be bound to make concessions in return for French and British agreement. Could he afford to take the risk?

Hitler's first reply showed uncertainty. He welcomed the idea of extending the original Locarno Pact to include attack from the air, while remaining evasive on the question of the proposed Eastern and Danubian Pacts. The German Government invited the British to continue discussions, and a visit to Berlin by the British Foreign Minister, Sir John Simon, was arranged for 7 March. On 4 March, however, the British Government published its own plans for increased armaments, basing this on 'the fact that Germany was . . . rearming openly on a large scale, despite the provisions of Part V of the Treaty of Versailles'. Great indignation was at once expressed in Germany, and Hitler contracted a 'chill' which made it necessary to postpone Sir John Simon's visit. On the 9th the German Government officially notified foreign governments that a German Air Force was already in existence. This seems to have been a kite with which to test the Western Powers' reaction. As Sir John Simon told the House of Commons that he and Mr. Eden were still proposing to go to Berlin and nothing else happened, it appeared safe to risk a more sensational announcement the next week-end. On 16 March 1935, the German Government proclaimed its intention of building up a peacetime army of five hundred and fifty thousand men.

Four days before, the French Government had doubled the period of Army service and reduced the age of enlistment to make good the reduced birth-rate of 1914–18. This served Hitler as a pretext for his own action. Germany, Hitler declared, was the one Power which had disarmed; now that the other Powers, far from disarming themselves, were actually beginning to increase their armaments, she had no option but to follow suit.

The announcement was received with enthusiasm in Germany, and on 17 March, Heroes Memorial Day, a brilliant military ceremony in the State Opera House celebrated the rebirth of the German Army. In this first open breach of the Treaty's provisions, Hitler had anticipated protests from abroad; what mattered was the action with which the other signatories of the Treaty proposed to support their protests.

The result more than justified the risks he had taken. The British Government, after making a solemn protest, proceeded to ask whether the Führer was still ready to receive Sir John Simon. The French appealed to the League, and an extraordinary session of the Council was at once summoned. But the French Note, too, spoke of searching for means of conciliation. This was not the language of men who intended to enforce their protests. When Sir John Simon and Mr. Eden at last visited Berlin at the end of March they found Hitler polite, even charming, but perfectly sure of himself and firm in his refusal to consider any pact of mutual assistance which included the Soviet Union. It was the Englishmen who had come to ask for cooperation and Hitler who was in the advantageous position of being able to say 'no', without having anything to ask in return. The very presence of the British representatives in Berlin, after the announcement of 16 March, was a triumph for his diplomacy.

In the weeks that followed, the Western Powers continued to make a display of European unity which, formally at least, was more impressive. At Stresa, on 11 April 1935, the British, French, and Italian Governments condemned Germany's action, reaffirmed their loyalty to the Locarno Treaty and repeated their declaration on Austrian independence. At Geneva the Council of the League duly censured Germany. In May, the French Government signed a pact with the Soviet Union by which each party undertook to come to the aid of the other in case of an unprovoked attack. This treaty was flanked by a similar pact, concluded at the same time, between Russia and France's most reliable ally, Czechoslovakia.

Yet, even if Hitler was taken aback by the strength of this belated reaction, and if the Franco-Russian and Czech-Russian treaties in particular faced him with awkward new possibilities, his confidence in his own tactics was never shaken. He proceeded to test the strength of this new-found unity; it did not take long to show its weaknesses.

On 21 May Hitler promulgated in secret the Reich Defence Law which placed Schacht in charge of economic preparations for war and reorganized the commands of the armed forces under himself as Supreme Commander of the Wehrmacht. But this was not the face that Hitler showed in public. On the evening of the same day, he appeared before the Reichstag to deliver a long and carefully prepared speech on foreign policy. In it are to be found most of the tricks with which Hitler lulled the suspicions and raised the hopes of the gullible. His answer to the censure of the Powers was not defiance, but redoubled assurances of peace, an appeal to reason, justice and conscience. The new Germany, he protested, was misunderstood, and his own attitude misrepresented.

No man ever spoke with greater feeling of the horror and stupidity of war than Adolf Hitler. Despite the failure of the other powers to disarm, Germany was still prepared to cooperate in the search for security. But she had rooted objections to the proposal of multilateral pacts, for this was the way to spread, not to localize, war. Moreover, Hitler declared, there was a special case, Bolshevik Russia, pledged to destroy the independence of Europe, a State with which a Nationalist Socialist Germany could never come to terms.

In place of the 'unrealistic' proposal of multilateral treaties, Hitler offered the signature of non-aggression pacts with Germany's neighbours. Germany's improved relations with Poland, he did not fail to add, showed how great a contribution such pacts could make to the cause of peace: this was the practical way in which Germany set about removing international misunderstandings.

Hitler supported his offer with the most convincing display of goodwill. The fact that Germany had repudiated the disarmament clauses of the Treaty of Versailles did not mean that she had anything but the strictest regard for the Treaty's other provisions —including the demilitarization of the Rhineland. She had no intention of annexing Austria and was perfectly ready to strengthen the Locarno Pact. She was ready to agree to the abolition of heavy arms; to limit the use of other weapons—such as the bomber and poison gas—by international convention; indeed, to accept an over-all limitation of armaments provided that it was to apply to all the Powers. Hitler laid particular stress on his willingness to limit German naval power to 35 per cent of the strength of the British Navy. He ended with a confession of his faith in peace. 'Whoever lights the torch of war in Europe can wish for nothing but chaos. We, however, live in the firm conviction that in our time will be fulfilled, not the decline, but the renaissance of the West. That Germany may make an imperishable contribution to this great work is our proud hope and our unshakable belief.'

Hitler's mastery of the language of Geneva was unequalled. He understood intuitively the longing for peace, the idealism of the pacifists, the uneasy conscience of the liberals, the reluctance of the great mass of their peoples to look beyond their own private affairs. At this stage these were greater assets than the uncompleted panzer divisions and bomber fleets he was still building, and Hitler used them with the same skill he had shown in playing on German grievances and illusions.

Although Hitler's attitude towards Britain was modified later by growing contempt for the weakness of her policy and the credulity of her governments, the idea of an alliance with her attracted him throughout his life. Such an alliance could only, in Hitler's view, be made on condition that Britain accepted the prospect of a German hegemony on the Continent and left Germany a free hand in attaining it. No British Government, even before the war, was prepared to go as far as an alliance on these terms, yet there was a section of British opinion which was sufficiently impressed by Hitler's arguments to be attracted to the idea, and Hitler was remarkably successful for a time in weakening the opposition of Great Britain to the realization of his aims. The policy of appeasement represented the acceptance by the British Government, at least in part, of Hitler's view of what British policy should be.

The speech of 21 May 1935 had been intended to influence opinion in Great Britain in Hitler's favour. The quickness of the British reaction was surprising. During his visit to Berlin in March Sir John Simon had been sufficiently impressed by a hint thrown out by the Führer to suggest that German representatives should come to London to discuss the possibility of a naval agreement between the two countries.

Early in June Joachim von Ribbentrop, whom Hitler now began to use for special missions, flew to London. He returned with the British signature of a naval pact which bound the Germans not to build beyond 35 per cent of Britain's naval strength, but tacitly recognized Germany's right to begin naval rearmament and specifically agreed that Germany should have the right to build up to 100 per cent of the submarine strength of the British Commonwealth. The affront to Britain's partners, France and Italy, both of whom were also naval powers, but neither of whom had been consulted, was open and much resented. The unanimity of the Powers' condemnation of German rearmament was destroyed. The British Government, in its eagerness to secure a private advantage, had given a disastrous impression of bad faith and had accepted Hitler's carefully calculated offer without a thought of its ultimate consequences.

In September the Führer attended the Party's rally at Nuremberg, where for the first time detachments of the new German Army took part in the parade. The Reichstag was summoned to Nuremberg for a special session, and Hitler presented for its unanimous approval the Nuremberg Laws directed against the Jews, the first depriving Germans of Jewish blood of their citizenship, the second forbidding marriages between Germans and Jews and the employment of German servants by Jews. These laws, Hitler declared, 'repay the debt of gratitude to the movement under whose symbol [the swastika, now the national emblem] Germany has recovered her freedom'.

The same month, while Hitler was making use of his power to gratify his hatred of the Jews, a quarrel began at Geneva which was to provide him with the opportunity to extend his power outside the German frontiers of 1914.

After the murder of Dollfuss, Mussolini had been outspoken in his dislike and contempt for the 'barbarians' north of the Alps, and he had cooperated with the other Powers in their condemnation of Germany's unilateral decision to rearm. Mussolini, however, had long been contemplating a showy success for his régime in Abyssinia, and he hoped to profit by French and British preoccupation with German rearmament to carry out his adventure on the cheap.

Abyssinia had appealed to the League in March. In September the British Government, having just made a sensational gesture of appeasement to Germany by the Naval Treaty of June, astonished the world for the second time by taking the lead in demanding the imposition of sanctions against Italy. The Baldwin Government thus made the worst of both worlds. By insisting on the imposition of sanctions Great Britain made an enemy of Mussolini and destroyed all hope of a united front against German aggression. By her refusal to drive home the policy of sanctions, in face of Mussolini's bluster, she dealt the authority of the League as well as her own prestige a fatal blow, and destroyed any hope of finding in collective security an effective alternative to the united front of the Great Powers against German aggression.

The ultimate beneficiary of these blunders was, not Mussolini, but Hitler. The advantages to be derived from the quarrel between Italy and the Western Powers did not escape Hitler, and the further development of the dispute only gave him greater cause for satisfaction. Not only was Italy driven into a position of isolation, in which Mussolini was bound to look more favourably on German offers of support, but the League of Nations suffered a fatal blow to its authority.

As Mussolini later acknowledged, it was in the autumn of 1935 that the idea of the Rome-Berlin Axis was born. No less important was the encouragement which the feebleness of the opposition to aggression gave Hitler to pursue his policy without regard to the risks.

By March 1936 Hitler judged the moment opportune for another *coup* in foreign policy. The proposed treaty between France and Russia had become a subject of bitter controversy in French politics, and in London there was no enthusiasm for France's latest commitment. On 27 February 1936, the Franco-Soviet treaty was finally ratified.

Hitler's reply was to march German troops into the demilitarized Rhineland. It was a proposal which thoroughly alarmed Hitler's generals. German rearmament had barely begun, and the first conscripts had been taken into the Army only a few months before. If the French and their allies marched, the Germans would be heavily outnumbered. Furthermore, reoccupation of the Rhineland represented not only a breach of the Treaty of Versailles but a *casus foederis* under the Locarno Pact. Hitler did not dispute these facts; he based his decision on the belief that the French would not march—and he was right.

On the morning of 7 March, as the German soldiers were marching into the Rhineland, greeted with flowers flung by wildly enthusiastic crowds, Neurath, German Foreign Minister, summoned the British, French, and Italian Ambassadors to the Wilhelmstrasse and presented them with a document which contained, in addition to Germany's grounds for denouncing the Locarno Pact, new and far-reaching peace proposals. As M. François-Poncet, the French Ambassador, described it, 'Hitler struck his adversary in the face, and as he did so declared: "I bring you proposals for peace!"' In place of the discarded Locarno Treaty, Hitler offered a pact of non-aggression to France and Belgium, valid for twenty-five years and supplemented by the air pact to which Britain attached so much importance.

At noon Hitler addressed the Reichstag. His speech was another masterpiece of reasonableness. Referring to the 'useless strife' between France and Germany, he asked: 'Why should it not be possible to lift the general problem of conflicting interests between the European states above the sphere of passion and unreason and consider it in the calm light of a higher vision?'

It was France, Hitler declared, who had betrayed Europe by her alliance with the Asiatic power of Bolshevism, pledged to destroy all the values of European civilization—just as it was France who, by the same action, had invalidated the Locarno Pact. Once again, reluctantly but without flinching, he must bow to the inevitable and take the necessary steps to defend Germany's national interests.

Hitler later admitted: 'The forty-eight hours after the march into the Rhineland were the most nerve-racking in my life. If the French had then marched into the Rhineland we would have had to withdraw with our tails between our legs, for the military resources at our disposal would have been wholly inadequate for even a moderate resistance.' Events, however, followed exactly the same pattern as the year before. There were anxious consultations between Paris and London, appeals for reason and calm, and much talk of the new opportunities for peace offered by Hitler's proposals. The Locarno Powers conferred; the Council of the League conferred; the International Court at The Hague was ready to confer. Germany's action was again solemnly condemned and the censure again rejected by Hitler. But no one marched —except the Germans.

Meanwhile Hitler dissolved the Reichstag and invited the German people to pass judgement on his policy. The election results, a reported 98.8 per cent in favor, left little doubt that a substantial majority of the German people approved Hitler's action, or that it raised the Führer to a new peak of popularity in Germany.

While European governments began to accommodate themselves to the new balance of power, Austria became more and more uneasy. The premise upon which Austrian independence was based, the unity and military superiority of Italy, France, and Great Britain in face of Germany, was being destroyed. Sooner or later Mussolini's 1934 guarantee of Italian divisions on the Brenner frontier would be withdrawn.

In 1936 Papen, whose aim was to undermine Austrian independence from within and to bring about the *Anschluss* peacefully, gained his first successes. On 13 May, Prince Starhemberg, the Austrian Vice-Chancellor and an outspoken opponent of the Austrian Nazis, was forced to resign. Three weeks after Starhemberg had gone, the Austrian Chancellor, Schuschnigg, informed Mussolini that the Austrian Government was about to sign an agreement with Germany. The Duce, though repeating his assurances of support for Austrian independence, gave his approval.

The Austro-German Agreement of 11 July 1936 was designed on the surface to ease and improve relations between the two countries. It recognized Austria's full sovereignty and promised mutual non-intervention in internal affairs. The most important of its secret clauses specified that the Austrian Government agreed to give representatives of the so-called

National Opposition in Austria, 'respectable' crypto-Nazis like Glaise-Horstenau and later Seyss-Inquart, a share in political responsibility.

Ostensibly, Austro-German relations were now placed on a level satisfactory to both sides. But, in fact, for the next eighteen months the Germans used the Agreement as a lever with which to exert increasing pressure on the Austrian Government and to extort further concessions. The Agreement thus marked a big step forward in the policy of capturing Austria by peaceful methods.

The Agreement also materially improved Hitler's prospects of a *rapprochement* with Italy. Here again he had extraordinary luck. On 4 July 1936, the League Powers tacitly admitted defeat and withdrew the sanctions they had tried to impose on Italy. Less than a fortnight later, on 17 July, civil war broke out in Spain and created a situation from which Hitler was able to draw no fewer advantages than from Mussolini's Abyssinian adventure.

Hitler was at Bayreuth on 22 July when he received a personal letter from General Franco. That night he decided to give active help to Franco. In the course of the next three years German aid to Franco was never sufficient to win the war for him, but Hitler's policy was to prolong the war. In return for aid, Germany secured economic advantages (valuable sources of raw materials in Spanish mines); useful experience in training her airmen and testing equipment such as tanks in battle conditions; above all, strategic and political advantages.

France, for geographical reasons alone, was more deeply interested in what happened in Spain than any other of the Great Powers, yet the ideological character of the Spanish Civil War divided, instead of uniting, French opinion. Many Frenchmen were prepared to support Franco as a way of hitting at their own Left-wing Government, led by Léon Blum. The Spanish Civil War exacerbated all those factors of disunity in France upon which Hitler had always hoped to play.

From the first Mussolini intervened openly in Spain, giving all the aid he could spare to bring about a victory for Franco. The common policy of Italy and Germany towards Spain created one of the main foundations on which the Rome-Berlin Axis was built, and the Spanish Civil War provided ample scope for such cooperation.

In September 1936, Hitler judged circumstances favourable for creating a closer relationship between Germany and Italy in order to exploit a situation in which the two countries had begun to follow parallel courses. The July Agreement between Germany and Austria removed the biggest obstacle to an understanding between Rome and Berlin, and on 29 June the German Ambassador conveyed to Ciano, the Italian Foreign Minister, an offer from Hitler to consider the recognition of the new Italian Empire—a point on which the Duce was notoriously touchy—whenever Mussolini wished. In September Hitler sent Hans Frank, his Minister of Justice, on an exploratory mission to Rome.

Frank brought a cordial invitation from the Führer for both Mussolini and Ciano to visit Germany. Germany, he said, had no claims in the Mediterranean: 'The interests of the Germans are turned towards the Baltic, which is their Mediterranean.' Frank declared that the Austrian question was now settled, and he suggested that Germany and Italy should pursue a common policy in presenting their colonial demands. Mussolini affected a certain disinterestedness, but a month later Ciano set out for Germany.

At Berchtesgaden on 24 October Hitler and Ciano met. Although Hitler monopolized the conversation he was obviously at pains to impress Ciano with his friendliness. The gist of his remarks was the need for Italy and Germany to create a common front against Bolshevism and the Western Powers. If England faced the formation of a strong German-Italian bloc, she might well seek to come to terms with it, and if not, Germany and Italy would have the power to defeat her.

A protocol prepared by the Italian and German Foreign Offices before Ciano's visit covered in some detail German-Italian cooperation on a number of issues. When Mussolini went to Milan on 1 November 1936 he spoke of an agreement between the two countries and for the first time used the famous simile of an axis, 'round which all those European states which are animated by a desire for collaboration and peace may work together.'

By the end of 1936 Hitler, by skilfully exploiting Mussolini's situation, had succeeded in establishing one of the two alliances on which he had counted in *Mein Kampf*. But the second condition, an understanding with Britain, still eluded him.

In August, Hitler, determined on a new approach to London, had appointed Ribbentrop as the German Ambassador to the Court of St James. An ambitious man, Ribbentrop succeeded in persuading the new Chancellor that he could provide him with more reliable information about what was happening abroad than reached him through the official channels of the Foreign Office. With Party funds he set up a Ribbentrop Bureau on the Wilhelmstrasse.

Arrogant, vain, humourless, and spiteful, Ribbentrop was one of the worst choices Hitler ever made for high office. But he shared many of Hitler's own social resentments (especially against the regular Foreign Service), he was prepared to prostrate himself before the Führer's genius, and his appointment enabled Hitler to take the conduct of relations with Great Britain much more closely into his own hands.

What puzzled Hitler and Ribbentrop was that although the British were disinclined to take any forceful action on the Continent and only too prepared to put off awkward decisions, they were wary of committing themselves to cooperate with Germany. Hitler was reluctant to take open action which would alienate them, in the hope that he might still win them over, yet he was tempted at times to regard Britain as 'finished' and her value either as an ally or an opponent as negligible. This alternation of moods never wholly disappeared from Hitler's ambivalent attitude towards Britain.

Hitler's best argument with the Conservative Government in Britain was the common interest of the European States in face of Communism. The Spanish Civil War sharpened the sense of ideological conflict in Western Europe, and many people in England as well as in France, who would have looked askance at a blatant German nationalism, were impressed by Hitler's anti-Communism.

Anti-Communism could also be used to provide the basis for a power-bloc which included Japan. In November Ribbentrop flew to Berlin from London for the signature of the Anti-Comintern Pact. The ideological objectives of the pact—the defeat of the Communist 'world-conspiracy'—gave it a universal character which a straightforward agreement aimed against Russia could not have had. It was expressly designed to secure the adherence of other States, and it was not long before Hitler began to collect new signatories. The public provisions of the pact dealt with no more than the exchange of information on Comintern activity, cooperation in preventive measures, and severity in dealing with Comintern agents. There was also a secret Protocol which dealt specifically with Russia and bound both parties to sign no political treaties with the U.S.S.R. In the event of an unprovoked attack or threat of attack by Russia on either Power, the Protocol added, each agreed to 'take no measures which would tend to ease the situation of the U.S.S.R.'

From every point of view, Hitler could feel satisfaction with his fourth year of power. The remilitarization of the Rhineland, German rearmament, and the contrast between his own self-confident leadership and the weakness of the Western Powers had greatly increased his prestige both abroad and at home. When the Olympic Games were held in Berlin in August 1936, thousands of foreigners crowded the capital, and the opportunity was used with great skill to put the Third Reich on show. At Nuremberg, in September, the Party Rally, which lasted a week, was on a scale which even Nazi pageantry had never before equalled.

Hitler rounded off his first four years of office by a long speech to the Reichstag on 30 January 1937, in which he formally withdrew Germany's signature from those clauses of the Treaty of Versailles which had denied her equality of rights and laid on her the responsibility for the war. 'Today,' Hitler added, 'I must humbly thank Providence, whose grace has enabled me, once an unknown soldier in the war, to bring to a successful issue the struggle for our honour and rights as a nation.'

It was an impressive record to which Hitler was able to point, not only in the raising of German prestige abroad, but in economic improvement and the recovery of national confidence at home. Between January 1933 and December 1934 the number of registered unemployed fell from six millions to two million six hundred thousand. Granted that some measure of economic recovery was general at this time, none the less in Germany it was more rapid and went further than elsewhere, largely as a result of heavy Government expenditure on improving the resources of the country and on public works.

It is natural, therefore, to ask whether there was not some point up to which the Nazi movement was a force for good, but after which its original idealism became corrupted. All the evidence indicates that from the first Hitler and the other Nazi leaders thought in terms solely of power, their own power, and the power of the nation over which they ruled.

In a secret memorandum of 3 May 1935, Dr. Schacht, the man who had the greatest responsibility for Germany's economic recovery, wrote: 'The accomplishment of the armament programme with speed and in quantity is *the* problem of German politics, and everything else should be subordinated to this purpose, as long as the main purpose is not imperilled by neglecting all other questions.' This view is repeated again and again through all the discussions on economic policy in these years. Hitler persistently refused to take into account any other economic or social objective besides the overriding need to provide him with the most efficient military machine possible in the shortest possible time.

The driving force behind German rearmament was Hitler. In August 1936, the period of conscription was extended to two years, while at Nuremberg, in September, impatient with the difficulties raised by the economic experts, Hitler proclaimed a Four-Year Plan and put Göring in charge fully armed with the powers to secure results whatever the cost. German economy was henceforward subordinated to one purpose, preparation for war. Although Germany made a remarkable economic recovery, and by the end of this period was one of the best-equipped industrial nations in the world, this was reflected, not in the standard of living of her people, but in her growing military strength.

Moreover, the biggest single factor in the recovery of confidence and faith in Germany was the sense of this power, expressed in an increasingly aggressive nationalism which had little use for the rights of other, less powerful nations. The psychology of Nazism, no less than Nazi economics, was one of preparation for war. Both depended for their continued success upon the maintenance of a national spirit and a national effort which in the end must find expression in aggressive action. War, the belief in violence and the right of the stronger, were not corruptions of Nazism, they were its essence. Recognition of the benefits which Hitler's rule brought to Germany in the first four years of his régime needs to be tempered therefore by the realization that for the Führer—and for a considerable section of the German people—these were the by-products of his true purpose, the creation of an instrument of power with which to realize a policy of expansion that was to admit no limits.

Throughout 1937, although Hitler was still at pains to protest his love of peace, there was a new note of impatience in his voice. The demand for the return of Germany's colonies was raised with increasing frequency in 1937.

There were two particular grounds for Hitler's confidence that the world would recognize his claim: the progress of German rearmament, and the consolidation of the Axis. Göring, now the economic dictator of Germany, had as little respect for economics as Hitler. In December 1936 Göring told a meeting of industrialists that it was no longer a question of producing economically, but simply of producing. It was quite immaterial in securing foreign exchange whether the provisions of the law were complied with or not, provided only that foreign exchange was brought in somehow to provide funds for armament. Hitler's and Göring's programme of autarky and the search for *ersatz* raw materials were criticized by Dr. Schacht, but his economic arguments fell on deaf ears. Göring continued to ride rough-shod over economic theories and economic facts alike.

The consolidation of the Rome-Berlin Axis was marked by increased consultation between the two parties and frequent exchanges of visits culminating in Mussolini's State Reception in Germany in September, and Italy's signature of the Anti-Comintern Pact in November. The initiative still came from Berlin, and Hitler watched with some anxiety the attempts of the British and French to renew friendly relations with the Duce.

On 2 January 1937, Ciano signed a 'gentlemen's agreement' with England in which each country recognized the other's vital interests in the freedom of the Mediterranean, and agreed that there should be no alteration in the *status quo* in that region. Shortly afterwards Hitler sent Göring to Rome on an exploratory mission. Each side regarded with some suspicion the other's attempts to reach an understanding with England. Above all, Austria was still a danger-point in German-Italian relations. Göring urged Mussolini to bring pressure to bear on the Austrian Government to observe the terms of the Austro-German Agreement, and although he made plain Germany's dislike of the Schuschnigg Government, he added the assurance that for Hitler's part there would be no surprises as far as Austria was concerned. Göring's tactful behaviour reassured the Italians.

The suspicions and difficulties were not easily removed, but the pull of events was too strong for Mussolini. His Mediterranean ambitions, his intervention in Spain, his anxiety to be on the winning side and to share in the plucking of the decadent democracies, not least his resentment over British and French policy in the past, were added to the vanity of a dictator with a bad inferiority complex in international relations, and pointed to the advantages of the partnership which Hitler persistently pressed on him. On 23 September the Duce set out for Germany in a new uniform specially designed for the occasion.

Hitler received the Duce at Munich, where the Nazi Party put on a superbly organized show. Mussolini had hardly recovered his breath when he was whisked away to a display of Germany's military power at the

Army manoeuvres in Mecklenburg, and of her industrial resources in the Krupp factories at Essen. The visit reached its climax in Berlin, where the two dictators stood side by side to address a crowd of eight hundred thousand. Mussolini returned from Germany bewitched by the display of power which had been carefully staged for him. Hitler had stamped on the Duce's mind an indelible impression of German might from which he was never able to set himself free.

Hitler laid himself out to charm as well as to impress, and publicly acclaimed the Duce as 'one of those lonely men of the ages on whom history is not tested, but who themselves are the makers of history'. Hitler's admiration for Mussolini was unfeigned. Mussolini, like himself—and like Stalin, whom Hitler also admired—was a man of the people; Hitler felt at ease with him as he never felt when with members of the traditional ruling classes.

Three weeks later Ribbentrop appeared in Rome to urge the Duce to put Italy's signature to the year-old Anti-Comintern Pact between Germany and Japan. Ribbentrop was disarmingly frank. He had failed in his mission to London, he told Mussolini, and had to recognize that the interests of Germany and Great Britain were irreconcilable. This was excellent hearing for the Duce, and he made little difficulty about signing the Pact. In return, Mussolini told Ribbentrop that he was tired of mounting guard over Austrian independence, especially if the Austrians no longer wanted their independence.

Hitler's exploitation of the quarrel between Italy and the Western Powers was beginning to yield dividends; in his cultivation cf Mussolini's friendship he had found the key to unlock the gate to Central Europe.

Hitler's interest in Italy did not lead him to neglect Poland. In 1936 the Poles, worried by the growth of Nazi influence in Danzig and still distrustful of Germany's fair words, tried to strengthen their ties with France. Well aware of the stiffening in the Polish attitude, Hitler and his aides gave the most convincing assurances. In Danzig, Hitler declared, Germany would act entirely by way of an understanding with Poland, and with respect for all her rights. Ribbentrop laid heavy emphasis on the common interests of Poland and Germany in face of the menace of Bolshevism. Göring assured the Poles that Germany would not attack them and told them in confidence that Germany needed a strong Poland; a weak Poland would be a standing invitation to Russian aggression, and for that reason Germany had no quarrel with the Franco-Polish alliance.

Hitler followed these reassurances by offering to negotiate a minorities treaty with Poland, which was signed in Berlin on 5 November. So long as Poland refused to cooperate against Germany, it was impossible to build up effective resistance to Hitler's eastern ambitions. If Italy's friendship was the key to Austria, Poland's was one of the keys to Czechoslovakia.

The German denunciation of the Locarno Pact had been followed by the reversion of Belgium to a professed policy of neutrality which, in King Leopold's words, 'should aim resolutely at placing us outside any dispute of our neighbours'. The withdrawal of Belgium was a further stage in the disintegration of the European Security system. Yet London and Paris still did not give up their attempts to reach some form of general agreement with Hitler.

A new impetus was given to these dragging negotiations with the replacement of Mr. Baldwin by Neville Chamberlain as Prime Minister at the end of May 1937. Baldwin has been characterized by Sir Winston Churchill as possessing a genius for waiting upon events, knowing little of Europe, and disliking what he knew. 'Neville Chamberlain, on the other hand, was alert, business-like, opinionated and self-confident in a very high degree. . . . His all-pervading hope was to go down in history as the great Peacemaker; and for this he was prepared to strive continually in the teeth of facts, and face great risks for himself and his country.'

The first fruits of Chamberlain's policy were the visit of Lord Halifax, then Lord President of the Council, to Germany in November 1937. Hitler, who agreed to receive Halifax at Berchtesgaden, showed himself both wilful and evasive. It was impossible, he declared, to make agreements with countries where political decisions were dictated by party considerations and were at the mercy of the Press. The British could not get used to the fact that Germany was no longer weak and divided; any proposal he made was automatically suspected, and so on.

Later, after Halifax had reported, Chamberlain wrote in his private journal: 'The German visit was from my point of view a great success because it achieved its object, that of creating an atmosphere in which it is possible to discuss with Germany the practical questions involved in a European settlement.' However sincere Chamberlain's desire to reach a settlement with Germany, in practice it amounted to an invitation to diplomatic blackmail which Hitler was not slow to exploit.

On 5 November, Hitler disclosed something of his real thoughts to a small group of men in a secret meeting at the Reich Chancellery. It was the same day the treaty with Poland was signed, and a fortnight before he listened to Chamberlain's well-meant messages. Only five others were present besides himself and Colonel Hossbach, the adjutant whose minutes are the source of our information. They were Field-Marshal von Blomberg, the German War Minister; Colonel-General von Fritsch, Commander-in-Chief of the Army; Admiral Raeder, Commander-in-

Chief of the Navy; Göring, Commander-in-Chief of the Air Force, and Neurath, the German Foreign Minister.

Hitler began by explaining that what he had to say was the fruit of his deliberation and experiences during the past four and a half years. Then he put the problem in the simplest terms: 'The aim of German policy was to make secure and to preserve the racial community and to enlarge it. It was therefore a question of space.' Germany's future, Hitler declared, could only be safeguarded by acquiring additional *Lebensraum* in Europe, and it could be found only at the risk of conflict.

Germany had to reckon with two 'hate-inspired antagonists'—not Russia despite all Hitler's talk of the Bolshevik menace, but Britain and France. Neither country was so strong as appeared. There were signs of disintegration in the British Empire—Ireland, India, the threat of Japanese power in the Far East and of Italian in the Mediterranean. In the long run, the Empire could not maintain its position. France's situation was more favourable than that of Britain, but she was confronted with internal political difficulties.

Germany's problem could be solved only by force, and there were several things to be considered. The peak of German power would be reached by 1943–5. After that, equipment would become obsolete, and the rearmament of the other Powers would reduce the German lead. Germany's problem of *Lebensraum* must be solved by 1943–5 at the latest.

However, the necessity for action could arise before that date. If internal strife in France reached such a pitch as to disable the French Army, Germany must immediately strike a blow against the Czechs. Another possibility was that France might become involved in war with another state and so could not take action against Germany. The diversion might emerge from the present tension in the Mediterranean, and Hitler was resolved to take advantage of any French or British military preoccupation to attack the Czechs.

In any case, the first objective must be to overrun Czechoslovakia and Austria and so secure Germany's eastern and southern flanks.

The significance of this meeting in November 1937 has been a subject of considerable controversy. It is surely wrong to suggest, as has William Shirer, that this was the occasion when 'the die was cast. Hitler had communicated his irrevocable decision to go to war.' Hitler was far too skilful a politician to make an irrevocable decision on a series of hypothetical assumptions. Far from working to a timetable, he was an opportunist, prepared to profit by whatever turned up, to wait for the mistakes made by others.

It is far more probable, therefore, that the reason for the meeting which Hossbach recorded was to override the doubts about the pace of rearmament expressed by General Fritsch, and earlier by Schacht, than to announce some newly conceived decision to commit Germany to a course deliberately aimed at war.

But to look for such a decision is to misunderstand the character of Hitler's foreign policy and his responsibility for the war which followed. For while Hitler's tactics were always those of an opportunist, the aim of his foreign policy never changed from its first definition in *Mein Kampf* in the 1920s to the attack on Russia in 1941: German expansion towards the East. Such a policy, as Hitler explicitly recognized in *Mein Kampf*, involved the use of force and the risk of war. He repeated this in November 1937: 'Germany's problem could only be solved by force and this was never without attendant risk.' What changed was not the objective or the means, but Hitler's judgement of the risks he could afford to run.

In his first four years of power, Hitler relied upon his skill as a politician to exploit the divisions, feebleness of purpose and bad conscience of the other Powers to win a series of diplomatic successes without even the display of force. In 1938–9, with German rearmament under way and his confidence fortified by success, he was prepared to take bigger risks and to invoke the threat of force if his claims were refused.

It is in this context that the meeting of November 1937 is to be seen. The harangue which Hitler delivered reflects the change of mood at the end of the first period and the opening of the second, a new phase in which Hitler was ready to increase the pressure and enlarge the risks of his foreign policy.

The picture he had formed of the immediate future was inaccurate. Events did not follow the course he predicted; war came at a date and as a result of a situation he had not foreseen. But the inaccuracy of the details matters little, for Hitler was ready to take advantage of any situation that emerged. The importance of the occasion lies in the changed tone in which Hitler spoke, in his readiness to run the risk of war and to annex Czechoslovakia and Austria whenever circumstances offered a favourable opportunity, 'even as early as 1938'.

The years of preparation and concealment were at an end: the Man of Peace was to give way to the Man of Destiny, a new role in which, by March 1939, Hitler was to achieve both the objectives of November 1937: the annexation of Austria and the conquest of Czechoslovakia.

CHAPTER SEVEN

THE DICTATOR

In the spring of 1938, on the eve of his greatest triumphs, Adolf Hitler entered his fiftieth year. His physical appearance was unimpressive, his bearing still awkward. The falling lock of hair and the smudge of his moustache added nothing to a coarse and curiously undistinguished face, in which the eyes alone attracted attention. The quality which his face possessed was mobility, an ability to express the most rapidly changing moods, at one moment smiling and charming, at another cold and imperious, cynical and sarcastic, or swollen and livid with rage.

Speech was the essential medium of his power. Hitler talked incessantly, often using words less to communicate his thoughts than to release the hidden spring of his own and others' emotions, whipping himself and his audience into anger or exaltation by the sound of his voice. Talk had another function, too. 'Words,' he once said, 'build bridges into unexplored regions.' As he talked, conviction would grow until certainty came and the problem was solved.

Hitler always showed a distrust of argument and criticism. Unable to argue coolly himself, since his early days in Vienna his one resort had been to shout his opponent down. The questioning of his assumptions or of his facts rattled him; the introduction of intellectual processes of criticism and analysis marked the intrusion of hostile elements which disturbed the exercise of manipulating emotion. Hence Hitler's hatred of the intellectual: in the masses 'instinct is supreme and from instinct comes faith. . . . While the healthy common folk instinctively close their ranks to form a community of the people, the intellectuals run this way and that, like hens in a poultry-yard. With them it is impossible to make history; they cannot be used as elements supporting a community.'

As an orator Hitler had obvious faults. The timbre of his voice was harsh; he spoke at too great length; he was often repetitive and verbose; he lacked lucidity. These shortcomings, however, mattered little beside the extraordinary impression of force, the immediacy of passion, the intensity of hatred, fury, and menace conveyed by the sound of the voice alone.

One of the secrets of his mastery over a great audience was his instinctive sensitivity to the mood of a crowd, a flair for divining the hidden passions, resentments and longings in their minds.

One of his most bitter critics, Otto Strasser, wrote: 'Hitler responds to the vibration of the human heart with the delicacy of a seismograph, or perhaps of a wireless receiving set, enabling him, with a certainty with which no conscious gift could endow him, to act as a loudspeaker proclaiming the most secret desires, the least admissible instincts, the sufferings, and personal revolts of a whole nation. . . . Adolf Hitler enters a hall. He sniffs the air. For a minute he gropes, feels his way, senses the atmosphere. Suddenly he bursts forth. His words go like an arrow to their target, he touches each private wound on the raw, liberating the mass unconscious, expressing its innermost aspirations, telling it what it most wants to hear.'

Hitler had another role, that of visionary and prophet. This was the mood in which he indulged, talking far into the night, in his house on the Obersalzberg, surrounded by the remote peaks and silent forests of the Bavarian Alps; or in the Eyrie he had built six thousand feet up above the Berghof, approached only by a mountain road blasted through the rock and a lift guarded by doors of bronze. There he would elaborate his fabulous schemes for a vast empire; his plans for breeding a new élite biologically preselected; his design for reducing whole nations to slavery in the foundation of his new empire. These themes had long fascinated Hitler, but he was to show that he was capable of translating his fantasies into a terrible reality. The invasion of Russia, the S.S. extermination squads, the planned elimination of the Jewish race; the treatment of the Poles and Russians, the Slav *Untermenschen*—these, too, were the fruits of Hitler's imagination.

All this combines to create a picture of which the best description is Hitler's own famous sentence: 'I go the way that Providence dictates with the assurance of a sleepwalker.' Hermann Rauschning writes: 'Dostoevsky might well have invented him, with the morbid derangement and the pseudo-creativeness of his hysteria.' With Hitler, indeed, one is uncomfortably aware of never being far from the realm of the irrational.

The baffling problem about this strange figure is to determine the degree to which he was swept along by a genuine belief in his own inspiration and the degree to which he deliberately exploited the irrational side of human nature, both in himself and others, with a shrewd calculation. For it is salutary to recall, before accepting the Hitler Myth at anything like its face value, that it was Hitler who invented the myth, assiduously cultivating and manipulating it for his own ends. So long as he did this he was brilliantly successful; it was when he began to believe in his own magic, and accept the myth of himself as true, that his flair faltered.

So much has been made of the charismatic nature of Hitler's leadership that it is easy to forget the astute and cynical politician in him. It is this mixture of calculation and fanaticism, with the difficulty of telling where one ends and the other begins, which is the peculiar characteristic of Hitler's personality:

to ignore or underestimate either element is to present a distorted picture.

The link between the different sides of Hitler's character was his extraordinary capacity for self-dramatization. Again and again one is struck by the way in which, having once decided rationally on a course of action, Hitler would whip himself into a passion which enabled him to bear down all opposition, and provided him with the motive power to enforce his will on others. He was capable of a calculated synthetic fury in which he worked himself into a frenzy of indignation, which he could assume or discard at will in order to gain concessions from foreign governments anxious for peace.

One of Hitler's most habitual devices was to place himself on the defensive, to accuse those who opposed him of aggression and malice, and to pass rapidly from a tone of outraged innocence to the full thunders of moral indignation. It was always the other side who were to blame, and in turn he denounced the Communists, the Jews, the Republican Government, or the Czechs, the Poles, and the Bolsheviks for their 'intolerable' behaviour which forced him to take drastic action in self-defence.

Hitler in a rage appeared to lose all control of himself. His face became mottled and swollen with fury, he screamed at the top of his voice, spitting out a stream of abuse, waving his arms wildly and drumming on the table or the wall with his fists. As suddenly as he had begun he would stop, smooth down his hair, straighten his collar and resume a more normal voice.

This skilful and deliberate exploitation of his own temperament extended to other moods. When he wanted to persuade or win someone over he could display great charm. Until the last days of his life he retained an uncanny gift of personal magnetism which defies analysis, but which many who met him have described. This was connected with the curious power of his eyes, which are persistently said to have had some sort of hypnotic quality. Similarly, when he wanted to frighten or shock, he showed himself a master of brutal and threatening language.

Yet another role was the impression of concentrated will-power and intelligence, the leader in complete command of the situation and with a dazzling knowledge of the facts. To sustain this part he drew on his remarkable memory, which enabled him to reel off complicated orders of battle, technical specifications and long lists of names and dates without a moment's hesitation. The fact that the details and figures which he cited were often found to contain inaccuracies did not matter: it was the immediate effect at which he aimed. The swiftness of the transition from one mood to another was startling: one moment his eyes would be filled with tears and pleading, the next blazing with fury, or glazed with the faraway look of the visionary.

Hitler, in fact, was a consummate actor, with the actor's and orator's facility for absorbing himself in a role and convincing himself of the truth of what he was saying at the time he said it. In his early years he was often awkward and unconvincing, but with practice the part became second nature to him, and with the immense prestige of success behind him, and the resources of a powerful state at his command, there were few who could resist the impression of the piercing eyes, the Napoleonic pose, and the 'historic' personality.

Hitler had the gift of all great politicians for grasping the possibilities of a situation more swiftly than his opponents. He saw, as no other politician did, how to play on the grievances and resentments of the German people, as later he was to play on French and British fear of war and fear of Communism. His insistence upon preserving the forms of legality in the struggle for power showed a brilliant understanding of the way to disarm opposition.

A German word, *Fingerspitzengefühl*—'finger-tip feeling'—which was often applied to Hitler, well describes his sense of opportunity and timing. Hitler knew how to wait. Clear enough about his objectives, he contrived to keep his plans flexible, and until he was convinced that the right moment had come, he would find a hundred excuses for procrastination. Once he had made up his mind to move, however, he would act boldly, taking considerable risks.

Surprise was a favourite gambit of Hitler's, in politics, diplomacy, and war: he gauged the psychological effect of sudden, unexpected hammer-blows in paralysing opposition. In war the psychological effect of the *Blitzkrieg* was just as important in Hitler's eyes as the strategic: it gave the impression that the German military machine was more than life-size, that it possessed some virtue of invincibility against which ordinary men could not defend themselves.

To attend one of Hitler's big meetings was to go through an emotional experience, not to listen to an argument or a programme. Every device for heightening the emotional intensity, every trick of the theatre was used. The Nuremberg rallies held every year in September were masterpieces of theatrical art, with the most carefully devised effects. To see the films of the rallies even today is to be recaptured by the hypnotic effect of thousands of men marching in perfect order, the music of the massed bands, the forest of standards and flags, the vast perspectives of the stadium, the smoking torches, the dome of searchlights. The sense of power, of force and unity was irresistible, and all converged with a mounting crescendo of excitement on the supreme moment when the Führer himself made his entry. Paradoxically, the man who was most affected by such spectacles was their originator, Hitler himself, and they played an indispensable part in the process of self-intoxication.

Hitler had grasped as no one before him what could be done with a combination of propaganda and terrorism. For the complement to the attractive power of the great spectacles was the compulsive power of the Gestapo, the S.S., and the concentration camp, heightened once again by skilful propaganda. Hitler was helped in this not only by his own perception of the sources of power in a modern urbanized mass-society, but also by possession of the technical means to manipulate them.

In making use of his formidable power Hitler had one supreme, and fortunately rare, advantage: he had neither scruples nor inhibitions. He was a man without roots or loyalties, and he felt respect for neither God nor man. Throughout his career he showed himself prepared to seize any advantage that was to be gained by lying, cunning, treachery, and unscrupulousness. He demanded the sacrifice of millions of German lives for the sacred cause of Germany, but in the last year of the war was ready to destroy Germany rather than surrender his power or admit defeat.

Wary and secretive, he entertained a universal distrust. He admitted no one to his counsels. 'He never', Schacht wrote, 'let slip an unconsidered word. He never said what he did not intend to say and he never blurted out a secret. Everything was the result of cold calculation.'

He had a particular and inveterate distrust of experts. He refused to be impressed by the complexity of problems, insisting until it became monotonous that if only the will was there any problem could be solved. Schacht, to whose advice he refused to listen and whose admiration was reluctant, says of him: 'Hitler often did find astonishingly simple solutions for problems which had seemed to others insoluble. He had a genius for invention. . . . His solutions were often brutal, but almost always effective.'

The crudest of Hitler's simplifications was the most effective: in almost any situation, he believed, force or the threat of force would settle matters—and in an astonishingly large number of cases he proved right.

In his Munich days Hitler always carried a heavy riding-whip, made of hippopotamus hide. The impression he wanted to convey was one of force, decision, will. Yet Hitler had nothing of the easy, assured toughness of a condottiere like Göring. His strength of personality, far from being natural to him, was the product of an exertion of will: from this sprang a harsh, jerky and over-emphatic manner which was very noticeable in his early days as a politician.

To say that Hitler was ambitious scarcely describes the intensity of the lust for power and the craving to dominate which consumed him. It was the will to power in its crudest and purest form, not identifying itself with the triumph of a principle as with Lenin or Robespierre—for the only principle of Nazism was power and domination for its own sake—nor finding satisfaction in the fruits of power, for, by comparison with other Nazi leaders like Göring, Hitler lived an ascetic life. For a long time Hitler succeeded in identifying his own power with the recovery of Germany's old position in the world, but as soon as the interests of Germany began to diverge from his own, his patriotism was seen at its true value—Germany, like everything else in the world, was only a means, a vehicle for his own power, which he would sacrifice with the same indifference as the lives of those he sent to the Eastern Front.

Although, looking backwards, it is possible to detect anticipations of this monstrous will to power in Hitler's early years, it remained latent until the end of the First World War and only began to appear noticeably when he reached his thirties. From the account in *Mein Kampf* it appears that the shock of defeat and the Revolution of November 1918 produced a crisis in which hitherto dormant faculties were awakened and directed towards the goal of becoming a politician and founding a new movement.

Resentment is marked in Hitler's attitude, and hatred intoxicated him. Many of his speeches are long diatribes of hate—against the Jews, the Marxists, the Czechs, the Poles, the intellectuals and the educated middle-classes who belonged to that comfortable bourgeois world which had once rejected him and which he was determined to destroy in revenge.

No less striking was his constant need of praise. The atmosphere of adulation in which he lived seems to have deadened the critical faculties of all who came into it. The most banal platitudes and the most grotesque errors of taste and judgement, if uttered by the Führer, were accepted as the words of inspired genius. It is to the credit of Röhm and Gregor Strasser, who had known Hitler for a long time, that they were irritated and totally unimpressed by this Byzantine attitude towards the Führer; no doubt, this was among the reasons why they were murdered.

A hundred years before Hitler became Chancellor, Hegel had pointed to the role of 'world-historical individuals' as the agents by which 'the Will of the World Spirit', the plan of Providence, is carried out.

Whether Hitler ever read Hegel or not, Hegel's writings about the Hero who is above conventional morality find an echo in Hitler's belief about himself. Cynical through he was, Hitler's cynicism stopped short of his own person: he came to believe that he was a man with a mission, marked out by Providence, and therefore exempt from the ordinary canons of human conduct.

Just before the occupation of Austria, in February

1938, he declared in the Reichstag: 'Above all, a man who feels it his duty at such an hour to assume the leadership of his people is not responsible to the laws of parliamentary usage or to a particular democratic conception, but solely to the mission placed upon him. And anyone who interferes with this mission is an enemy of the people.'

It was in this sense of mission that Hitler, a man who believed neither in God nor in conscience ('a Jewish invention, a blemish like circumcision'), found both justification and absolution. He was the Siegfried come to reawaken Germany to greatness, for whom morality and suffering were irrelevant. So long as this sense of mission was balanced by the cynical calculations of the politician, it represented a source of strength, but success was fatal. When half Europe lay at his feet and all need of restraint was removed, Hitler abandoned himself entirely to megalomania. Ironically, failure sprang from the same capacity which brought him success, his power of self-dramatization, his ability to convince himself. Hitler played out his 'world-historical' role to the bitter end. But it was this same belief which curtained him in illusion and blinded him to what was actually happening, leading him into that arrogant overestimate of his own genius which brought him to defeat. The sin which Hitler committed was that which the ancient Greeks called *hybris*, the sin of overweening pride, of believing himself to be more than a man. No man was ever more surely destroyed by the image he had created than Adolf Hitler.

After he became Chancellor Hitler had to submit to a certain degree of routine. This was against his natural inclination. He hated systematic work, hated to submit to any discipline, even self-imposed. Administration bored him and he habitually left as much as he could to others, an important fact in explaining the power of men like Hess and Martin

Bormann, who relieved him of much of his paperwork.

When he had a big speech to prepare he would put off beginning work on it until the last moment. Once he could bring himself to begin dictating he worked himself into a passion, rehearsing the whole performance and shouting so loudly that his voice echoed through the neighbouring rooms. The speech composed, he would invite his secretaries to lunch, praising and flattering them.

Most North Germans regarded such *Schlamperei,* slovenliness, and lack of discipline as a typical Austrian trait. In Hitler's eyes it was part of his artist nature: he should have been a great painter or architect, he complained, and not a statesman at all.

He passionately hated all forms of modern art, a term in which he included most painting since the Impressionists. His taste was for the Classical models of Greece and Rome, and for the Romantic, and he had a particular fondness for nineteenth-century painting of the more sentimental type. He admired painstaking craftsmanship, and habitually kept a pile of paper on his desk for sketching in idle moments.

Architecture appealed strongly to him—especially Baroque—and he had grandiose plans for the rebuilding of the big German cities. The qualities which attracted him were the monumental and the massive; the architecture of the Third Reich, like the Pyramids, was to reflect the power of its rulers.

Hitler looked upon himself not only as a connoisseur of painting and an authority on architecture, but as highly musical. In fact, his liking for music did not extend very much further than Wagner, some of Beethoven, and Bruckner, light opera and operettas. Hitler never missed a Wagner festival at Bayreuth and he claimed to have seen such operas as *Die Meistersinger* and *Götterdämmerung* more than a hundred times. He was equally fond of the cinema, and at the height of the political struggle in 1932 he and Goebbels would slip into a picture-house to see *Mädchen in Uniform*, or Greta Garbo. When the Chancellery was rebuilt he had projectors and a screen installed on which he frequently watched films in the evening, including many of the foreign films he had forbidden in Germany.

Hitler rebuilt both the Chancellery and his house on the Obersalzberg after he came to power, the original Haus Wachenfeld becoming the famous Berghof. He had a passion for big rooms, thick carpets, and tapestries. Rauschning speaks of 'the familiar blend of *petit bourgeois* pleasures and revolutionary talk'. He liked to be driven fast in a powerful car; he liked cream cakes and sweets (specially supplied by a Berlin firm); he liked flowers in his rooms, and dogs; he liked the company of pretty—but not clever—women; he liked to be at home up in the Bavarian mountains.

It was in the evenings that Hitler's vitality rose. He hated to go to bed—for he found it hard to sleep—and after dinner he would gather his guests and his household and sit talking with them until two or three o'clock in the morning, often later. For long periods the conversation would lapse into a monologue, but to yawn or whisper was to incur immediate disfavour.

There was little ceremony about life at the Berghof. Hitler had no fondness for formality or for big social occasions. Although he lived in considerable luxury, he was indifferent to the clothes he wore, ate very little, never touched meat, and neither smoked nor drank.

The chief reason for Hitler's abstinence seems to have been anxiety about his health. He lived an unhealthy life; he took part in no sport, never rode or swam, and suffered a good deal from stomach disorders as well as from insomnia. He was depressed at the thought of dying early, before he had had time to complete his schemes, and he hoped to add years to his life by careful dieting and avoiding stimulants. He became a crank as well as a hypochondriac, and preached the virtues of vegetarianism to his guests at table with the same insistence as he showed in talking politics.

Hitler had been brought up as a Catholic and was impressed by the organization and power of the Church. For the Protestant clergy he felt only contempt: 'They are insignificant little people, submissive as dogs, and they sweat with embarrassment when you talk to them. They have neither a religion they can take seriously nor a great position to defend like Rome.' It was 'the great position' of the Church that he respected; towards its teaching he showed the sharpest hostility. In Hitler's eyes Christianity was a religion fit only for slaves; he detested its ethics in particular. Its teaching, he declared, was a rebellion against the natural law of selection by struggle and the survival of the fittest. 'Taken to its logical extreme, Christianity would mean the systematic cultivation of the human failure.' From political considerations he restrained his anti-clericalism, seeing clearly the dangers of strengthening the Church by persecution. Once the war was over, he promised himself, he would root out and destroy the influence of the Christian Churches, but until then he would be circumspect.

There is no evidence to substantiate the once popular belief that Hitler resorted to astrology. He was a rationalist and a materialist, with no feeling or understanding for either the spiritual side of human life or its emotional, affective side. Emotion to him was the raw material of power. The pursuit of power cast its harsh shadow like a blight over the whole of his life. Everything was sacrificed to the 'world-historical' image; hence the poverty of his private life and of his human relationships.

After his early days in Munich, Hitler made few,

if any, friends. In a nostalgic mood he would talk regretfully of the *Kampfzeit*, the Years of the Struggle, and of the comradeship he had shared with the *Alte Kämpfer*, the Old Fighters. With almost no exceptions, Hitler's familiars belonged to the Nazi Old Guard. It was in this intimate circle, talking over the old days, that Hitler was most at his ease. He never lost his distrust of those who came from the bourgeois world. It was on the Old Guard alone that he believed he could rely, for they were dependent on him. More than that, he found such company, however rough, more congenial than that of the bankers and generals, high officials and diplomats, whose stiff manners and 'educated' talk roused all his old class resentment and the suspicion that they sneered at him behind his back—as they did.

Hitler enjoyed the company of women. Many women were fascinated by his hypnotic powers; there are well-attested accounts of the hysteria which affected them at his big meetings. If ladies were present at table he knew how to be attentive and charming, as long as they had no intellectual pretensions and did not try to argue with him. Gossip connected his name with that of a number of women in whose company he had been frequently seen, and speculated eagerly on his relations with them.

Much has been written, on the flimsiest evidence, about Hitler's sex life. Amongst the mass of conjecture, two hypotheses are worth serious consideration. The first is that Hitler was affected by syphilis.

There are several passages in *Mein Kampf* in which Hitler speaks with surprising emphasis of 'the scourge of venereal disease' and its effects. According to certain reports, Hitler contracted syphilis while he was a young man in Vienna. More than one medical specialist has suggested that Hitler's later symptoms—psychological as well as physical—could be those of a man suffering from the tertiary stage of syphilis. Unless, however, a medical report on Hitler should some day come to light this must remain an open question.

A second hypothesis is that Hitler was incapable of normal sexual intercourse. Putzi Hanfstängl, who knew Hitler well in his Bavarian days and later, says plainly that he was impotent. The gallantry, the hand-kissing and flowers, were an expression of admiration but led to nothing more. 'A bit of petting may have gone on, but that, it became clear to me, was all that Hitler was capable of. My wife summed him up very quickly: "Putzi," she said, "I tell you he is a neuter." '

This too must remain a hypothesis, but it is not inconsistent with what is known of Hitler's relations with the only two women in whom he showed more than a passing interest—his niece, Geli Raubal, and the woman he married on the day before he took his life, Eva Braun.

Geli and Friedl Raubal, the daughters of Hitler's widowed half-sister, Angela Raubal, accompanied their mother when she came to keep house for Hitler on the Obersalzberg in 1925. Geli was then seventeen, simple and attractive, with a pleasant voice which she wanted to have trained for singing. During the next six years she became Hitler's constant companion. This period in Munich Hitler later described as the happiest in his life; he idolized this girl, who was twenty years younger than himself, took her with him whenever he could—in short, he fell in love with her. Whether Geli was ever in love with him is uncertain. She was flattered and impressed by her now famous uncle, but she suffered from his hypersensitive jealousy. Hitler refused to let her have any life of her own; he refused to let her go to Vienna to have her voice trained; he was beside himself with fury when he discovered that she had allowed Emil Maurice, his chauffeur, to make love to her, and forbade her to have anything to do with any other man. Geli resented and was made unhappy by Hitler's possessiveness and domestic tyranny.

On the morning of 17 September 1931, Hitler left Munich with Hoffmann, his photographer and friend, after saying good-bye to Geli. He was bound for Hamburg, but had only got beyond Nuremberg when he was informed that Geli had shot herself in his flat shortly after his departure. Whatever the reason for Geli's suicide, her death dealt Hitler a greater blow than any other event in his life. For days he was inconsolable and his friends feared that he would take his own life. He never spoke of Geli again without tears coming into his eyes; according to his own statement to a number of witnesses, she was the only woman he ever loved. Her room at the Berghof was kept exactly as she had left it. Her photograph hung in his room in Munich and Berlin, and flowers were always placed before it on the anniversary of her birth and death. There are mysteries in everyone's personality, not least in that strange, contradictory, and distorted character which was Adolf Hitler, and it is best to leave it as a mystery.

Hitler's relations with Eva Braun were on a different level. As Speer later remarked, 'For all writers of history, Eva Braun is going to be a disappointment.'

Eva, the daughter of a Bavarian craftsman, was a pretty, empty-headed blonde who worked in Hoffmann's photographer's shop. Hitler occasionally invited her to be one of his party on an outing, but the initiative was on Eva's side.

Less than a year after Geli's death, Eva Braun, then twenty-one, attempted to commit suicide. Hitler was understandably sensitive to such a threat at a time when he was anxious to avoid any scandal and, according to Hoffmann, 'it was in this manner that Eva Braun got her way and became Hitler's *chère amie*'.

Eva was kept very much in the background. She stayed at Hitler's flat or the Berghof, but was rarely allowed in public with Hitler and had to stay in her room during big receptions or dinners. When Hitler's half-sister, Frau Raubal, who still kept house at the Berghof after Geli's death and hated the upstart Eva, left for good in 1936, Eva took her place as *Hausfrau* and sat on Hitler's left hand when he presided at lunch. Only after Eva's sister, Gretl, married Fegelein, Himmler's personal representative with the Führer, during the war, was Eva allowed to appear more freely in public. She could then be introduced as Frau Fegelein's sister and the Führer's reputation preserved untarnished.

Eva made no pretensions to intellectual gifts or to any understanding of politics. Her interests in life were sport—she was an excellent skier and swimmer—animals, the cinema, sex, and clothes. In return for her privileged position she had to submit to the same petty tyranny that Hitler had attempted to establish over Geli. She only in secret dared to dance or smoke, because the Führer disapproved of both; she lived in constant terror lest a chance photograph or remark should rouse Hitler's anger at her being in the company of other men.

After the beginning of the war Eva's position became more secure. Hitler cut himself off from all social life and was wholly absorbed in the war. She had no more rivals to fear, and the liaison had now lasted so long that Hitler accepted her as a matter of course. In time, Hitler became genuinely fond of Eva. Her empty-headedness did not disturb him; on the contrary, he detested women with views of their own. It was her loyalty which won his affection and it was as a reward for her loyalty that, after more than twelve years of a relationship which was more domestic than erotic in character, Hitler finally gave way and on the last day of his life married her.

It may well be doubted whether Hitler, absorbed in

the dream of his own greatness, ever had the capacity to love anyone deeply. At the best of times he was never an easy man to live with: his moods were too incalculable, his distrust too easily aroused. He was quick to imagine and slow to forget a slight; there was a strong strain of vindictiveness in him which often found expression in a mean and petty spite.

There is no doubt that Hitler, if he was in the right mood, could be an attractive, indeed a fascinating companion. On the outings in which he delighted he not only showed great capacity for enjoyment himself, but put others at their ease. He could talk well and he had the actor's gift of mimicry to amuse his companions. On the other hand, his sense of humour was strongly tinged with a malicious pleasure in other people's misfortunes or stupidities. He would laugh delightedly at the description by Goebbels of the indignities the Jews had suffered at the hands of the Berlin S.A. Indifferent towards the sufferings of others, he lacked any feeling of sympathy, was intolerant and callous, and filled with contempt for the common run of humanity. Pity and mercy he regarded as humanitarian clap-trap and signs of weakness. The more absorbed he became by the arrogant belief in his mission and infallibility the more complete became his loneliness, until in the last years of his life he was cut off from all human contact and lost in a world of inhuman fantasy where the only thing that was real or mattered was his own will.

'A man who has no sense of history,' Hitler declared, 'is like a man who has no ears or eyes.' He claimed to have had a passionate interest in history since his schooldays and he displayed considerable familiarity with the course of European history. His conversation was studded with historical references and parallels. More than that: Hitler's whole cast of thought was historical, and his sense of mission derived from his sense of history.

Like his contemporary Spengler, Hitler was fascinated by the rise and fall of civilizations. He saw himself born at a similar critical moment in European history when the liberal bourgeois world of the nineteenth century was disintegrating. What would take its place? The future lay with the 'Jewish-Bolshevik' ideology of the masses unless Europe could be saved by the Nazi racist ideology of the élite. This was his mission and he drew upon history to fortify him in it.

To his *Weltanschauung*, however repellent, Hitler remained remarkably consistent. Once formed, it was rigid and inflexible. Hitler's was a closed mind, violently rejecting any alternative view, refusing to criticize or allow others to criticize his assumptions. Of historical study as a critical discipline, or of the rich fields of human history beside the quest for power, war, and the construction of empires, he was invincibly ignorant.

If Hitler's views on politics were dogmatic and intolerant, his opinions on every other topic were, in addition, banal, narrow-minded, and totally unoriginal as well as harsh and brutal. What he had to say about marriage, women, education, religion, bore the indelible stamp of an innate vulgarity and coarseness of spirit. He was not only cut off from the richest experiences of ordinary human life—love, marriage, family, human sympathy, friendship—but the whole imaginative and speculative world of European literature was closed to him. Everything that spoke of the human spirit and of the thousand forms in which it has flowered, from mysticism to science, was alien to him.

The basis of Hitler's political beliefs was a crude Darwinism. 'Man has become great through struggle. . . . Whatever goal man has reached is due to his originality plus his brutality. . . . All life is bound up in three theses: Struggle is the father of all things, virtue lies in blood, leadership is primary and decisive.' It followed from this that 'through all the centuries force and power are the determining factors. . . . Only force rules. Force is the first law.' Force was more than the decisive factor in any situation; it was force which alone created right. 'Always before God and the world, the stronger has the right to carry through what he wills.'

The ability to seize and hold a decisive superiority in the struggle for existence Hitler expressed in the idea of race, the role of which is as central in Nazi mythology as that of class in Marxist. All that mankind has achieved, Hitler declared in *Mein Kampf*, has been the work of the Aryan race. Although Hitler frequently talked as if he regarded the whole German nation as of pure Aryan stock (whatever that may mean) his real view was rather different. It was only a part of any nation which could be regarded as Aryan. These constituted an élite within the nation which stamped its ideas upon the development of the whole people, and by its leadership gave this racial agglomeration an Aryan character which in origin belonged only to a section. Thus Hitler's belief in race could be used to justify both the right of the German people to ride roughshod over 'inferior' peoples, and the right of the Nazis, representing an élite, to rule over the German people.

This explains why Hitler often referred to the Nazi capture of power in Germany as a racial revolution, since it represented the replacement of one ruling caste by another. As Hitler told Otto Strasser in May 1930: 'We want to make a selection from the new dominating caste which is not moved, as you are, by any ethic of pity, but is quite clear in its own mind that it has the right to dominate others because it represents a better race.'

What Hitler was seeking to express in his use of the word 'race' was his belief in inequality—both between peoples and individuals—as another of the iron laws of Nature. He had a passionate dislike of the egalitarian doctrines of democracy in every field, economic, political and international.

Just as he opposed the concept of 'race' to the democratic belief in equality, so to the idea of personal liberty Hitler opposed the superior claims of the *Volk,* the primitive, instinctive tribal community of blood and soil.

The *Volk* not only gave meaning and purpose to the individual's life, it provided the standard by which all other institutions and claims were to be judged. Justice, truth and the freedom to criticize must all be subordinated to the overriding claims of the *Volk* and its preservation.

As soon as Hitler began to think and talk about the organization of the State, the metaphor which dominated his mind was that of an army. He saw the State as an instrument of power in which the qualities to be valued were discipline, unity and sacrifice. It was from the Army that he took the *Führerprinzip,* the leadership principle, upon which first the Nazi Party, and later the National Socialist State, were built.

In Hitler's eyes the weakness of democracy was that it bred irresponsibility by leaving decisions always to anonymous majorities, thus putting a premium on the avoidance of difficult and unpopular decisions. At the same time, the Party system, freedom of discussion and freedom of the Press sapped the unity of the nation. From this, he told the Hitler Youth, 'we have to learn our lesson: one will must dominate us, we must form a single unity; one discipline must weld us together; one obedience, one subordination must fill us all, for above us stands the nation.'

'Our Constitution,' wrote Nazi Germany's leading lawyer, Dr. Hans Frank, 'is the will of the Führer.' This was in fact literally true. The Weimar Constitution was never replaced; it was simply suspended by the Enabling Law, which was renewed periodically and placed all power in Hitler's hands.

Yet Hitler was equally careful to insist that his power was rooted in the people; his was a plebiscitary and popular dictatorship, a democratic Caesarism. It is obvious that Hitler felt—and not without justification—that his power, despite the Gestapo and the concentration camps, was founded on popular support to a large degree.

Like all revolutionary movements, Nazism drew much of its strength from the formation of a new leadership drawn from other than the traditional classes. The Party's fourteen years of struggle served as a process of natural selection, and even before coming to power, the Party created the cadres of leadership to take over the State.

Once in power the Party remained the guarantor of the National Socialist character of the State. 'Our Government is supported by two organizations: politically by the community of the *Volk* organized in the National Socialist movement, and in the military sphere by the Army.' These, to use another phrase of Hitler's, were the two pillars of the State. The Party was a power held in reserve to act, if the State should fail to safeguard the interests of the *Volk;* it was the link between the Führer and his *Volk;* finally it was the agent for the education of the people in the Nazi *Weltanschauung*. Education is an ambiguous word in this context; on another occasion Hitler spoke of 'stamping the Nazi *Weltanschauung* on the German people.' 'The main plank in the Nationalist Socialist programme,' Hitler declared in 1937, 'is to abolish the liberalistic concept of the individual and the Marxist concept of humanity and to substitute for them the *Volk* community, rooted in the soil and bound together by the bond of its common blood.'

While Hitler's attitude towards liberalism was

one of contempt, towards Marxism he showed an implacable hostility. The difference is significant. Liberalism had lost its attraction in the age of mass-politics, especially in Germany, where it had never had deep roots. Marxism, however, whether represented by revisionist Social Democracy or revolutionary Communism, was a rival *Weltanschauung* able to exert a powerful attractive force over the masses comparable with that of Nazism. Ignoring the profound differences between Communism and Social Democracy in practice and the bitter hostility between the rival working-class parties, he saw in their common ideology the embodiment of all that he detested—mass democracy and a levelling egalitarianism as opposed to the authoritarian state and the rule of an élite; equality and friendship among peoples as opposed to racial inequality and the domination of the strong; class solidarity versus national unity; internationalism versus nationalism.

The Marxist conception of class war and of class solidarity cutting across frontiers was a particular threat to Hitler's own exaltation of national unity founded on the community of the *Volk*. His single-minded concept of the national interest was to be embodied in, and guaranteed by, the absolutism of the State.

Just as Hitler ascribed to the 'Aryan' all the qualities and achievements which he admired, so all that he hated is embodied in another mythological figure, that of the Jew. There can be little doubt that Hitler believed what he said about the Jews; from first to last his anti-Semitism is one of the most consistent themes in his career, the master idea which embraces the whole span of his thought. In whatever direction one follows Hitler's train of thought, sooner or later one encounters the satanic figure of the Jew, the universal scapegoat. One of Hitler's favourite phrases, which he claimed—very unfairly

—to have taken from Mommsen, was: 'The Jew is the ferment of decomposition in peoples.'

In 1922 Hitler said: 'The Jew has never founded any civilization, though he has destroyed hundreds. He possesses nothing of his own creation to which he can point. Everything he has stolen. . . . In the last resort it is the Aryan alone who can form States and set them on their path to future greatness.' From this early speech through the Nuremberg Laws of 1935 and the pogrom of November 1938 to the destruction of the Warsaw Ghetto and the death camps of Mauthausen and Auschwitz, Hitler's purpose was plain and unwavering. He meant to carry out the extermination of the Jewish race in Europe, using the word 'extermination' not in a metaphorical but in a precise and literal sense as the deliberate policy of the German State—and he very largely succeeded. On a conservative estimate, between four and four and a half million Jews perished in Europe under Hitler's rule—apart from the number driven from their homes who succeeded in finding refuge abroad.

Stripped of their romantic trimmings, all Hitler's ideas can be reduced to a simple claim for power which recognizes only one relationship, that of domination, and only one argument, that of force.

Hitler was not original in this view. Every single one of his ideas—from the exaltation of the heroic leader, the racial myth, anti-Semitism, the community of the *Volk,* and the attack on the intellect, to the idea of a ruling élite, the subordination of the individual and the doctrine that might is right—is to be found in anti-rational and racist writers in most European countries during the hundred years which separate the Romantic movement from the foundation of the Third Reich. By 1914 they had become the commonplaces of radical, anti-Semitic and pan-German journalism in every city in Central Europe, including Vienna and Munich, where Hitler picked them up.

Hitler's originality lay not in his ideas, but in the terrifyingly literal way in which he set to work to translate them into reality, and his unequalled grasp of the means by which to do this. To read Hitler's speeches and table talk is to be struck again and again by the lack of magnanimity or of any trace of moral greatness. His comments on everything except politics display a cocksure ignorance and an ineradicable vulgarity. Yet this vulgarity of mind, like the insignificance of his appearance, was perfectly compatible with brilliant political gifts. Accustomed to associate such gifts with the qualities of intellect which Napoleon possessed, or with the strength of character of a Cromwell or a Lincoln, we are astonished and offended by this combination. Yet to underestimate Hitler as a politician, to dismiss him as an ignorant demagogue, is to make precisely the mistake that so many Germans made in the early 1930s.

It was not a mistake which those who worked closely with him made. Whatever they felt about the man, however much they disagreed with the rightness of this or that decision, they never underrated the ascendancy which he was able to establish over all who came into frequent contact with him. Generals who arrived at his headquarters determined to insist on the hopelessness of the situation not only failed to make any protest when they stood face to face with the Führer, but returned shaken in their judgement and half convinced that he was right after all.

The final test of this ascendancy belongs to the later stages of this history when, with the prestige of success destroyed, the German cities reduced to ruins, and the greater part of the country occupied, this figure, whom his people no longer saw or heard, was still able to prolong the war long past the stage of hopelessness until the enemy was in the streets of Berlin and he himself decided to break the spell. But the events of these earlier years cannot be understood unless it is recognized that, however much in retrospect Hitler may seem to fall short of the stature of greatness, in the years 1938 to 1941, at the height of his success, he had succeeded in persuading a great part of the German nation that in him they had found a ruler of more than human qualities, a man of genius raised up by Providence to lead them into the Promised Land.

CHAPTER EIGHT

FROM VIENNA TO PRAGUE
1938–9

The winter of 1937–8 marks the turning-point in Hitler's policy from the restricted purpose of removing the limitations imposed on Germany by the Treaty of Versailles to the bolder course which brought the spectacular triumphs of the years 1938–41. It was not so much a change in the direction or character of his foreign policy as the opening of a new phase in its development. The time was ripe, he judged, for the realization of aims he had long nurtured.

Hitler had no cut-and-dried views about how he was to proceed, but he was revolving certain possibilities in his mind and, granted favourable circumstances, he was prepared to move against Austria and Czechoslovakia as early as the new year, 1938.

The prospects Hitler had unfolded at the meeting of 5 November 1937 alarmed at least some of those who were present. The Army leaders, Blomberg and Fritsch, feared risking war with Great Britain and France, and were anxious about such material points as the incomplete state of Germany's western fortifications, France's military power and the strength of the Czech defences.

These doubts were not removed by Hitler's irritable assurances that he was convinced Britain would never fight and that he did not believe France would go to war on her own. On 9 November, Fritsch renewed his objections to Hitler. Neurath, the Foreign Minister, also attempted to see Hitler and dissuade him from the course he proposed to follow. Hitler hated to have his intuition subjected to analysis.

Within less than three months of the meeting of 5 November, Blomberg, Fritsch, and Neurath were removed from office, while those who remained were the two who had silenced whatever doubts they felt—Göring and Raeder.

Hitler's most persistent critic, Dr. Hjalmar Schacht, had already gone at the end of 1937. He did not oppose German rearmament: it was Schacht who, by his device of the Mefo-bills, enabled Hitler to finance his big programme of rearmament and public works without an excessive inflation. It was Schacht again who set up the elaborate network of control over German imports, exports, and foreign exchange transactions, and who provided a new basis for Germany's foreign trade by his barter trading, blocked-mark accounts and clearing agreements, manipulating these with such skill as to secure great advantages for Germany in trade negotiations.

There were limits, however, beyond which Schacht as an economist felt it dangerous to make increasing demands on the economy for rearmament, and in 1935–6 he several times warned Hitler that these limits were being approached. Hitler was irritated. In the long run, if he got the arms, he believed that he would be able to solve Germany's economic problems by other than economic means. On the other hand, he needed Schacht, with his unrivalled grasp of finance and foreign trade, to steer Germany through the first difficult years until she was strong enough to take what she wanted.

Ironically, it was Schacht who persuaded Hitler, in April 1936, to appoint Göring as Commissioner for Raw Materials and Foreign Exchange, in the hope that this would put a stop to the extravagant waste of Germany's foreign exchange assets and her limited supplies of raw materials by Party agencies, such as the Ministry of Propaganda. Göring, having once entered the field of economic policy, began to take an interest in what was going on and to amass power.

In September 1936 Hitler named him Plenipotentiary for the Four-Year Plan, a scheme to make Germany self-sufficient which Schacht regarded as impossible. To Schacht's contempt for Göring's ignorance of economics was added the pique of a vain and ambitious man at the rise of a rival power. On 5 September Schacht went on leave of absence from the Ministry of Economics, and his resignation was reluctantly accepted on 8 December 1937. In order to preserve appearances he remained Minister without Portfolio, and for the time being President of the Reichsbank as well. Schacht's successor as Minister of Economics was Walther Funk, but the post was shorn of the greater part of its powers, being wholly subordinated to Göring.

Two principal institutions of the State had so far escaped the process of *Gleichschaltung*—the Foreign Service and the Army. Both were strongholds of that upper-class conservatism which roused all Hitler's suspicion and dislike. Hitler had at first accepted the view that the cooperation of the professional diplomats and the generals was indispensable to him, but he rapidly came to feel contempt for the advice he received from the Foreign Office, whose political as well as social traditions he regarded as too respectable and too limited for the novel, half-revolutionary, half-gangster tactics with which he meant to conduct his foreign policy. Neurath, the Foreign Minister, one of President Hindenburg's appointments, still retained some independence of position, and would have to be replaced.

But the critical relationship was that with the Army. So far the bargain of 1934 had worked well, but not without signs of trouble, which were ominous for the future. The generals, although delighted with the rearmament of Germany, were critical of the speed with which it had been rushed through. The flood of conscripts which began to pour into the depots was more than the four thousand officers of the small Regular Army could train satisfactorily. The figure of thirty-six divisions for the peacetime force which Hitler announced in 1935 had been arbitrarily fixed without the agreement of the General Staff, who would have preferred twenty-one divisions. According to Manstein, who was at that time Chief of Staff at the headquarters of the important Military Area III (in Berlin), he and his commanding officer learned of Hitler's decision for the first time over the radio.

In 1936, again, Hitler had sprung his decision to reoccupy the Rhineland on the Army High Command with the least possible notice. On this occasion Blomberg and Fritsch protested to Hitler at the risks he was running. Hitler did not forget their opposition, more especially as events justified his judgement and not theirs.

If Hitler found it difficult to get on with the stiff, buttoned-up hierarchy of the Army, the generals had their grievances as well. They knew little of what was discussed in the circle round the Führer. The Army had lost its old independent position; there was a new master in Germany, one with whose foreign and internal politics the generals were far from being in agreement. They disliked the Pact with Poland; they were inclined to be friendly with Russia and China, while they had little use for an alliance with Japan and nothing but scorn for Italy. They were alarmed at the prospect of a two-front war after the Franco-Russian Pact, and felt a traditional respect for France as a great military Power, which was at variance with Hitler's contemptuous dismissal of the French as a divided nation. The Party's attitude towards the Churches—in particular the arrest of Pastor Martin Niemöller in July 1937—roused considerable opposition in the Officer Corps. This was reinforced by dislike of the S.S. and S.A., whose ideas began to penetrate the Army as the inevitable result of conscription. The S.S. leader, Himmler, and his chief lieutenant, Heydrich—who had been expelled from the Naval Officer Corps for scandalous conduct in 1930—now entertained the ambition of humbling the proud independence of the Officer Corps, which treated the S.S. and its 'officers' with icy contempt.

The man to whom the Army looked to defend its interests was not the man they now considered Hitler's puppet, Blomberg, who was Minister of War and Commander-in-Chief of the Wehrmacht, but Colonel-General von Fritsch, Commander-in-Chief of the Army (Reichswehr). Fritsch was to prove unequal to the hopes placed in him.

The key to the relationship between the Army and the Nazi régime was Hitler's own attitude. Closer acquaintance had reduced the exaggerated respect he originally felt for the generals. After 1935–6 he saw them as no more than a group of men who, with few exceptions, lacked understanding of anything outside their own highly important but narrow field of specialization, a caste whose pretensions were unsupported by political ability or, when put to the test, by solidarity in face of an appeal to their self-interest. By the beginning of 1938, with his power securely established and with the foundations of German military rearmament solidly laid, he no longer felt the same need to buy the Army's support on its own terms. Thus when Blomberg and Fritsch, on the grounds of Germany's unpreparedness, attempted to apply a brake to the development of his foreign policy, he disregarded their judgement and felt no need to appease their doubts.

Hitler was still smarting with irritation at the opposition Fritsch had expressed in November 1937, when an apparently unconnected series of events provided him with the chance to end once and for all the pretensions of the High Command to independent views.

The trouble began with Blomberg's eagerness to get married to a certain Fräulein Erna Grühn, who,

Blomberg admitted, was a lady with a 'past'. Aware of the rigid views of the Officer Corps on the social suitability of the wife of a Field-Marshal and a Minister of War, Blomberg consulted Göring as a brother-officer. Göring actively encouraged him, and when the marriage took place—very quietly—on 12 January, Hitler and Göring were the two principal witnesses.

At this stage, however, complications arose. A police dossier disclosed that the wife of the Field-Marshal had a police record as a prostitute. Blomberg had dishonoured the Officer Corps, and the generals saw no reason to spare a man whom they had long disliked for his attitude towards Hitler. Supported by Göring, and urged on by the Chief of Staff, General Beck, Fritsch requested an interview with Hitler and presented the Army's protest: Blomberg must go. Hitler eventually agreed, and the question then arose who was to succeed Blomberg.

Fritsch was the obvious candidate, but Himmler opposed him as the man responsible for defeating his attempts to extend the power of the S.S. to the Army, and Göring was ambitious to get Blomberg's place for himself. Indeed, the part played by Göring throughout the whole affair—his encouragement of Blomberg, and the fact that it was he who informed Hitler of the information which had come to light about Erna Grühn after the marriage—invites suspicion. Finally, Hitler too must have hesitated to appoint a man who had shown himself so lukewarm and unconvinced by the Führer's genius at the secret conference on 5 November.

Himmler and Göring settled the matter by again producing a police dossier, this time to show that General von Fritsch had been guilty of homosexual practices. They went further: when Hitler summoned Fritsch to the Chancellery at noon on 26 January and faced him with the charges in the dossier, they arranged for the Commander-in-Chief to be confronted with Hans Schmidt, a young man who made his living by spying on and blackmailing well-to-do homosexuals.

Schmidt identified Fritsch as one of those from whom he had extorted money. The identification was worthless. The officer in question, it later emerged, was not Fritsch at all, but a retired cavalry officer by the name of Frisch. Eventually Schmidt confessed in court that he had been threatened by the Gestapo if he did not agree to their demands and incriminate Fritsch, as Himmler ordered. Schmidt paid for this indiscretion with his life. But by then the trick had served its purpose. In face of Hitler's angry charges, General von Fritsch maintained an indignant silence, and when Hitler sent him on indefinite leave, he refused to take any action to defend himself.

For a few days it looked as if the affair might lead to a major crisis. A furious battle developed behind the scenes round the question whether there should be a court of inquiry into the charges against Fritsch and, if so, who should conduct it, the Army or the Party. Behind this loomed the far bigger question of who was to be given command of the Army, for now not only Blomberg's office of Minister of War and Commander-in-Chief of the Armed Forces, but also Fritsch's as Commander-in-Chief of the Army, had to be filled.

Hitler's solution to the problem was presented to the German Cabinet, when it met for the last time during the Third Reich on Monday, 4 February. After announcing Blomberg's resignation, Hitler added that Fritsch, too, had asked to be relieved of his duties because of ill health. Blomberg's successor as C.-in-C. of the Armed Forces was to be Hitler himself. The old post of War Minister, which Blomberg had held as well, was to be abolished. The work of the War Ministry was henceforward to be done by a separate High Command of the Armed Forces (Oberkommando der Wehrmacht, the familiar O.K.W.

of the war communiqués), which in fact became Hitler's personal staff. To the head of the O.K.W. he appointed a man who was to prove quite incapable of withstanding him, even if he had wanted to—General Wilhelm Keitel.

In General von Brauchitsch Hitler found a man acceptable to the Officer Corps in Fritsch's post of Commander-in-Chief of the Army. At the same time, however, he took the chance to retire sixteen of the senior generals and to transfer forty-four others to different commands. To console Göring, Hitler promoted him to the rank of Field-Marshal, which made him the senior German officer.

The purge extended further than the Army. Neurath was relieved of his office as Foreign Minister, and replaced at the Wilhelmstrasse by the subservient Ribbentrop. The three ambassadors in the key posts of Vienna (Papen), Rome (Hassell), and Tokyo (Herbert von Dirksen) were simultaneously replaced.

Thus Hitler succeeded in removing the few checks which remained upon his freedom of action by using a situation not of his making to establish a still firmer grip upon the control of policy and the machinery of the State. He had replaced Blomberg and Fritsch, Neurath and Schacht, with creatures of his own will—Keitel, Ribbentrop, and Funk—and added to the power in his own hands by assuming direct control of the Armed Forces. When put to the test, the claim of the Army to stand apart from the process of *Gleichschaltung* in the totalitarian state had proved to be hollow.

This last stage in the revolution after power marked the end of the Conservatives' hopes that they still preserved some slender guarantees against the recklessness of the Nazis. It was the prelude to the new era in foreign policy.

About nine o'clock on the evening of 4 February (the day of the Cabinet Meeting in Berlin) Franz von Papen, the Führer's special representative in Vienna, was sitting in his study at the Legation when the telephone bell rang. It was the State Secretary, Lammers, speaking from the Reich Chancellery, with the brief announcement that Papen's mission in Vienna was at an end and he had been recalled.

Papen's surprise was considerable, for he had seen Hitler personally only the week before and nothing had then been said about his recall or transfer. The question everyone asked in Vienna was whether Papen's recall would be followed by a change in German policy towards Austria.

For the past eighteen months, Austro-German relations had been governed by the Agreement of July 1936. The interpretation of that Agreement by the two parties, however, showed wide differences of opinion. The Austrian Government had originally, though not very strongly, hoped that the Agreement might serve as a final settlement of the differences between Germany and themselves. The Germans, on the other hand, clearly regarded it as a lever with which to exercise pressure on the Austrians in such a way that, in the end, Austrian independence would be undermined and the *Anschluss* carried out peacefully, on the initiative of a Nazi-controlled government in Vienna. Chancellor Schuschnigg's room for manoeuvre in resisting such pressure was limited. For there was a strong underground Nazi party in Austria which bitterly resented the policy of peaceful penetration represented by Papen and itched to seize power by a putsch. Hitler had so far refused to give the Austrian Nazis a free hand, but this was an alternative which he could use to bring additional pressure on the Austrian Government if Schuschnigg proved too obstinate or evasive in meeting German demands.

On 25 January 1938 a police raid on the Austrian Nazis' headquarters in Vienna brought to light plans for a rising in the spring of 1938 and for an appeal to Hitler to intervene. Schuschnigg was uneasy and only too well aware of Austria's increased isolation with the establishment of the Rome-Berlin Axis. At the time of Papen's sudden recall, the Austrian Chancellor was already revolving in his mind the advantage of a personal meeting with Hitler to put German-Austrian relations on a clearer footing.

Papen now had very good reasons of his own for advocating such a meeting. Here might be a chance to reinstate himself, and the day after his dismissal he hurried to Berchtesgaden to put the proposal to Hitler personally. The suggestion that Schuschnigg might come to see him aroused Hitler's interest. Ignoring the fact that he had just recalled Papen from his post, he ordered him to return at once to Vienna and make the necessary arrangements.

Schuschnigg tried to safeguard himself by asking for assurances that the 1936 Agreement should remain the basis of Austro-German relations in the future. Papen was perfectly ready to give such assurances, and on the evening of 11 February Schuschnigg and his Secretary of State for Foreign Affairs, Guido Schmidt, quietly left Vienna by train. The next morning they were driven up the mountain roads from Salzburg to the Obersalzberg. Hitler was waiting on the steps of the Berghof when they arrived, and at once conducted the Austrian Chancellor into his study for a private talk before lunch.

Scarcely had they sat down than Hitler, brushing aside Schuschnigg's polite remarks about the view from his window, launched into an angry tirade against the whole course of Austrian policy. Schuschnigg's attempts to interrupt and defend himself were shouted down as Hitler rapidly worked himself into a towering rage. He talked excitedly of his mission: 'I have achieved everything that I set out to do, and have thus become perhaps the greatest German in history.' Characteristically, he began to abuse Schuschnigg for ordering defence works to be constructed on the border. This was an open affront to Germany, which was perfectly capable of exerting 'just revenge'.

Austria, Hitler sneered, was alone: neither France, nor Britain, nor Italy would lift a finger to save her. And now his patience was exhausted. Unless Schuschnigg was prepared to agree, at once, to all that he demanded, he would settle matters by force. For the moment, however, Hitler said nothing of what his demands were. After this two-hour tirade, he suddenly broke off and led his guest in to lunch. Throughout the meal he was charm itself, but the presence of German generals at the table did not escape the Austrians' notice.

In the middle of the afternoon Schuschnigg was taken to see Ribbentrop and Papen. They presented him with a draft of Hitler's demands. The Austrian Government was to lift the ban on the Nazi Party and to recognize that National Socialism was perfectly compatible with loyalty to Austria. Seyss-Inquart, a 'respectable' crypto-Nazi, was to be appointed Minister of the Interior, with control of the police and the right to see that Nazi activity was allowed to develop along the lines indicated. An amnesty for all imprisoned Nazis was to be proclaimed, and Nazi officials and officers who had been dismissed were to be reinstated. Another pro-Nazi, Glaise-Horstenau, was to be appointed Minister of War, and to assure close relations between the German and Austrian Armies there was to be a systematic exchange of officers. Finally, the Austrian economic system was to be assimilated to that of Germany.

Schuschnigg's efforts to secure alterations in the draft, other than minor changes, were unavailing. Hitler told him: 'You will either sign as it is and fulfil my demands within three days, or I will order the march into Austria.' When Schuschnigg explained that, although willing to sign, he could not, by the Austrian Constitution, guarantee ratification, or the observance of the time limit for the amnesty, Hitler flung open the door, and, turning Schuschnigg out, shouted for General Keitel. The effect of the summons was well calculated, and Schmidt remarked to the Chancellor that he would not be surprised if they were arrested in the next five minutes. Half an hour later, however, Hitler again sent for Schuschnigg. 'I have decided to change my mind,' he told him. 'For the first time in my life. But I warn you—this is your very last chance. I have given you three more days before the Agreement goes into effect.'

This was the limit of Hitler's concessions, and the Austrian Chancellor had little option but to sign. His one anxiety now was to get away. Just twenty-four hours after they had arrived at Salzburg the Austrian Chancellor's train set out again for Vienna. Up at the Berghof the Führer relaxed: it had been a highly successful day.

The next day, the 13th, General Alfred Jodl recorded a meeting with Keitel: 'He tells us that the Führer's order is that military pressure by shamming military action should be kept up until the 15th.' On the 14th Jodl added: 'The effect is quick and strong. In Austria the impression is created that Germany is undertaking serious military preparations.' On the 16th, the Austrian Government announced a general amnesty for Nazis and the reorganization of the cabinet, with Seyss-Inquart as Minister of the Interior. The Austrian Nazis were already boasting that within a matter of weeks, if not of days, they would be in the saddle and give Austria a taste of the whip.

At the end of the first week of March Schuschnigg resolved upon a desperate expedient which, he hoped, would destroy the strongest argument Hitler had so far used—that a majority of the Austrian people were in favour of an *Anschluss* with Germany. He determined to hold a plebiscite on Sunday 13 March, in which the Austrian people should be invited to . . . declare whether they were in favour of an Austria which was free and independent, German and Christian.

Hitler does not appear to have been informed of the plebiscite until the afternoon of 9 March. He was furious that Schuschnigg should try to obstruct him in this way, but he had been taken by surprise and nothing was prepared for such an eventuality. German and Austrian Nazi leaders were hastily summoned, and military plans were drawn up.

The question that most preoccupied Hitler was Mussolini's reaction, and he sent off Prince Philip of Hesse to the Duce with a letter which he had instructions to deliver personally. It began with the unconvincing argument that Austria had been conspiring with the Czechs, in order to restore the Hapsburgs and 'to throw the weight of a mass of at least twenty million men against Germany if necessary'. Hitler then continued that the Germans in Austria were being oppressed by their own government. Now, at last, the Austrian people were rising against their oppressors, and Austria was being brought to a state of anarchy. He could no longer remain passive in face of his responsibilities as Führer of the German Reich and as a son of Austrian soil.

In the course of the night 10–11 March Hitler gave his orders to Glaise-Horstenau, who was still waiting in Berlin, and packed him off to Vienna by plane. At two o'clock in the morning Directive No. 1 for Operation Otto, the invasion of Austria, was issued. By the time Hitler went to bed early on the morning of 11 March the Army trucks and tanks were already beginning to roll south towards the frontier.

The same morning, in the Chancellor's flat in Vienna, the telephone woke Schuschnigg from his sleep at half past five. The Chief of Police, Skubl, was on the line: the German border at Salzburg had been closed an hour before, and all rail traffic between the two countries stopped. The Chancellor dressed and drove through the still dark streets to early Mass at St Stephen's Cathedral.

There were other early risers in Vienna that morning. At dawn Papen left by air for Berlin, and not long afterwards the Austrian Minister of the Interior, Seyss-Inquart, was to be seen walking up and down at the airport waiting for the promised message from Hitler. It was brought by Glaise-Horstenau, and at half past nine the two Ministers called on the Chancellor to present Hitler's demands: the original plebiscite must be cancelled and replaced by another to be held in three weeks' time. For two hours the three men argued to and fro: not until after lunch did Schuschnigg, having obtained President Miklas's approval, finally agree. But by now the German demands were being raised. At 2.45 p.m. Göring, on Hitler's instructions, began a series of telephone calls from Berlin to Vienna, demanding first the resignation of Schuschnigg, then the appointment of Seyss-Inquart. Schuschnigg resigned, but President Miklas stubbornly refused to make Seyss-Inquart Chancellor.

At half past five an angry Göring, roaring down the wire from Berlin, demanded to speak to Seyss-Inquart: 'Look here, you go immediately . . . and tell the Federal President that, if the conditions which are known to you are not accepted immediately, the troops already stationed at the frontier will move in tonight along the whole line, and Austria will cease to exist. . . .'

Shortly after half past seven Schuschnigg broadcast to the nation the news that Germany had delivered an ultimatum, and rebutted the lie that civil war had broken out in Austria. The President still refused to appoint Seyss-Inquart as Chancellor.

Some time after eight o'clock Göring rang up again. If Schuschnigg had resigned, Seyss-Inquart should regard himself as still in office and entitled to carry out necessary measures in the name of the Government. Anyone who objected would have to face a German court-martial by the invading troops. With Seyss-Inquart technically in office the façade of legality could be preserved. When Keppler telephoned to Berlin shortly before 9 p.m. to report that Seyss-Inquart was acting as instructed, Göring replied: 'Listen. You are the Government now. Listen carefully and take notes. The following telegram should be sent here to Berlin by Seyss-Inquart: "The provisional Austrian Government, which, after the dismissal of the Schuschnigg Government, considers it its task to establish peace and order in Austria, sends to the German Government the urgent request to support it in its task and to help it prevent bloodshed. For this purpose it asks the German Government to send German troops as soon as possible."' Göring added a few more instructions. Seyss-Inquart was to form a government from the names on the list sent from Berlin; the frontiers were to be watched, to prevent people getting away.

Throughout the country the local Nazis were already seizing town halls and government offices. The threat of a seizure of power by force, which was implicit in the noisy mob filling the street outside the Chancellery, contributed to the atmosphere of compulsion before which in the end President Miklas had to yield, and a little before midnight he capitulated. To avoid bloodshed and in the hope of securing at least the shadow of Austrian independence, he nominated Seyss-Inquart as Federal Chancellor of Austria. At two o'clock in the morning Seyss-Inquart's appeal to Hitler to halt German troops at the frontier was turned down: the occupation must go on.

The hour for crossing the frontier was fixed at daybreak on Saturday, the 12th. Before then Hitler had received the message he had been waiting for all day —news from Rome. When Prince Philip rang up at 10.25 p.m. on the night of 11 March, it was Hitler, not Göring, who came to the telephone to hear the reassuring words: 'The Duce accepted the whole thing in a very friendly manner. He sends you his regards. . . .' Hitler's response was effusive: 'Please tell Mussolini I will never forget him for this. . . . I will never forget, whatever may happen. If he should ever need any help or be in any danger, he can be convinced that I shall stick to him, whatever may happen, even if the whole world were against him.'

With that load off his mind Hitler was content to leave Göring to take over the direction of affairs in Berlin. He was not much worried about the risk of British or French intervention, but Göring took steps to secure a promise that the Czech Army would not come to Austria's aid.

Hitler himself crossed the frontier on the 12th, and drove through the decorated villages into the crowded, cheering streets of Linz. He was in an excited mood: he had come home at last. The next day he went out to lay a wreath on his parents' grave at Leonding. To Mussolini he sent a telegram: 'I shall never forget this. Adolf Hitler.' To the crowds he declared: 'If Providence once called me forth from this town to be the leader of the Reich, it must, in so doing, have charged me with a mission, and that mission could be only to restore my dear homeland to the German Reich.'

That night, perhaps under the influence of the enthusiastic reception he had met in his homeland, Hitler decided not to set up a satellite government under Seyss-Inquart but to incorporate Austria directly into the Reich.

Next morning, the Sunday on which the ill-fated plebiscite was to have been held, one of Hitler's State Secretaries, Stuckart, flew to Vienna to place Hitler's plan before the new Austrian Government. The terms in which the suggestion was framed admitted of only one answer. A Cabinet meeting was hurriedly summoned, and late on the night of the 13th Seyss-Inquart was able to present the Führer with the text of a law already promulgated, the first article of which read: 'Austria is a province of the German Reich.' Hitler was deeply moved. Tears ran down his cheeks, and he turned to his companions with the remark: 'Yes, a good political action saves blood.' The same night the arrests began: in Vienna alone they were to total seventy-six thousand.

Austria was to have its plebiscite after all, a plebiscite in which not only Austria but the whole of Greater Germany was to take part, this time under Nazi auspices. When he presented his report to the Reichstag on 18 March Hitler announced the dissolution of the Reichstag and new elections for 10 April, appealing for another four years of power to consolidate the gains of the new Grossdeutschland.

In the course of the electoral campaign Hitler travelled from end to end of Germany, with a closing demonstration at Vienna on the 9th. To the Burgomaster of Vienna he said: 'Be assured that this city is in my eyes a pearl. I will bring it into that setting which is worthy of it and I will entrust it to the care of the whole German nation.'

Under the Nazi system of voting there was no room for surprises—99.08 per cent voted their approval of his actions; in Austria the figure was higher still: 99.75. 'For me,' Hitler told the Press, 'this is the proudest hour of my life.'

The union of Austria with Germany was the fulfilment of a German dream older than the Treaty of Versailles, which had specifically forbidden it, or even than the unification of Germany, from which Bismarck had deliberately excluded Austria. With the dissolution of the Hapsburg Monarchy at the end of the war many Austrians saw in such a union with Germany the only future for a State which, shorn of the non-German provinces of the old Empire, appeared to be left hanging in the air. Austria's none too happy experience in the post-war world, including grave economic problems like unemployment, added force to this argument. If the rise of Nazism in Germany diminished Austrian enthusiasm for an *Anschluss*, yet the pull of sentiment, language and history, reinforced by the material advantages offered by becoming part of a big nation, was strong enough to awaken genuine welcome when the frontier barriers went down and the German troops marched, in, garlanded with flowers. For months, even years, Austria had been living in a state of insecurity, and there was a widespread sense of relief that the tension was at an end, and that what had appeared inevitable had happened at last, peacefully. Moreover, the Austrian Nazi Party had attracted a considerable following before 1938. Vienna, where the Jews had played a more brilliant part than in almost any other European city, was an old centre of anti-Semitism, and in provinces like Styria Nazism made a powerful appeal.

Dissillusionment was not slow to come. Some of the worst anti-Semitic excesses took place in Vienna, and many who had welcomed the *Anschluss* were shocked by the characteristic Nazi mixture of arrogance and ignorance, and of terrorism tempered by corruption. Even Austrian Nazis were soon to complain at the shameless way in which the new province was plundered. Vienna was relegated to the position of a provincial town and the historical traditions of Austria obliterated.

None the less, in 1938 Hitler had a plausible case to argue when he claimed that the *Anschluss* was only the application of the Wilsonian principle of self-determination. Those outside Austria who wanted to lull their anxieties to sleep again could shrug their shoulders and say it was inevitable—after all, the Austrians were Germans, and Hitler himself an Austrian. In Rome Mussolini made the best of a bad job and shouted down Italian doubts by loudly proclaiming the value and strength of the Axis. In London and Paris there was uneasiness, but reluctance to draw too harsh conclusions. The French reaffirmed their obligations to the Czechs, but Chamberlain, who still hoped for a settlement with the Dictators, refused to consider a British guarantee to Czechoslovakia.

Yet those who, like Churchill, saw in the annexation of Austria a decisive change in the European balance of power were to be proven right. The acquisition of Vienna placed the German Army on the edge of the Hungarian plain and at the threshold of the Balkans. To the south, Germany now had a common frontier with Italy and Yugoslavia, no more than fifty miles from the Adriatic. To the north, Hitler was in a position to press Czechoslovakia from three directions. Germany's strategic position had been immeasurably improved. Nor was the contribution of Austria's resources in iron, steel, and magnesite to be disregarded.

The execution of the *Anschluss*, it is true, had been hastily improvised, in answer to a situation Hitler had not foreseen, but such a step had always been one of his first objectives in foreign policy, and the ease with which it had been accomplished was bound to tempt him to move on more rapidly to the achievement of the next. Every step Hitler had taken in foreign policy since 1933 had borne an increased risk, and every time he had been successful in his gamble. To his astonishment and delight this time there had not even been a special session of the League of Nations to rebuke him. The door to further successful adventures appeared to be already half-open, needing only a vigorous kick to swing it right back.

German armaments were now increasing at a much more rapid rate. On 1 April 1938, twenty-seven or twenty-eight divisions were ready; by the late autumn of 1938, including reserve divisions, this figure had grown to fifty-five divisions. Expenditure on rearmament was mounting by leaps and bounds.

This increase in German military strength did not yet amount to military supremacy in Europe, but taken with the fears, disunity and weakness of the opposition—those psychological factors to which he always attached the greatest importance—Hitler calculated that by the autumn of 1938 he would be in a position to press home his next demands with an even greater chance of success. Ten days after the result of the plebiscite on Austria had been announced, therefore, on 21 April, Hitler sent for General Keitel and set his Staff to work out new plans for aggression.

There was no doubt where Hitler would turn next. He had hated the Czechs since his Vienna days, when they had appeared to him as the very type of those Slav *Untermenschen*—'subhumans'—who were challenging the supremacy of the Germans in the Hapsburg Monarchy. The Czechoslovak State was the symbol of Versailles—democratic in character, a strong supporter of the League of Nations, the ally of France and of Russia. The Czech Army, a first-class force, backed by the famous Skoda armament works and provided with defences comparable with the Maginot Line in strength, was a factor which had to be eliminated before he could move eastwards. For the Bohemian quadrilateral was a natural defensive position of almost unequalled strategic value in the heart of Central Europe, within less than an hour's flying time from Berlin, and a base from which, in the event of war, heavy blows could be dealt at some of the most important German industrial centres. The annexation of Czechoslovakia was the second necessary step in the development of his programme for securing Germany's future.

The Czechs had few illusions about their German neighbours. They had done their best to build up their own defences and to buttress their independence by alliances with France and Russia. On paper this meant that any attack on Czechoslovakia must inevitably lead to general war; in actuality it meant that France and Great Britain, in their anxiety to avoid war, were prepared to go to great lengths to prevent the Czechs invoking the guarantees they had been given. The lever with which Hitler planned to undermine the Czech Republic was the existence inside its frontiers of a German minority of some three and a quarter millions, former subjects of the old Hapsburg Empire. The grievances of the Sudeten Germans had been a persistent source of trouble in Czech politics since the foundation of the Republic. The rise of the Nazis to power across the frontier and the growing strength of Germany had been followed by sharpened demands from the Sudeten Germans for a greater measure of autonomy from Prague and by the spread of Nazi ideas and Nazi organization among the German minority. From 1935 the German Foreign Office secretly subsidized the Nazi Sudeten German Party under the leadership of Konrad Henlein, and during 1938 Henlein succeeded in ousting the rival parties among the German minority from the field.

Aware of the dangers represented by this Trojan Horse within their walls, the Czech Government made a renewed effort towards the end of 1937 to reach a satisfactory settlement with the Sudeten German leaders. Ostensibly this remained the issue throughout the whole crisis—a square deal for the German minority in Czechoslovakia. By presenting the issue in this way Hitler succeeded in confusing public opinion in the rest of the world, and mobilizing sympathy for the wrongs of an oppressed minority.

After the annexation of Austria, however, the rights or wrongs of the German minority ceased to be anything more than the excuse with which Hitler was pushing his foot into the door. On 28 March 1938 he delegated Henlein to be his representative in

Sudetenland and gave him instructions to put forward continuing demands which would be unacceptable to the Czech Government. In this way Hitler planned to create a situation of permanent unrest in Czechoslovakia which could be progressively intensified until it reached a pitch where he could plausibly represent himself as forced, once again, to intervene in order to prevent civil war.

The events in Austria had already led to big demonstrations in the Sudetenland, much wild talk of 'going home to the Reich' and the intimidation of Czechs living near the frontier. Rumours were current of troop movements on the German side of the frontier and an imminent invasion. But events must not outstrip Germany's preparations. Henlein was told to keep his supporters in hand, and Hitler laid it down that plans to breach the Czech fortifications must be prepared to the smallest detail.

Hitler set out for Rome on 2 May on a State visit. Every Party boss and Nazi hanger-on tried to squeeze into the four special trains which were needed to carry the German delegation and its cumbrous equipment of special uniforms. The competition to share, at Italian expense, in the endless galas, receptions and banquets, the expensive presents and imposing decorations, was intense. Nothing appealed to the gutter-élite of Germany so much as a free trip south of the Alps.

Hitler was delighted with Italy. He was less pleased by the fact that protocol required him, as Head of the State, to stay with the King and endure the formality of a reception at the Palace.

Since the two dictators had last met in September the Axis had been subjected to considerable strain. For the first time they met as neighbours. The *Anschluss* was not forgotten in Italy, where anxiety about the South Tyrol had revived, while the Anglo-

Italian Agreement of April had been noted without enthusiasm in Berlin. When Ribbentrop produced a draft German-Italian treaty of alliance, Mussolini and Ciano were evasive. But Hitler's references to the new Italy were in a generous vein, and by the time he left he had succeeded in restoring cordiality between the two régimes.

On Hitler's return to Germany he found the diplomatic situation developing even more favourably than he had expected. The French and English had joined in urging the Czechs to make the utmost concessions possible to the Sudeten Germans. Not content with this, Lord Halifax instructed the British Ambassador in Berlin to tell Ribbentrop that Britain was pressing the Czechs to reach a settlement with Henlein, and to ask for German cooperation. Ribbentrop's reply was that the British and French action was warmly welcomed by the Führer, who must have been delighted at the way in which the other Powers were doing his work for him. A day or two later the General Staff reported that a dozen German divisions were stationed on the Czech frontiers ready to march at twelve hours' notice.

On 20 May Keitel sent Hitler a draft of the military directive for Operation Green (the code name for Czechoslovakia). On the same day, however, there was an unexpected turn to events. The Czech Government, alarmed by German troop concentrations near the frontier, ordered a partial mobilization of its forces. The British and French Governments at once warned Hitler of the grave danger of general war if the Germans made any aggressive move against the Czechs, and the French, supported by the Russians, reaffirmed their promise of immediate aid to Czechoslovakia. This display of solidarity left Hitler with no option but to call a retreat.

On 23 May the Czech Ambassador in Berlin was assured that Germany had no aggressive intentions towards his country, and indignant denials were made by the German Foreign Office of the reports of troop concentration.

The alarm felt in London and Paris at the prospect of German military action, however, rapidly turned into irritation with the Czechs. The advocates of appeasement blamed President Benes for his 'provocative' action, while Chamberlain determined never to run so grave a risk of war again.

For a week Hitler remained at the Berghof in a black rage, which was not softened by the crowing of the foreign Press at the way in which he had been forced to climb down. Then, on 28 May, he suddenly appeared in Berlin and summoned a conference of military advisers at the Reich Chancellery. Hitler sketched with angry gestures on a map exactly how he meant to eliminate the State which had dared to inflict this humiliation on him.

Hitler hoped by the swiftness of his action to forestall effective intervention. In any case, thirty-six divisions were to be immediately mobilized, with X-day set for not later than 1 October 1938.

Hitler's tactics for the next three months, June, July, and August of 1938, were devoted to preparing the ground for intervention. In their anxiety to avoid war London and Paris urged the Czechs to make more and more concessions to the Sudeten Germans. Hitler noted with satisfaction the strain to which the Czechs were being subjected, their feeling of being pushed and hurried by their friends, their sense of isolation.

To the east Hitler pressed the Rumanians not to make transit facilities available to the Russians, who might come to Czechoslovakia's aid. He kept a watchful eye on Poland, where opinion was inflexibly opposed to any idea of Soviet troops marching across its territory; the Polish Government was preoccupied with the possibility of taking advantage of their difficult position to secure the valuable district of Teschen in Czech Silesia.

The Poles were not the only people whose appetite for territory might be encouraged at the expense of Czechoslovakia. At the end of the war Hungary had lost the whole of Slovakia to the new Czechoslovak State. Budapest's demand for the return at least of the districts inhabited by Magyars, better still of the whole province, had never wavered. To safeguard Slovakia against the claims of the Hungarian revisionists, Benes had concluded the Little Entente with Rumania and Yugoslavia, who were affected by similar Hungarian claims. The Hungarians were eager enough to take advantage of the Czechs' difficulties, but were worried lest in doing so they should provide a *casus foederis* for the Little Entente and commit themselves too completely to the German side. The tortuous efforts of the Hungarians to sit on the fence to the last moment met with little appreciation in Berlin.

With Russia Hitler confined himself to putting obstacles in the way of her giving aid to the Czechs, and played heavily on British and French dislike of inviting Russian cooperation in order to keep any United Front from coming into existence. So successful had been Hitler's anti-Bolshevik propaganda that the Franco-Russian and Russo-Czech pacts of 1935 were regarded by the British and French Governments as liabilities rather than as assets. On the Russian side there was equal distrust of the Western Powers, and a determination not to go one step ahead of France and Britain in risking war with Germany. Hitler's remark to Otto Strasser—'There is no solidarity in Europe'—was still true.

Meanwhile Henlein continued negotiations with the Czech Government in a desultory way, taking care always to find fresh objections to the successive Czech offers of a greater measure of Home Rule for the Sudeten districts. The Sudeten German Party kept feeling against the Czechs in the frontier districts at fever-pitch: by the end of the summer the tension between the Sudeten population and the Czech officials was reaching snapping-point.

On the German side of the frontier the military preparations were systematically continued. Hitler, however, now began to encounter some resistance in the Army High Command. The risk of a general war alarmed his staff officers, and not all were convinced by Hitler's declaration that intervention by France and Britain could be discounted. The opposition was led by General Ludwig Beck, Chief of Staff of the Army, who tried to persuade the Commander-in-Chief of the Army, General von Brauchitsch, to make a stand against Hitler.

Brauchitsch, although he agreed with Beck's argument, tried to avoid taking action. In the first week of August, however, at Beck's insistence, a meeting of the leading commanders was held in Berlin under the chairmanship of Brauchitsch. Only two generals dissented from Beck's views. This time Brauchitsch went so far as to submit Beck's memorandum to Hitler.

News of the generals' conference had already reached Hitler. After a stormy argument with Brauchitsch, on 10 August Hitler summoned another conference, this time to the Berghof. The senior generals were excluded; Hitler appealed to the younger generation of the Army and Air Force leaders, and for three hours he used all his skill to set before them the political and military assumptions on which his plans were based. Then, for the first—and last—time at a meeting of this sort, he invited discussion. The result was disconcerting. General von Wietersheim, Chief of Staff to the Army Group Commander at Wiesbaden, General Adam, got up and said bluntly that it was his own, and General Adam's, view that the western fortifications against France could be held for only three weeks. A furious scene followed,

Hitler cursing the Army as good-for-nothing and shouting: 'I assure you, General, the position will not only be held for three weeks, but for three years.'

When Beck resigned, and demanded that the Commander-in-Chief resign with him, Brauchitsch refused. General Halder took over Beck's duties as Chief of Staff from 1 September. Hitler rejected any alteration in his policy, yet he was conscious of the fact that the opposition, if it had been silenced, had not been convinced.

On 3 September Hitler summoned Keitel, Chief of the O.K.W., and Brauchitsch to the Berghof to go over the final arrangements. Field units were to be moved up on 28 September and X-day fixed by noon on 27 September. The next day President Benes put an end to the Sudeten leaders' game by inviting them to set down on paper their full demands, with the promise to grant them immediately whatever they might be. In their embarrassment at having no more cause for agitation, the Sudeten leaders used the pretext of incidents at Moravska-Ostrava to break off the negotiations with Prague and send an ultimatum demanding the punishment of those responsible before they could be resumed. Benes had knocked the bottom out of the argument that the issue was Sudeten grievances and immediately afterwards, Henlein left for Germany.

For some time past the foreign embassies in Berlin had been reporting that the opening of the final stage of the Czech crisis would coincide with the Nuremberg Party Rally, due to begin on 6 September. Hitler's closing speech on 12 September, it was said, would show which way the wind was blowing and perhaps decide the issue of peace or war.

Meanwhile a small group of conspirators was discussing the possibility of seizing Hitler by force as soon as he gave the order to attack the Czechs, and putting him on trial before the People's Court. Amongst those involved were Beck, Karl Goerdeler,

Ulrich von Hassell, General von Witzleben, General Erich Höppner, Colonel Hans Oster, and Ewald von Kleist. Everything depended upon the conspirators' ability to persuade Brauchitsch, Halder, and the other generals of the certainty of a general war if Czechoslovakia was attacked. Soundings were made in London in the hope of securing incontestable proof that Britain and France would support the Czechs, in the event of a German attack. Such evidence as they were able to get, however—including a letter from Churchill—failed to convince Brauchitsch or Halder, and the conspiracy hung fire. On 9 September Hitler held another military conference, attended by Halder, Brauchitsch and Keitel. Hitler was highly critical of the Army's plans, condemning them for failure to provide the concentration of forces which alone would secure the quick, decisive success he needed. His aim was to drive at once right into the heart of Czechoslovakia and leave the Czech Army in the rear. X-day was now fixed for 30 September and was to be preceded by a rising in Sudetenland.

Hitler still held to the belief that Britain and France would not go to war over Czechoslovakia. As if to confirm his view, *The Times*, on 7 September, published its famous leader suggesting the possibility of Czechoslovakia ceding the Sudetenland to Germany. When Hitler stood in the spotlights at the huge stadium on the final night of the Rally, all the world was waiting to hear what he would say—and they were not disappointed. His speech was remarkable for a brutal attack on another State and its President such as had rarely, if ever, been heard in peacetime before.

Hitler made no attempt to disguise his anger at the humiliation of 21–2 May, which, he declared, had been deliberately planned by Benes, who spread the lie that Germany had mobilized. 'You will understand, my comrades, that a Great Power cannot for a second time suffer such an infamous encroachment upon its rights. . . .' Yet, for all his tone of menace Hitler was careful not to pin himself down; he did not commit himself to precise demands—only 'justice' for the Sudeten Germans—nor to the course of action he would follow if the demands were not made.

The speech was the signal for a rising in the Sudetenland, and several people were shot. But the Czech Government proclaimed martial law, put down the rising, and by the 15th had the situation well in hand. Henlein fled to Bavaria with several thousand of his followers.

The failure of nerve came, not in Prague, but in Paris. Twenty-four hours after Hitler's speech at Nuremberg the French Government reached a point where it was so divided on the question of its obligations to Czechoslovakia that the French Prime Minister, Daladier, appealed to Chamberlain to make the best bargain he could with Hitler. Chamberlain at once sent a message to Hitler proposing an interview.

Hitler was delighted. His vanity was gratified by the prospect of the Prime Minister of Great Britain making his first flight at the age of sixty-nine in order to come and plead with him. Hitler did not even offer to meet him half-way, but awaited him at the Berghof. They were accompanied to Hitler's study only by the interpreter.

Hitler began by a long, rambling account of all that he had done for Germany in foreign policy, how he had restored the equality of rights denied by Versailles, yet at the same time had signed the Pact with Poland, followed by the Naval Treaty with Britain, and had renounced Alsace-Lorraine. The question of the return of the Sudeten Germans, however, was different, he declared, since this affected race, which was the basis of his ideas. These Germans must come into the German Reich.

Chamberlain, who had spent most of the interview so far listening and watching Hitler, interrupted to ask if this was all he wanted, or whether he was aiming at the dismemberment of Czechoslovakia. Hitler replied that there were Polish and Hungarian demands to be met as well—what was left would not interest him.

Chamberlain attempted to narrow the problem down to practical considerations: if the Sudetenland was to be ceded to Germany, how was this to be done, what about the areas of mixed nationality, was there to be a transfer of populations as well as a change of frontiers? Hitler became excited. That was academic. 'Three hundred Sudetens have been killed, and things of that kind cannot go on; the thing has got to be settled at once. I am determined to settle it; I do not care whether there is a world war or not.'

Outside the autumn day was dying, the wind howled and the rain ran down the window-panes. Up in this house among the mountains two men were discussing the issue of war or peace, an issue that must affect millions of people they had never seen or heard of. It was this thought which preoccupied Chamberlain, and now he too began to grow angry.

'If the Führer is determined to settle this matter by force,' he retorted, 'without waiting even for a discussion between ourselves to take place, what did he let me come here for? I have wasted my time.'

With perfect timing, Hitler hesitated and let his mood change. 'Well, if the British Government were prepared to accept the idea of secession in principle, and to say so, there might be a chance then to have a talk.' Chamberlain declined to commit himself until he had consulted the British Cabinet; but if Hitler was prepared to consider a peaceful separation of the Sudeten Germans from Czechoslovakia, then he believed there was a way out, and would return for a second meeting. Meanwhile he asked Hitler for an assurance that he would not take precipitate action until he had received an answer. With all the appearance of making a great concession, Hitler agreed, knowing perfectly well that X-day was still a fortnight away.

Hitler never supposed that the British Prime Minister would be able to secure the Czechs' agreement to a voluntary surrender of the Sudetenland to Germany. He saw the interview at Berchtesgaden as a further means of ensuring that Britain and France would not intervene. He had drawn the British Prime Minister into advocating the cession of the Sudetenland on the grounds of self-determination; if, as he anticipated, this was rejected by Prague, in Chamberlain's eyes the responsibility of war would rest on the unreasonable Czechs, and Britain would be less likely than ever to go to Czechoslovakia's aid, or to encourage the French to do so.

During the week that followed, therefore, Hitler continued his preparations to attack Czechoslovakia. On the 18th the Army reported its plans for the deployment of five armies against the Czechs, a total of thirty-six divisions, including three armoured divisions.

Political preparations matched the military. On 20 September the Slovak People's Party, at Henlein's prompting, put forward a claim to autonomy for the Slovaks. On the same day Hitler sharply urged the Hungarians to present their demands to Czechoslovakia for the return of the districts claimed by Hungary. On the 21st the Poles delivered a Note in Prague, asking for a plebiscite in the Teschen district. By the 22nd the Sudeten Freikorps, armed and equipped in Germany, had seized control of the Czech towns of Eger and Asch.

Chamberlain again flew to Germany on the 22nd, and was received by Hitler at Godesberg, on the Rhine. Chamberlain was in an excellent temper. He had succeeded in getting the agreement of both the British and French Governments to the terms with which he now returned and in forcing the Czechs to accept them by an Anglo-French ultimatum. He was prepared to present a plan for the transfer of the Sudeten districts of Czechoslovakia to Germany without a plebiscite, leaving a commission to settle the details including the transfer of populations where no satisfactory line could be drawn. The existing alliances which the Czechs had with Russia and France were to be dissolved, while Britain would join in an international guarantee of Czechoslovakia's independence and neutrality.

When Chamberlain had finished speaking, Hitler inquired whether his proposals had been submitted to the Czech Government, and accepted by them. The Prime Minister replied: 'Yes.' There was a brief pause. Then Hitler said, quite quietly: 'I'm exceedingly sorry, but after the events of the last few days this solution is no longer any use.'

Chamberlain was both angry and puzzled; he had, he declared, taken his political life in his hands to secure agreement to Hitler's demands—only for Hitler to turn them down. Nor did Hitler give any clear indication of the reasons why he had changed his mind. He talked of the demands of the Poles and Hungarians, the unreliability and treachery of the Czechs; he argued himself into a fury over the wrongs and sufferings of the Sudeten Germans; above all, he insisted on the urgency of the situation and the need for speed. The whole problem, he shouted, must be settled by 1 October. If war was to be avoided, the Czechs must at once withdraw from the main areas to be ceded and allow them to be occupied by German forces. Afterwards a plebiscite could be held to settle the detailed line of the frontier, but the essential condition was a German occupation of the Sudetenland, at once.

Chamberlain took note of Hitler's new demands, but he refused to commit himself. He had assumed that Hitler's demands for the cession of the Sudetenland had been sincerely meant, but Hitler's real intention, as defined in his directive after the May crisis, was 'to destroy Czechoslovakia by military action'. The Sudeten question was used simply as a means to create favourable political conditions for carrying out his objective.

The next day Chamberlain tried to persuade Hitler to accept a compromise. Instead Hitler added a new time limit requiring the Czechs to begin evacuation of the territory to be ceded by 26 September, and to complete it by 28 September, in four days' time. The exchanges between the two men had already become tart when Ribbentrop brought the news that the Czechs had ordered mobilization. That, Hitler declared, settled matters. The argument went to and fro, Hitler excitedly denouncing the Czechs, Chamberlain not concealing his indignation at Hitler's impatience and his anger at the way he had been treated.

Then, once again, with perfect timing, Hitler produced his final 'concession'. 'You are one of the few men for whom I have ever done such a thing,' he declared. 'I am prepared to set one single date for the Czech evacuation—1 October—if that will facilitate your task.' He once more promised that the Czech problem was the last territorial demand which he had to make in Europe. This remark, Chamberlain told the House of Commons, was made 'with great earnestness'.

On 25 September the British Cabinet decided that it could not accept the terms Hitler had offered in the Godesberg Memorandum. On the 26th the British Government at last gave assurances of support to France if she became involved in war with Germany as a result of fulfiling her treaty obligations, and preparations for war were expedited in both Britain and France. The British Prime Minister, however, still refused to give up hope and resolved to make one last appeal to Hitler.

In the interval Hitler had swung back into his most intransigent mood, working his resentment, hatred and impatience of opposition up to the pitch where they would provide him with the necessary stimulus for his speech at the Berlin Sportpalast on the evening of the 26th. This was the mood in which Sir Horace Wilson and the British Ambassador found him when they arrived to present the British Prime Minister's letter three hours before the meeting in the Sportpalast was due to begin. The utmost the British representatives were able to get out of him was agreement to conduct negotiations with the Czechs, on the basis of their acceptance of the Godesberg Memorandum. If he was to hold back his troops, Hitler demanded an affirmative reply within less than forty-eight hours, by 2 p.m. on Wednesday 28 September.

Hitler's speech at the Sportpalast was a masterpiece of invective which even he never surpassed. He began with a survey of his own efforts to arrive at a settlement with the other Powers in the past five years, and instanced the familiar catalogue of problems that had already been solved. 'And now before us stands the last problem that must be solved and will be solved. It is the last territorial claim which I have to make in Europe, but it is the claim from which I will not recede and which, God willing, I will make good.' The origin of the Czech problem, he declared, was the refusal of the Peacemakers to apply their own principle of self-determination.

From this point Hitler's account became more and more grotesque in its inaccuracy. Having established a rule of terror over the subject peoples, the Slovaks, Germans, Magyars, and Poles, Benes had set out systematically to destroy the German minority; they were to be shot as traitors if they refused to fire on their fellow Germans. The Germans were so persecuted that hundreds of thousands fled into exile; thousands more were butchered by the Czechs. Meanwhile Benes put his country at the service of the Bolsheviks as an advanced air base from which to bomb Germany.

Hitler briefly explained the Godesberg proposals, brushing aside the Czech objection that these constituted a new situation. The Czechs had already agreed to the transfer of the districts demanded; the only difference was the German occupation. 'My patience is now at an end. I have made Herr Benes an offer which is nothing but the execution of what he himself has promised. The decision now lies in his hands: Peace or War. He will either accept this offer and now at last give the Germans their freedom, or we will go and fetch this freedom for ourselves.'

The next morning Hitler was still in the same exalted mood; the process of self-intoxication with his own words was still at work. When Sir Horace Wilson asked where the conflict would end, if the Czechs rejected the German demands, Hitler retorted that the first end would be the total destruction of Czechoslovakia. Wilson then added that the war could scarcely be confined to Czechoslovakia, and that Britain would feel obliged to support France if she went to the aid of the Czechs, and Hitler replied, 'If France and England strike, let them do so. It is a matter of complete indifference to me. I am prepared for every eventuality. It is Tuesday today, and by next Monday we shall all be at war.'

As soon as Wilson had gone Hitler ordered the movement of the assault units, twenty-one reinforced regiments, totalling seven divisions, up to their action stations. They must be ready to go into action on 30 September. A concealed mobilization was put into operation, including that of five further divisions in Western Germany.

Yet Hitler had not slammed the door. Even at the height of his frenzy in the Sportpalast the night before he had still left open the alternative to war which he had put forward at Godesberg. While Ribbentrop and Himmler were in favour of war, there were others in Hitler's entourage who pressed him to make a settlement, among them Göring. Events began to support their arguments. In particular, the news that Great Britain and France were taking active steps in preparation for war, and Sir Horace Wilson's warning that Britain would support France impressed Hitler more than all Chamberlain's appeals.

At this moment the group of conspirators who planned to carry out a *coup d'état* and seize Hitler by force were making renewed preparations. Hitler's decision, of course, cannot have been influenced by a plot of which he remained ignorant, but the conspiracy reveals something of the dismay that was felt in the Army High Command at the risk of a general war—and of this Hitler was perfectly well aware.

During the late afternoon of the 27th, a mechanized division in full field equipment rumbled through the main streets of Berlin, and was greeted with almost complete silence by the crowds, who turned their backs rather than look on. For a long time, Hitler stood at the window to watch, and the total lack of enthusiasm—in contrast to the scenes of 1914—is reported to have made a singularly deep impression on him. At ten o'clock the Commander-in-Chief of the Navy, Raeder, arrived to reinforce the Army's arguments that their fortifications and forces were inadequate—an appeal that was given weight by the news, received during the night, that the British fleet was being mobilized.

Hitler was sufficiently interested in keeping open the line to London to send a letter to Chamberlain, which was delivered at 10.30 p.m. on the 27th. It was skilfully phrased to appeal to Chamberlain, and the closing sentence—'I leave it to your judgement whether . . . you consider you should continue your effort . . . to bring the Government in Prague to reason at the very last hour'—spurred the British Prime Minister to make one final effort.

With the morning of 28 September—'Black Wednesday'—all hope of avoiding war seemed to have gone. A sense of gloom hung over Berlin, no less than over Prague and Paris and London.

Shortly after eleven o'clock the French Ambassador, François-Poncet, brought an offer which went a good way to meet the Godesberg demands, providing for the immediate occupation of part of the Sudetenland by 1 October and the occupation of the rest in a series of stages up to 10 October. The plan had not yet been accepted by the Czechs, but if it was agreed to by Hitler France would demand Czech acceptance.

The decisive move, however, appears to have been an appeal which the British Government addressed to Mussolini. The Duce was now thoroughly alarmed at the prospect of a European war for which Italy was ill-prepared. At the British request he agreed to send Attolico, the Italian Ambassador in Berlin, to Hitler.

Breathless and hatless, Attolico arrived at the Chancellery while Hitler was still engaged with François-Poncet. Mussolini, Attolico began, sent assurances of full support whatever the Führer decided, but he asked him to delay a final decision for twenty-four hours in order to examine the new proposals put forward by Paris and London. Mussolini's appeal made an impression on Hitler; after a slight hesitation he agreed.

Next, Sir Nevile Henderson arrived, bringing Chamberlain's reply to Hitler's letter of the night before. In a last appeal the British Prime Minister put forward the suggestion of an international conference to discuss the necessary arrangements to give Hitler what he wanted. Mussolini, meanwhile, was again on the telephone to Attolico, who was instructed to inform Hitler of the Duce's support for

Chamberlain's proposals of a conference. At the same time, Mussolini pointed out, the new proposals brought by François-Poncet, which would form the basis of any discussions, would allow Hitler to march his troops into the Sudetenland by 1 October, the date to which he had publicly committed himself.

Between one and two o'clock that afternoon Hitler agreed on condition that Mussolini should be present in person and the conference held at once, either in Munich or Frankfurt. Mussolini accepted and chose Munich. The same afternoon invitations were sent to London and Paris.

Hitler was eager to see Mussolini before the conference began, and early next day, 29 September, he boarded the Duce's train at Kufstein on the old German–Austrian frontier. According to Italian accounts, Hitler greeted Mussolini with an elaborate exposition, illustrated with a map, of his plans for a lightning attack on Czechoslovakia, followed by a campaign against France, and it was Mussolini who persuaded Hitler to give the conference a chance and not to assume that it would fail from the beginning.

The meeting of the two dictators with the British and French Prime Ministers began in the newly built Führerhaus at 12.45 p.m. Hitler, pale, excited and handicapped by his inability to speak any other language but German, leaned a good deal on Mussolini. Hitler, however, left the meeting in little doubt of what was required of it. Attempts by Chamberlain and Daladier to secure representation for the Czechs produced no results: Hitler refused categorically to admit them to the conference. Either the problem was one between Germany and Czechoslovakia, which could be settled by force in a fortnight; or it was a problem for the Great Powers, in which case they must take the responsibility and impose their settlement on the Czechs.

The conference had been so hastily improvised that it lacked any organization. There were constant interruptions while members of one delegation or another went in and out to prepare alternative drafts. Finally, in the early hours of the morning of 30 September, agreement was reached, and the two dictators left to the British and French the odious task of communicating to the Czechs the terms for the partition of their country.

The Munich Agreement contained few substantial variations from the proposals of the Godesberg Memorandum. On 1 October German troops marched into the Sudetenland, as Hitler had demanded. The promised plebiscite was never held and the frontiers when finally drawn followed strategic much more than ethnographical lines, leaving two hundred and fifty thousand Germans in Czechoslovakia, and including eight hundred thousand Czechs in the lands ceded to the Reich. Czechoslovakia lost her system of fortifications together with eleven thousand square miles of territory. To this must be added crippling industrial losses and the disruption of the Czech railway system. President Benes was forced to go into exile and one of the first acts of the new régime was to denounce the alliance with Russia. On 10 October the Czechs ceded the Teschen district to Poland, and on 2 November Ribbentrop and Ciano dictated the new Czech-Hungarian frontier at a ceremony in Vienna from which the two other signatories of the Munich Agreement and the guarantors of Czechoslovakia were blatantly excluded.

Hitler's prestige rose to new heights in Germany, where relief that war had been avoided was combined with delight in the gains that had been won on the cheap.

Abroad the effect was equally startling, and Churchill described the results of the Munich settlement in a famous speech on 5 October 1938: 'At Berchtesgaden . . . £1 was demanded at the pistol's point. When it was given (at Godesberg), £2 was demanded at the pistol's point. Finally the Dictator consented to take £1 17s. 6d. and the rest in promises of goodwill for the future. . . . We are in the presence of a disaster of the first magnitude.'

Austria and the Sudetenland within six months represented the triumph of Hitler's methods of political warfare. His diagnosis of the weakness of the Western democracies, and of the international divisions which prevented the formation of a united front against him, had been brilliantly vindicated. Five years after coming to power he had raised Germany from one of the lowest points of her history to the position of the leading Power in Europe—and this not only without war, but with the agreement of Great Britain and France. He can scarcely have failed to appreciate the fact that, twenty years after the end of the First World War, he had dictated terms to the victorious Powers of 1918 in the very city in the back streets of which he had begun his career as an unknown agitator.

Yet Hitler was more irritated than elated by his triumph. The indignation against the Czechs into which he had lashed himself, his pride and sensitivity to prestige, his ingrained dislike of negotiation, his preference for violence, his desire for a sensational success for the new German Army he had created, the recurrent thought of the original plan to secure the whole of Czechoslovakia—all these were factors liable to start up once more the conflict in his mind. By allowing himself to be persuaded into accepting a negotiated settlement, Hitler came to believe that he had been baulked of the triumph he had really wanted, the German armoured divisions storming across Bohemia and a conqueror's entry into the Czech capital. This was still his objective.

A new directive for the Armed Forces was issued on 21 October which listed, immediately after measures to defend Germany, preparations to liquidate the remainder of Czechoslovakia.

The well-meaning efforts of the appeasers during the next six months, far from mollifying, only irritated Hitler further. He objected to this attempt to put him on his best behaviour, to treat him as a governess treats a difficult child, appealing to sweet reason and his better instincts. After Munich he was more determined than ever not to be drawn into the kind of general settlement which was the object of Chamberlain's policy. He refused to make any concessions in return for the gestures of appeasement offered to him; he was interested in appeasement only in so far as it was the equivalent of capitulation and the complete abandonment of British interest in continental affairs.

On the night of 9–10 November a carefully organized pogrom against the Jewish population throughout Germany was carried out as revenge for the murder of a Nazi diplomat by a young Jew in Paris. Horror and indignation were immediately expressed in both Great Britain and the United States. President Roosevelt recalled the American Ambassador in Berlin, and the British Press was unanimous in its condemnation of the Nazi outrages. Hitler flew into a rage. Hatred of the Jews was perhaps the most sincere emotion of which he was capable. To his resentment against Britain was added the fury that the British should dare to express concern for the fate of the German Jews. He now saw London as the centre of that Jewish world conspiracy with which he had long inflamed his imagination, and Great Britain as the major obstacle in his path.

One of Hitler's diplomatic objectives during the winter of 1938–9 seems to have been to detach France from Great Britain. In mid-October he suggested to François-Poncet a joint declaration guaranteeing the existing Franco-German frontier, thus confirming Germany's abandonment of any claim to Alsace-Lorraine, together with an agreement to hold consultations on all questions likely to affect mutual relations. The French Government proved amenable, and the proposed Declaration was signed on 6 December.

In the meantime Hitler was at pains to strengthen his relations with Italy. At Munich Ribbentrop had produced a draft for a defensive military alliance between Germany, Italy, and Japan, and at the end of October the German Foreign Minister visited Rome to urge the Duce to put his signature to the treaty. Mussolini was wary of committing himself to an outright military alliance, and he resented Ribbentrop's visit to Paris in December and the German–French Declaration at precisely the moment when Italy was raising her own claims to Tunisia, Corsica, and Nice. Yet the very fact that Mussolini was beginning a new quarrel with France forced him back into his old position of dependence on Germany. At the beginning of the New Year, after two months of hesitation, the Duce suppressed his doubts and said he was willing to accept the suggested treaty.

In his speech of 30 January Hitler was lavish in his praise of Fascist Italy and her great leader: he had Mussolini where he wanted him and the working partnership of the Axis had been reaffirmed.

What would Hitler's next move be? No question more absorbed the attention of every diplomat and foreign correspondent in the winter of 1938–9. Every rumour was caught at and diligently reported.

Immediately after Munich, Dr. Funk went on a tour of the Balkans. His visit underlined the state of economic dependence upon Germany in which all these countries—Hungary, Rumania, Bulgaria, and Yugoslavia—now found themselves. Their political docility was secured by other means besides preferential trade and currency agreements—by the organization of the German minorities; by subsidies to local parties on the Nazi model, like the Iron Guard in Rumania; by playing on internal divisions between different peoples and different classes in the same country; by encouraging the territorial claims of one country, like Hungary, and rousing the fears of another, like Rumania. After the annexation of Austria and the capitulation of Munich these countries had to recognize that they were in the German sphere of influence and must shape their policy accordingly.

After Munich Hitler did not hesitate to express his disappointment with the Hungarians for failing to press their claims on the Czechs more pertinaciously. He showed considerable impatience with the Hungarian demands on Slovakia, and refused to agree to their annexation of Ruthenia, with a common Hungarian-Polish frontier. Instead an autonomous Ruthenia was set up within the new Czechoslovakia, in the same relationship towards its German patrons as the autonomous Slovakia. The little town of Chust, which was the capital of Ruthenia, soon became the centre of a Ukrainian national movement, eager to bring freedom to the oppressed Ukrainian populations of Poland and of the Soviet Union.

Poland and Russia took the threat sufficiently seriously to discover a common interest, despite their inveterate hostility, and the Pact of Nonaggression between the two countries was reaffirmed.

This Polish–Soviet *rapprochement* may have played a part in persuading the Germans to postpone their plans for a Greater Ukraine. At any rate, by the end of January it was a different set of anxieties which was beginning to occupy the Foreign Ministries in London and Paris.

On 24 January Lord Halifax, the British Foreign Secretary, wrote an appreciation of the European situation, which was to be laid before President Roosevelt and which was subsequently sent to the French Government as well. Lord Halifax began by saying that he had received a large number of reports that indicated Hitler was resentful of the Munich

Agreement and considered Great Britain primarily responsible for his humiliation. 'As early as November there were indications which gradually became more definite that Hitler was planning a further foreign adventure for the spring of 1939.' Lord Halifax continued that possibly Hitler was 'considering an attack on the Western Powers as a preliminary to subsequent action in the east.'

The directives Hitler issued to the Armed Forces are the surest indication of the way in which his mind was moving. On 21 October, the Army and the other Forces were ordered to be prepared at all times for three eventualities: the defence of Germany, the liquidation of Czechoslovakia, and the occupation of Memel. On 24 November Hitler added a fourth eventuality, the occupation of Danzig, and on 17 December he instructed the Army to make its preparations to occupy the rest of Czechoslovakia.

In point of time, the preliminary moves for securing Danzig overlap the preparations for the liquidation of Czechoslovakia, but it will be convenient to treat relations with Poland separately and to conclude this chapter with the occupation of Prague.

After the cession to Poland and Hungary of a further 5,000 square miles of territory, with a population of well over a million souls, the Government in Prague was obliged to grant far-reaching autonomy to the two eastern provinces of Slovakia and Ruthenia, each of which had its own Cabinet and Parliament and maintained only the most shadowy relation with the Central Government. Even this did not satisfy the Germans, who made a long series of further demands upon the unfortunate Czechs.

The German documents in fact leave the clear impression that Hitler was only seeking a favourable opportunity to carry out the destruction of the Czechoslovak State of which he had been baulked at Munich, and for this reason the Germans steadily refused to give the guarantee for which the Czechs anxiously pleaded.

There were considerable practical advantages to be derived from such a move. The German Army was anxious to replace the long, straggling German–Czech frontier, which still represented a deep enclave in German territory, with a short easily-held line straight across Moravia from Silesia to Austria. The German Air Force was eager to acquire new air bases in Moravia and Bohemia. The seizure of Czech Army stocks and of the Skoda arms works, second only to Krupps, would represent a major reinforcement of German strength. The rearmament of Germany was beginning to impose a severe strain on the German economy and standard of living. The occupation of Bohemia and Moravia would help to alleviate this strain. Czech reserves of gold and foreign currency, Czech investments abroad and the agricultural and manpower resources of the country could be put to good use. At the same time, another cheap success in foreign policy would distract attention from any shortages at home and add to the prestige of the régime.

The role of fifth column, which had been played by the Sudeten Germans in 1938, was now assigned to the Slovaks, assisted by the German minority left within the frontiers of the new State. The demand of the Slovak extremists for complete independence was carefully cultivated by the Germans. The Prague Government had either to act or watch the separatist intrigues in Slovakia and Ruthenia break up the state. On 6 March the President of Czechoslovakia, Emil Hacha, dismissed the Ruthenian Government from office, and, on the night of 9–10 March, the Slovak as well. It did not take Hitler long to grasp that here was the opportunity for which he had been waiting.

Some of the Slovak leaders seem to have shown a

last-minute reluctance to play the part for which they were cast, but they were prodded in the back by the well-organized German minority. Durcansky, one of the dismissed Ministers, was hurried across the border in the German Consul's car. Over the Vienna radio he denounced the new Slovak Government formed by Karol Sidor, and called on the Slovak Hlinka Guard to rise. Arms were brought across the river from Austria and distributed to the Germans, who occupied the Government buildings in Bratislava. The British Consul in the city reported that the enthusiasm of the Slovak population was lukewarm, and continued so even after the declaration of Slovak independence. But Hitler was not interested in what the Slovaks thought; all he wanted was the declaration, and he took drastic measures to get it.

It had already been announced in Berlin on the morning of 11 March that Tiso, the deposed Slovak Premier, had appealed to Hitler. That night the two chief German representatives in Vienna, Bürckel and Seyss-Inquart, accompanied by five German generals, arrived in Bratislava and pushed their way into a meeting of the Slovak Government. Bürckel is reported to have told the new Premier, Sidor, that they must proclaim the independence of Slovakia at once, or Hitler—who had decided to settle the fate of Czechoslovakia—would disinterest himself in the Slovaks' future.

Early the next morning, Sunday the 12th, at a further Cabinet meeting, Tiso said that Bürckel had brought him an invitation from Hitler to go at once to Berlin. This he had been obliged to accept, for the consequences of refusal, Bürckel had added, would be the occupation of Bratislava by German troops and of Eastern Slovakia by the Hungarians.

In the early evening of Monday, 13 March, Hitler received Tiso and Durcansky in the Reich Chancellery. Tiso, a Catholic priest, had first to listen to a long, angry speech in which Hitler denounced the

Czechs and expressed astonishment at his own forbearance.

The attitude of the Slovaks, Hitler continued, had also been disappointing. After Munich he had prevented Hungary from occupying Slovakia in the belief that the Slovaks wanted independence, and had thereby risked offending his Hungarian friends. The new Slovak Premier, Sidor, however, now declared that he would oppose the separation of Slovakia from Czechoslovakia. If Slovakia wished to make herself independent, Hitler said, he would support this endeavour. If she hesitated, he was no longer responsible for the events which might follow. To add point to Hitler's remarks, Ribbentrop conveniently produced a message reporting Hungarian troop movements on the Slovak frontiers.

Tiso was then allowed to go. When the Deputies of the Slovak Parliament met the next morning, Tiso read out a proclamation of independence for Slovakia which Ribbentrop had already had drafted in Slovak. Whether they liked it or not, the Deputies had no option but to accept the independence thrust upon them, and by noon on Tuesday, 14 March, the break-up of Czechoslovakia had begun.

Hitler was now ready to deal with the Czechs. By 13 March the news from Slovakia had been crowded out of the front pages of the German Press by violent stories of a Czech 'reign of terror' directed against the German minority in Bohemia and Moravia. There was little truth in any of these atrocity stories, but they served their purpose, not least in helping Hitler to whip up his own indignation.

On Monday the 13th the Czech Government made a last effort to avert German action by a direct appeal to Hitler, and the following day President Hacha and the Foreign Minister, Chvalkovsky, set out for the German capital by train. An hour before they left, the Hungarians presented an ultimatum demanding the withdrawal of all Czech troops from Ruthenia, and the President had not yet crossed the Czech frontier when news reached Prague that German troops had already occupied the important industrial centre of Moravska Ostrava.

In Berlin President Hacha was received with all the honours due to a Head of State and, when he reached the Chancellery, found an S.S. Guard of Honour drawn up in the courtyard. The irony was barely concealed. Not until after 1 a.m. was the President admitted to Hitler's presence. Ill at ease, politically inexperienced, old, tired, and without a card in his hand, Hacha tried to soften Hitler's mood by ingratiating himself. He had no grounds for complaint over what had happened in Slovakia, he declared, but he pleaded with the Führer for the right of the Czechs to continue to live their own national life.

When Hacha had finished his abject plea Hitler began to speak. Once again he reviewed the course of his dealings with the Czechs; once again he repeated the charge that they had failed to break with the old régime of Benes and Masaryk. He no longer had confidence in the Czech Government, and was invading Czechoslovakia this very morning. The Czechs could resist and be destroyed, or allow the troops to enter peacefully.

Hacha and Chvalkovsky were then taken into another room for further talks with Göring and Ribbentrop. During this interlude Göring threatened to destroy Prague by bombing and Hacha fainted. After he was revived by an injection from Hitler's doctor, Morell, Hacha was put through to Prague by telephone, and the Czech Government undertook to order no resistance to the German advance. In the meantime a draft communiqué had been prepared for Hacha's signature. Its smooth terms were a masterpiece of understatement. The Führer had received President Hacha at the latter's request, and the President 'confidently placed the fate of the Czech people in the hands of the Führer'. Not a word was said of threats or invasion.

Hitler could hardly contain himself. He burst into his secretaries' room and invited them to kiss him. 'Children,' he declared, 'this is the greatest day of my life. I shall go down to history as the greatest German.'

Two hours later German troops crossed the frontier. When the British and French Ambassadors called at the Wilhelmstrasse to deliver their inevitable protests they were met with the argument that the Führer had acted only at the request of the Czech President.

By the afternoon of 15 March Hitler was on his way to Prague. His proclamation to the German people revived the stories of an 'intolerable' reign of terror, which had forced him to intervene to prevent the 'complete destruction of all order in a territory . . . which for over a thousand years belonged to the German Reich.' That night he spent in the palace of the Kings of Bohemia with the swastika waving from its battlements. Hitler had paid off another of the historic grudges of the old Hapsburg Monarchy, the resentment of the Germans of the Empire in face of the Czech claim to equality.

The next day, Hitler reaffirmed the claim of the Germans to the territories of Bohemia and Moravia in which the upstart Czechs had dared to establish their own national state. As a contemptuous sop to the Western Powers, Hitler recalled the 'moderate' Neurath and named him the first Protector of Bohemia and Moravia. But the real power over the Czechs lay in the hands of the Head of the Civil Administration and the Secretary of State. To these two offices Hitler appointed Henlein and Karl Hermann Frank, the leaders of the Sudeten German Party, as a suitable reward for their services.

On the 16th Tiso sent Hitler a telegram asking him

to take Slovakia under his protection, which had been drafted with German help during his visit to Berlin on the 13th. Hitler graciously responded and German troops promptly moved in to guarantee Slovakia's newly won independence. Ruthenia, however, which was no longer of interest to the Germans, was abandoned to the Hungarians, whose troops marched in, overrunning all opposition, and soon reached the Polish border, establishing a common frontier between Hungary and Poland.

Hitler was back in Vienna on the 18th for the drafting of the Treaty of Protection between Germany and Slovakia. Slovakia granted to Germany the right to maintain garrisons in its territory and promised to conduct its foreign policy in the closest agreement with its Protector. A secret protocol allowed Germany the fullest rights in the economic exploitation of the country.

As the year before in Austria, so now in Czechoslovakia, the speed of the operation staggered the world.

Nowhere was the German action more resented than in Rome. Attolico, the Italian Ambassador in Berlin, had only been informed of the German intentions on 14 March, and the arrival of Philip of Hesse with his usual message of thanks for Italy's unshakeable support scarcely mollified Mussolini. 'The Italians will laugh at me,' he told Ciano; 'every time Hitler occupies a country he sends me a message.' The Duce was gloomy and worried at the prospect of the expansion of German influence down the Danube and in the Balkans, his own chosen sphere of interest, but anger was tempered by the calculation that Hitler was now too powerful to oppose and that it was best to be on the winning side.

Mussolini was the prisoner of his own policy. Hitler's success roused his envy, yet the more he attempted to imitate the Führer the more dependent he became upon him. A personal letter from Hitler on the twentieth anniversary of the Fascist movement helped to smooth the Duce's ruffled feathers, and by 26 March he was making a speech full of aggressive loyalty to the Axis. If he was to find compensation it must be within the framework of the Axis, not outside it. The Duce's eyes began to wander towards Albania.

In London the effects of Hitler's *coup* were more far-reaching. Prague has rightly been taken as the turning point in British foreign policy, the stage at which the British Government set to work, however ineptly, to organize resistance to any further aggressive move by the German dictator.

Much has been made of the fact that by his seizure of lands inhabited by Czechs, not by Germans, Hitler had now departed from the principle of self-determination to which he had hitherto appealed. But Hitler's adoption of the phraseology of national self-determination was no more to be taken seriously than his use of the language of the League when that suited his purpose.

Those who had been taken in by the arguments of self-determination made the same mistake as those who had imagined that Hitler would be bound by the Nazi Party Programme or by the policy of 'legality' after he became Chancellor. They failed to recognize that Hitler had only one programme: power, first his own power in Germany, and then the expansion of German power in Europe. The rest was window-dressing. This had been his programme before Prague, as it remained his programme afterwards. The only question was whether the other Powers would let him achieve the German domination of Europe without taking effective action to stop him.

CHAPTER NINE

HITLER'S WAR

1939

After the annexation of Austria, Hans Dieckhoff, one of the senior men in the German Foreign Office, remarked to Ribbentrop that Bismarck would have taken years to consolidate his position before making another move. 'Then,' Ribbentrop retorted, 'you have no conception of the dynamics of National Socialism.'

The obvious weakness of Hitler's policy, the fault which destroyed him as surely as it had destroyed Napoleon, was his inability to stop. By the end of 1938 Hitler had everything to gain by waiting for a year or two before taking another step, sitting back to profit from the divisions and hesitations of the other European Powers, instead of driving them, by the fears he aroused, into reluctant combination. Moreover, a temporary relaxation of the rearmament drive would have had considerable economic benefits for Germany.

At a meeting of the Reich Defence Council on 18 November, Göring spoke at length on the need to concentrate all the resources of the nation on raising the level of rearmament from a current index of 100 to one of 300. Everything was to be subordinated to this single task, regardless of the fact, which Göring frankly admitted, that the German economy was already showing strain.

Alarmed at the financial consequences of Göring's new measures, Schacht, as President of the Reichsbank, presented a memorandum on 7 January 1939, signed by the directors of the Bank, protesting against the Government's reckless expenditure. Hitler sent for Schacht and handed him his dismissal. 'You don't fit into the National Socialist picture,' he told him. Schacht was replaced by the docile Funk, and a secret decree placed the Bank under the direct orders of the Führer, with the obligation to provide the Government with whatever credits it demanded.

If any country had good reason to fear Germany's intentions it was Poland. At the Peace Settlement of 1919, and afterwards when Germany was weak, the Poles had acquired territory, the loss of which was more resented by the Germans than perhaps any other part of the Versailles Settlement. In order to provide Poland with access to the sea, Danzig was separated from Germany and made into a Free City, where the Poles enjoyed special privileges, while East Prussia was divided from the rest of the Reich by the Polish Corridor. There was justice on both sides. Much of the land regained by Poland had been first seized by Prussia at the time of the Partitions, and was inhabited by Poles. But the Poles had in turn taken more than they could legitimately claim. German public opinion was as solid in demanding that the eastern frontiers should be redrawn as Polish opinion was in refusing it. The rise of the Nazis to power in Germany had been followed by the steady growth of Nazi influence in Danzig, and it appeared only a question of time before the city was reunited to the Reich.

Yet the first country with which Hitler had signed a Pact of Non-Aggression had been Poland. The obvious reason for Hitler's friendliness was his need to placate the most suspicious of Germany's neighbours and France's principal ally in Eastern Europe, until he was strong enough to risk her hostility with equanimity. But there is also evidence to show that Hitler was anxious to secure Polish support against Russia. If the Poles accepted the restoration of Danzig to Germany and other changes in their western frontiers, in return they could eventually find compensation eastwards against Russia.

The question was whether the Polish Government would accept the role of Germany's ally in Hitler's schemes of eastward expansion. For five years Göring had been trying to negotiate an agreement of German–Polish cooperation against Russia. Now, in October 1938, Ribbentrop made some additional proposals to Colonel Beck, the Polish Foreign Minister: the return of Danzig to the Reich, and the construction of a German road and railway across the Polish Corridor to link East Prussia with the rest of Germany. In return Poland's economic interests in Danzig would be safeguarded, and Germany would guarantee Poland's existing frontiers. Beck could have little doubt that Ribbentrop's proposals were being made with a sense of immediacy which had been lacking from Göring's expansive gestures.

The success of Beck's policy of independence had so far rested on the fact that neither Germany nor the Soviet Union had wished to bring pressure to bear on the Poles. He did not want Poland to become a German satellite, but at the same time he wanted to avoid becoming too dependent on Russia. The most he could do was to try to persuade the Germans not to press their demands, at the same time making it clear to them that any attempt to annex Danzig by force would lead to war. He was willing to discuss a German–Polish agreement about Danzig to replace the existing League of Nations régime, but not to consider the return of Danzig to the Reich. He was prepared to improve German communications with East Prussia, but not to agree to an extra-territorial road across the Corridor. Ribbentrop's persistence entirely failed to alter Beck's attitude.

It was unlikely that Hitler would allow this situation to continue indefinitely. On 24 November, he revised his secret Directive to the Armed Forces to provide for a further contingency: a lightning occupation of Danzig by German forces. But, for the moment, it suited Hitler not to press his demands on Poland. He had first to complete the liquidation of Czechoslovakia. At Berchtesgaden on 19 November he assured Beck that there would be no *fait accompli* in Danzig, and in his speech of 30 January he made his customary friendly reference to Poland and the five-year-old German–Polish Agreement. After that there was silence for the next seven weeks.

The occupation of Bohemia–Moravia, in March 1939, and the German assumption of a protectorate over Slovakia, at once transformed the situation. German garrisons in Slovakia on the southern flank of Poland (already German troops were on Poland's northern and western frontiers) were taken in Warsaw as a step explicitly directed against Polish security. The deliberate neglect of the Germans to inform them in advance increased the Polish Government's alarm.

Nor was this the end of German surprises. Again without informing Poland beforehand, Ribbentrop presented an ultimatum to the Lithuanian Government demanding the return of the Memelland, a strip of territory on the northern frontier of East Prussia which Germany had lost by the Treaty of Versailles. There was no question of Lithuanian resistance to Germany, and a week after reviewing his troops in Prague Hitler arrived in Memel by sea to acclaim the return of the city to the Reich.

For the moment the focus of Western anxieties was Rumania, where a far-reaching economic agreement on behalf of the Reich was signed on 23 March. Since the Treaty established a dominant position for Germany in the development of Rumania's very considerable agricultural and mineral resources, it appeared unlikely that Hitler would make a further move in that direction. An uneasy peace descended on the Balkans.

Hitler's attention was in fact now fixed on Poland. On 21 March, Ribbentrop asked the Polish Ambassador, Lipski, to call on him. Lipski complained of the German action in Slovakia, which had been undertaken without a word to the Poles. Ribbentrop replied that, if the matter of Danzig and the extra-territorial road and railway to East Prussia could be settled to Germany's satisfaction, the Slovak question could doubtless be dealt with in such a way as to remove Polish anxieties. The Führer had been disagreeably surprised at the failure of the Poles to make any constructive reply to his proposals, and it was important that he should not come to the conclusion that Poland was rejecting his offer. For that reason it would be advisable for Colonel Beck to come to Berlin as soon as possible, and for the Ambassador to report to Warsaw at once.

Hitler was now determined to press the Poles hard, but he was still thinking in terms of a peaceful settlement which would bind the Poles more closely to Germany. He would have been glad enough to recover Danzig without making enemies of the Poles. On the other hand, he was not prepared to wait indefinitely, and, if he had to use force, then characteristically he would impose a really drastic settlement, the outline of which he had ready at the back of his mind. Everything depended upon the answer the Poles gave him in the next few days.

On Sunday 26 March, Lipski informed Ribbentrop that the Polish Government was prepared to discuss Danzig as well as German communications with East Prussia, and was anxious to reach a settlement with Germany on both; but the German demands, the return of Danzig and the extra-territorial road and railway, were once again rejected. Ribbentrop received Lipski coldly and began to threaten.

The next day Ribbentrop told Lipski that the Polish counter-proposals were wholly unsatisfactory, began blustering about Polish outrages against the German minority in Poland and warned him that the German Press could hardly be restrained much longer from answering Polish attacks. On 28 March Beck announced that the Polish Government would equally regard any attempt by the Germans, or by the Danzig Senate, to change the *status quo* in the Free City as an act of aggression against Poland.

The next stage was to be the application of pressure to the recalcitrant Poles, but the situation was suddenly complicated by the unexpected intervention of the British Government, which, to Hitler's anger, refused to mind its own business. Alarming reports were reaching London of German preparations for immediate action against Danzig and Poland. This time Chamberlain announced that Great Britain would risk war to protect Polish independence, and the French associated themselves with the British in these assurances.

Hitler was both surprised and angered. The day after Chamberlain's announcement, he insisted excitedly that he was not to be turned from the path he had chosen. Germany could not submit to intimidation or encirclement. In the same mood Hitler issued a new directive to his commanders on 3 April which listed as the three contingencies for which they were to prepare, the defence of the frontiers, Operation White (war with Poland), and the seizure of Danzig. Plans for the smashing of the Polish armed forces were to be ready by 1 September 1939, the actual date of the German invasion of Poland. German policy, the directive stated, continued to be based on the avoidance of trouble with Poland. But 'should Poland reverse her policy towards Germany and adopt a threatening attitude towards the Reich, we may be driven to a final settlement, notwithstanding the existing Pact with Poland'. For the moment, however, Hitler seems to have been at a loss how to proceed. This time the British Prime Minister showed no disposition to fly to Germany, and the Polish Foreign Minister, instead of coming to Berlin as Ribbentrop had demanded, visited London, where agreement was announced on the preparation of a pact of mutual assistance between Great Britain and Poland.

Some of the anger which Hitler felt at this check came out in his speech to the Reichstag at the end of the month. On 14 April, following the Italian invasion of Albania on the 7th, President Roosevelt had addressed a message to Mussolini and Hitler, asking if they were willing to give assurances against aggression to a list of thirty countries. It was announced that Hitler would reply to the President on 28 April.

Hitler began his speech with a lengthy and elaborate defence of his foreign policy, up to the present. This is so frequent a feature of his speeches on foreign policy as to suggest that this act of self-justification was psychologically necessary, in order to kindle the indignation and conviction with which he could defend the most blatant acts of aggression.

After describing his action in Czechoslovakia as a service to peace, Hitler declared that he had nothing but feelings of friendship and admiration for Great Britain, but friendship could survive only if it was based on mutual regard for each other's interests and achievements. The British, by beginning their old game of encircling Germany, had destroyed the basis for the Anglo-German Naval Treaty of 1935, and he had therefore decided formally to denounce it.

With Poland, too, Hitler declared, he had been only too anxious to reach a settlement. He had never ceased to uphold the necessity for Poland to have access to the sea. But Germany also had legitimate demands, for access to East Prussia and for the return of the German city of Danzig to the Reich. To solve the problem Hitler had made an unprecedented offer to Poland, the terms of which he now repeated, with the careful omission of the German invitation to join in a bloc directed against Russia. The Poles, however, had not only rejected his offer, but had begun to lend themselves—like the Czechs the year before—to the international campaign of lies against Germany. In these circumstances the German–Polish Agreement of 1934 had no longer any validity, and Hitler had therefore decided to denounce this too. He was careful to add, however, that the door to a fresh agreement between Germany and Poland was still open.

Throughout this speech Hitler spoke in violent terms of the 'international warmongers' in the Democracies, whose one aim was to misrepresent German aims and to stir up trouble. He exalted the strength of the Axis, congratulating Mussolini on the occupation of Albania and the establishment of order in a territory which naturally belonged to Italy's *Lebansraum.*

Hitler had by now worked himself into the state of mind in which to answer President Roosevelt. The second half of his speech was marked by a display of sarcasm which produced roars of applause from the Reichstag. It is not difficult to pull holes in Hitler's argument when it is set down in cold print, or to point to the cheapness of his retorts, but to see and hear it brought to life on the film is to be struck once again by Hitler's mastery of irony and every other trick of the orator.

The American President had read him a lesson on the wickedness and futility of war: who should know better, Hitler retorted, than the German people, who had suffered from the oppression of an unjust peace treaty for twenty years? Mr. Roosevelt believed that all problems could be solved round the conference table: yet the first nation to express distrust of the

League of Nations was the U.S.A. 'The freedom of North America was not achieved at the conference table, any more than the conflict between the North and South.'

President Roosevelt pleaded for disarmament: the German people, trusting in the promises of another American President, had laid down their arms once before, only to see the other States repudiate their promises: the German people had had enough of unilateral disarmament.

Mr. Roosevelt was much concerned about German intentions in Europe. If Germany inquired about American policy in Central and South America she would be referred to the Monroe Doctrine and told to mind her own business. None the less, Hitler had approached each of the States mentioned by the President and had asked them if they felt threatened by Germany, and if they had asked the American President to request guarantees on their behalf. The reply in all cases had been negative. Not all the States mentioned by the President, however, had been able to reply. In Syria and Palestine the views of the inhabitants could not be ascertained owing to the occupation by French and British—not German—troops. The German Government was still willing to give assurances against aggression to any of the States referred to by the President, provided only that they came forward and asked for such assurances themselves.

This speech of 28 April 1939 is one of the most effective defences Hitler ever made of his use of the power he had secured in 1933. Yet his real skill lay in his evasion of the simple question which President Roosevelt has posed: was Nazi Germany entertaining further schemes of aggression?

To this question Hitler, for obvious reasons, preferred to give no clear answer. Instead, he confused the issue by repeating his own highly selected and exaggerated account of the history of Germany since

1918, by pointing to the inconsistencies and shortcomings of his critics, and by playing upon that historical pariah-complex of self-pity and self-justification which the German people had developed after the defeat of 1918. Even to question his intentions was made to appear as part of that denial of equal rights to the German people upon which he had fed his own and his audiences' indignation for so long.

The speech marked the close of a period of activity on Hitler's part. Having answered Polish obduracy and the British guarantees by the denunciation of the Naval Treaty and the German–Polish Pact, Hitler was content to sit back and wait. The door had been left open to further negotiations with the Poles, if they should change their minds.

Throughout the summer of 1939, from the end of April to August, Hitler was little to be seen in public. He could have devised no better tactics for the occasion. The measures hastily improvised by the Chamberlain Government had been due to the sense of urgency created by his action in Czechoslovakia and the Memelland. Once the tension relaxed, there was a good chance that London and Paris might be tempted to reduce rather than extend the obligations they had assumed. The British Government was still eager to avoid war and find a settlement, while in Paris the currents of appeasement and defeatism ran strongly beneath the surface. Nor was the guarantee which had been given to Poland of practical value, unless the Russians could be brought in as well.

Although the British opened negotiations in Moscow, they pursued them without conviction. The Russians looked upon Western policy with distrust and were determined not to accept any obligations towards Poland except as part of a mutual assistance pact binding Britain and France as well as themselves to the defence of Eastern Europe including the Soviet Union. In any case, the Poles and the other peoples living between Germany and Russia were not likely to accept help from a country they regarded with as much suspicion as the Germans.

Here was a situation from which Hitler did not find it hard to extract profit. Throughout the summer the remilitarization of Danzig, the training of the local S.S. and S.A. with arms smuggled across the frontier, and a series of incidents designed to provoke the Poles continued with little remission.

While pressure was kept up on Poland, German propaganda through the radio and the Press hammered home the argument that Danzig was not worth a war, and that war was only likely because of the obstinacy of the Poles. In their turn the Poles were warned not to trust their new friends, the British, who would soon tire of their energetic attitude, and sell them down the river, as they had sold the Czechs at Munich.

Meanwhile, relations with the other States in the German sphere of influence were strengthened, in order to keep Poland isolated. The visit of the Hungarian Prime Minister and Foreign Minister to Berlin, at the end of April, was followed by the State visit of Prince Paul, the Regent of Yugoslavia, at the beginning of June. Hitler laid himself out to please, and there was an impressive military review. A month later the Bulgarian Prime Minister was fêted in the German capital. The Germans showed an equal interest in the Baltic. The Pact of Non-Aggression signed with Lithuania after the cession of Memelland was matched by similar pacts with Latvia and Estonia.

Hitler's greatest success, however, was his alliance with Italy. The invasion of Albania in April, which Mussolini and Ciano regarded as the assertion of Italian independence, only bound them more closely to the Axis. The Germans now began to press for the signature of the military alliance which Mussolini had so far evaded. Ciano, who was uneasy about the state of German–Polish relations, invited Ribbentrop to come to Italy, in the hope of learning what was in Hitler's mind, and set off for their meeting at Milan (on 6 May) with a memorandum in which the Duce laid great stress on Italy's need of peace for a period of not less than three years.

Ribbentrop did not disguise Hitler's determination to recover Danzig and secure his motor-road to East Prussia, but he was sympathetic to Mussolini's insistence on the need to defer war. After dinner Ciano telephoned to Mussolini and reported that the conversations were going well. Thereupon the Duce, on the spur of the moment, apparently swayed by a gust of anti-British irritation after a year's hesitations and doubts, ordered Ciano to publish the news that an Italo-German alliance had been agreed upon. Ribbentrop would have preferred to wait until he could bring in Japan as well, but when he telephoned to Hitler he found the Führer eager to seize the chance offered by Mussolini's sudden change of mind. The announcement of the alliance, Hitler was convinced, would further weaken the British and French resolution to stand by Poland. On 21 May Ciano arrived in Berlin for the formal signature of the Pact of Steel.

The preamble of the Treaty announced that 'the German and Italian nations are determined to act side by side and with united forces for the securing of their living-space and the maintenance of peace'. It pledged Germany and Italy to full military support of each other in the event of war, and mutual agreement regarding any armistice.

Mussolini's chief anxiety, however, continued to be the possibility of war. On 30 May the Duce sent a secret memorandum for Hitler which emphasized Italy's need of a preparatory period of peace extending as far as the end of 1942. Hitler's reply was a vague suggestion that he should meet the Duce for a discussion some time in the near future. Beyond that he made no comment, and his silence seems to have been accepted by the Italians as assent.

Had Mussolini been present when Hitler met his senior Army, Navy, and Air Force officers on 23 May, the day after the Pact of Steel had been signed, his anxiety would have been vastly increased. Hitler began from the same premises as in the meeting in November 1937: the problem of *Lebensraum*, and the need to solve it by expansion eastwards. This time, however, he passed rapidly over the assumptions on which his views were based, and came almost at once to the military problem.

War, Hitler told his officers, was inevitable. 'Danzig is not the object of our activities. It is a question of expanding our living-space in the east, of securing our food-supplies, and of settling the Baltic problem. . . . There is no question of sparing Poland and we are left with the decision: To attack Poland at the first suitable opportunity.'

Hitler then went on to discuss the character of a war with the British, whom he described as the driving force against Germany, and of whose strength he spoke with appreciation. It would, he declared, be a life-and-death struggle and probably of long duration. Hitler saw the Ruhr as Germany's vital point, and so made the occupation of Holland and Belgium, to protect the Ruhr, a first objective. Because Britain's weakness was her need of sea-borne supplies, the Army's task must be to overrun the Low Countries and France in order to provide bases from which to blockade Britain. Analysing the lessons of the First World War, Hitler argued that a wheeling movement by the German Army towards the Channel ports at the outbreak of war—instead of towards Paris—might well have been decisive. This was in fact to be his own strategy in 1940.

Hitler's object was still, as it had always been, continental expansion eastwards, but the chief obstacle to his plans he now saw as England. War between Germany and Great Britain, Hitler was convinced, could not be avoided.

The German Army had now been expanded to a peacetime strength of thirty infantry divisions, five newly equipped Panzer divisions, four light divisions, and twenty-two machine-gun battalions. In four years its strength had been increased from seven to fifty-one divisions. Behind these forces stood a steadily increasing number of reserves and the most powerful armaments industry in the world. Nor had the Army been built up at the expense of the other branches of the armed forces. Since 1933 the German Navy had put into service two battleships of 26,000 tons; two armoured cruisers of 10,000 tons; seventeen destroyers, and forty-seven U-boats. It was engaged on completing two more battleships of 35,000 tons (actually much larger, for one of them was the *Bismarck*); four heavy cruisers of 10,000 tons; an aircraft carrier; five destroyers, and seven U-boats. The German Air Force, which had been entirely built up since 1933, now had a strength of 260,000 men, with twenty-one squadrons consisting of 240 echelons. Its anti-aircraft forces numbered close on three hundred batteries.

If, numerically, the number of divisions Germany could put in the field was still smaller than that of the French or Russian Armies, in quality, leadership, and equipment it was almost certainly an instrument without equal—and this time Hitler was determined to put it to use. By 15 June, he had the Army's plan for operations in Poland. Brauchitsch defined the object of the operation as 'to destroy the Polish armed forces. The political leadership demands that the war should be begun by heavy surprise blows and lead to quick successes.' On 27 July, the order was drafted for the occupation of Danzig, only the date being left blank for the Führer to write in.

Yet Hitler was in no hurry to move. For three months the resolution of the Poles, and the determination of the British and French to stand by their guarantees, had been subjected to strains, but had not been weakened. Only in one direction could Hitler see an opening, in Moscow. Talks between the British, French, and Russians were making little progress owing to suspicions on both sides. Hitler began toying with the possibility of a Russo-German agreement, which would guarantee Russia's neutrality in the event of a war between Germany and Poland. Britain and France would then be forced to recognize the impossibility of coming to the aid of Poland.

The basis for a deal was obvious. In the long run war between Germany and Russia was inevitable, so long as Hitler persisted in looking for Germany's living-space in the east. But in the short run, the last thing Hitler wanted was to become involved with Russia while he was still occupied with Poland. Stalin, for his part, was eager to postpone any clash with Germany as long as possible. He put forward various schemes for collective security, but he had an inveterate distrust of the Western Powers, doubted their determination to resist Hitler if put to the test and suspected them of trying to embroil the Soviet Union with Germany as a way of weakening both régimes.

In such a frame of mind, Stalin was likely to look on a separate deal with Germany as very much to Russia's advantage. It would enable him to buy time at the expense of Poland, and possibly secure important territorial and strategic advantages in Eastern Europe as part of his price. These could be used to strengthen the Soviet Union against the day when Hitler might feel free to put his designs against Russia into operation.

The obstacles in the way were twofold: the extreme distrust each side entertained towards the other, and the public commitments each had undertaken against the other. Hitler had made anti-Bolshevism a principal item of his propaganda stock-in-trade for twenty years; he had built up his foreign relations around the Anti-Comintern Pact, and, quite apart from propaganda, he had always looked towards Russia as the direction of Germany's future expansion. However opportunist his attitude, and however cynical his intentions towards Russia once the Pact had served its purpose, Hitler was bound to hesitate in face of such a repudiation of his own past.

But most people would be far more impressed by his astuteness in getting the Russians to sign than troubled about his inconsistency. A *rapprochement* with Russia would even be welcomed in certain quarters. Some members of the German Army, always preoccupied with the danger of a war on two fronts, favoured active collaboration with Russia. In the German Foreign Office, too, a policy of friendship with Russia did not lack advocates. To those who were struggling with the economic problems of the Four-Year Plan, the ability to draw upon Russian resources in raw materials would be a godsend. Finally, there was the argument which weighed most of all with Hitler—this was the surest way to isolate Poland, to deter the Western Powers from interfering, and to keep any conflict localized.

Trade talks with Russia began in June, but the Germans found the Russians tough and suspicious negotiators. Nevertheless, economic conversations were soon steered by Karl Schnurre, the German negotiator, into talk of a political alliance.

On 28 July France and Britain agreed to military staff talks in Moscow. But the undistinguished composition of the Western missions, and the fact that they were to proceed to Moscow not by plane but by slow boat, suggested that the British and French approached the talks with little conviction. By contrast the Germans appeared at last to have made up their minds.

The German press campaign against Poland, which had been damped down since June, was now renewed, and its scope was enlarged to include German claims not merely to Danzig, but to the whole of the Corridor, and even Posen and Upper Silesia, claims which were supported by a steady stream of reports describing Polish oppression of the German minority in these provinces.

At the same time the interference of the Germans in Danzig with the Polish customs and frontier guards in the Free Territory, and Polish economic reprisals, led to the most serious crisis so far in the dispute over the city. When the Danzig authorities notified a number of the Polish customs officers that they would be prevented from carrying out their duties, the Polish Government, alarmed at the undermining of their rights, sent an ultimatum demanding the withdrawal of the order under a time limit. The reply was a denial that any such order had been issued. The affair did not rest there. On 7 August, Forster, the Gauleiter of Danzig, was summoned to the Obersalzberg, and on his return told Carl Burckhardt, the League High Commissioner, that Hitler had reached the extreme limits of his patience. This was followed by a sharp exchange of notes between the Polish and German Governments, each warning the other of the consequences of further intervention. On the 11th, when Burckhardt himself visited Hitler, the Führer threatened that 'if the slightest thing was attempted by the Poles he would fall upon them like lightning with all the powerful arms at his disposal. . . .'

In Rome, meanwhile, the danger of war aroused increasing anxiety. Alarmed at the Germans' silence about their plans (the usual sign that they meant to spring a surprise on their allies) Ciano urgently requested Ribbentrop to meet him at Salzberg on 11 August.

The two Foreign Ministers spent altogether ten

hours together. Ciano pleaded with all the eloquence at his command for a peaceful settlement of the dispute with Poland; he found himself up against a brick wall. Ciano described the atmosphere of his talks with Ribbentrop as icy. Hitler, whom he was taken to see the following day at the Berghof, was more cordial, but appeared equally implacable in his decision. He was already lost in military calculations.

'I return to Rome,' Ciano wrote in his diary, 'completely disgusted with the Germans, with their leader, and with their way of doing things. They have betrayed us and lied to us. Now they are dragging us into an adventure which we do not want and which may compromise the régime and the country as a whole.'

There is no need to look far for an explanation of the indifference Hitler and Ribbentrop showed towards the ally they had welcomed with such effusiveness three months before. It was the prospect of an agreement with Russia which now dazzled the Führer and his Foreign Minister.

Yet the agreement had still to be negotiated, and the Russians did not seem eager. Hitler, with his eyes on the deadline he had fixed for the attack on Poland, began to press for the negotiations to start at once. On 14 August Ribbentrop telegraphed to the German Ambassador in Moscow, instructing him to see Molotov—if possible, Stalin—and to suggest that he should himself come to Moscow for direct discussions with the Russian Government. Molotov was not to be hurried. Such a visit, he said, 'required adequate preparation in order that the exchange of opinions might lead to results', to 'making concrete decisions'.

Ribbentrop at once accepted Molotov's suggestion and added that he was ready to come at any time after Friday 18 August with full powers to conclude a treaty.

Russia's military talks with the British and French

had reached a deadlock and were adjourned on 17 August; they were never seriously resumed. The Russians, however, meant to take full advantage of the Germans' eagerness. When Molotov received the German Ambassador the same day, the 17th, he told him that, first, they must conclude the trade agreement which had been hanging fire for several months; then they could turn to a non-aggression pact. Molotov made no reference to Ribbentrop's specific proposal of a date for his visit.

Molotov's tactics only added to Hitler's impatience. The trade treaty would be signed at once, Ribbentrop telegraphed back; what mattered was his immediate departure for Moscow.

On the 19th the Russian Government agreed to Ribbentrop's meeting—but only a week after the signature of the trade treaty, on 26 or 27 August. Faced with a delay which might disrupt the timetable he had laid down, Hitler swallowed his pride and brought himself to ask a personal favor of Stalin, the head of the Bolshevik state which he had proclaimed the irreconcilable enemy. On Sunday the 20th Hitler telegraphed Stalin, urging him to receive Ribbentrop on the 22nd, or the 23rd at the latest.

Hitler waited on tenterhooks for Stalin's reply. On Monday morning, Stalin indicated he would see Ribbentrop on the 23rd.

Hitler had already accepted Molotov's draft of the German-Soviet Non-Aggression Pact, but the Russians had added a postscript to the text: 'The present pact shall be valid only if a special protocol is signed simultaneously covering the points in which the High Contracting Parties are interested in the field of foreign policy.'

To put it in crude terms, the Soviet Government did not propose to sign until it learned what its share of the spoils was going to be, and how Eastern Europe was to be parcelled up. It was to complete this process of horse-trading that Ribbentrop was now to fly to Moscow. But to Hitler the one thing that mattered was the Pact; whatever its precise wording, the Pact meant the neutrality of Russia, an end to any threat of a British-French-Russian agreement to block German designs in Eastern Europe, and the isolation of Poland. To get Stalin's signature Hitler was prepared to promise anything.

Ribbentrop left for Moscow on the morning of the 22nd. Later that day one after another of the senior commanders of the Armed Services drove up the mountain road to the Berghof for a conference specially summoned by the Führer. No discussion was permitted; they were there to listen.

Hitler recited at length the reasons why a conflict with Poland was inevitable and ranged over the various possibilities of French and British reaction. He spoke of the need for iron resolution, 'no shrinking back from anything.' With the exhortation: 'Close your hearts to pity. Act brutally', Hitler dismissed his generals. The order for the start of hostilities, he added, would begin later: the probable time would be dawn on Saturday, 26 August.

Hitler stayed on at the Berghof, waiting for news from Moscow. It was here that Henderson, the British Ambassador, found him on the afternoon of the 23rd. On the previous day the British Cabinet had met to discuss the announcement that Germany and Russia were about to conclude a Pact of Non-Aggression. The Cabinet concluded that they were determined to fulfil their obligations to Poland. The British Government began to call up reservists and ordered their Ambassador in Berlin to deliver to Hitler personally a letter from the Prime Minister which stated that, if necessary, the British Government was 'resolved, and prepared, to employ without delay all the forces at their command, and it is impossible to foresee the end of hostilities once engaged.'

If the warning Chamberlain's letter contained took Hitler by surprise, its effect on him was at once relieved by the repetition of Chamberlain's familiar offer to continue his search for a peaceful settlement. Far from being impressed, he had no sooner received the Ambassador than he launched into a violent tirade against the British upon whom he laid the main blame for the crisis: it was the British guarantee to Poland, he declared, which had prevented the whole affair being settled long ago. He gave the wildest account of Polish excesses against the German minority in Poland, refused even to consider the suggestion of negotiations, and bitterly reproached the British for the way in which they had rejected his offers of friendship.

In his reply to Chamberlain, Hitler took note of the British intention to fulfil their guarantee to Poland, but this, he added, could make no difference to his determination: 'Germany, if attacked by England, will be found prepared and determined.'

It is unlikely that Hitler believed the British would put his determination to the test. Confident that the actual conclusion of the Pact between Germany and Russia would shake the resolution of the Western Powers, on the evening of his interview with Henderson (the 23rd) Hitler fixed the date for the attack on Poland at 4.30 a.m. on Saturday the 26th.

The news from Moscow was good. Ribbentrop reached the Russian capital at noon on the 23rd. He drove almost immediately to the Kremlin for his talks with Stalin and Molotov. That evening, while copies of the Agreement were being prepared for signature, toasts were drunk in the most cordial atmosphere.

The public document was a straightforward Pact of Non-Aggression, but to it was appended a protocol of the utmost secrecy, which only became known after the war. By this Germany and Russia agreed to divide the whole of Eastern Europe into spheres of influence. Finland, Estonia, and Latvia were recognized to lie in the Soviet sphere, Lithuania and Vilna in the German. A partition of Poland was foreshadowed along the line of the Rivers Narew, Vistula, and San. Russia stated her interest in the Rumanian province of Bessarabia. Ribbentrop, with an eye to German economic interests in that part of the world, was content to declare Germany's 'complete political disinterestedness in these areas.'

In the early hours of the 24th both documents were signed, and by one o'clock in the afternoon Ribbentrop was on his way back to Berlin, filled with enthusiasm for Germany's new friends. Stalin was less easily carried away. As the German delegation left the Kremlin he took Ribbentrop by the arm and repeated: 'The Soviet Government takes the new Pact very seriously. I can guarantee on my word of honour that the Soviet Union will not betray its partner.' The doubts were barely concealed.

Ribbentrop returned to Berlin in the firm belief that he brought with him an agreement which gave Hitler a free hand to deal the Poles a blow from which they would not recover for fifty years. To get this Ribbentrop had been prepared to risk straining Germany's relations with the Italians and the Japanese, to make nonsense of the Anti-Comintern Pact, and to grant sweeping concessions to the Russians in Eastern Europe. Immediately, these appeared a small price to pay for one of the most dramatic diplomatic *coups* in history.

Hitler was waiting in Berlin to greet his triumphant Foreign Minister as 'a second Bismarck'. That evening he spent with Göring and Weizsäcker, the State Secretary in the Foreign Office, listening to Ribbentrop's account of his reception in Moscow.

The news from London, however, was disappointing. Parliament had met on the 24th and had unanimously applauded firm statements by the Prime Minister and Foreign Secretary. Chamberlain's government had not fallen and the news of the Nazi-Soviet Pact had so far failed to produce the expected results.

The attack on Poland was now due to start in thirty-six hours. Hitler let the orders stand but determined to make one further effort to detach the Western Powers from the Poles.

On the morning of the 25th, after dispatching a long and somewhat embarrassed letter to Mussolini, Hitler sent an unofficial envoy, Herr Birger Dahlerus, to London. What Dahlerus took to Lord Halifax, however, was no new offer for a solution of the crisis over Poland, but simply an assurance of Hitler's willingness to come to an agreement with England.

While Dahlerus was acting in good faith, in hope of averting war, Hitler was seeing the British Ambassador. He began by recalling Henderson's hope, expressed at the end of their last conversation, that an understanding between Britain and Germany might still be possible. He spoke with regret of the general war to which it seemed the British attitude must now lead.

After naming certain conditions—the eventual fulfilment of Germany's colonial demands, the preservation of his special obligations to Italy, and his refusal ever to enter into a conflict with Russia—Hitler then made his offer: a German guarantee of the existence of the British Empire, with the assurance of German assistance, irrespective of where such assistance might be required. With this he coupled his willingness to accept a reasonable limitation of armaments, and to regard Germany's western frontiers as final.

His offer, reduced to simple terms, was a bribe in return for looking the other way while he strangled Poland. Henderson, after some argument, agreed to transmit the offer to London. Hitler urged him to spare no time and offered to put a German plane at his disposal.

Hitler next received Attolico, the Italian Ambassador. When he learned that no reply to his letter to Mussolini had yet reached Berlin, Hitler sent Ribbentrop to telephone Ciano.

A message was now brought to Hitler announcing the signature in London of the Pact of Mutual Assistance between Britain and Poland. The final signature of the Agreement had been held up for months by one delay or another: the fact that it should be signed on this very day, after Hitler had made his final offer to Great Britain, meant that his latest attempt to drive a wedge between the British and the Poles had failed.

A second piece of unwelcome news followed, this time from Rome. Hitler's message of the 25th, with its hint of imminent action against Poland, had found Mussolini still in a state of painful hesitation. He could not ignore the poor state of the Italian Army and the lack of preparations for a major war. The message which Attolico now brought to the Reich Chancellery expressed Mussolini's satisfaction at the agreement with Russia, but stated that the Duce declined going to war at this stage, unless Germany sent immediate military supplies.

Added to the news from London the effect of Mussolini's reply was to convince Hitler that he must give himself more time. Less than twelve hours before zero hour the invasion was halted.

The strain on Hitler at this point was intense. But, haggard and preoccupied though he appeared, he had lost neither his nerve nor his skill in political calculation. He went on probing the strength of the opposition he had encountered, searching for a weak point, trying to gauge how far he could go, exploring every possibility that turned up, revealing his inner mind to no one, waiting for the opportunity and the intuitive impulse to carry him forward.

Less than twenty-four hours after he had postponed the attack on Poland, Hitler fixed a new zero hour and thereafter did not depart from it.

Confused negotiations with Rome, with Paris, and with London followed. Hitler's immediate reply to Mussolini had been to ask him what he needed to complete his preparations, in order to see whether Germany could supply the deficiencies. The Italian answer was a list so extensive as to rule out any hope of this. Hitler replied that some of the Italian demands could be met, but not before the outbreak of hostilities.

The Duce, Ciano reports, was beside himself at the poor figure he was obliged to cut, and he telegraphed Hitler, expressing his regret at being unable 'to afford you real solidarity at the moment of action.'

The tone of Hitler's answer was resigned; he contented himself with three further requests to his Italian ally: support for Germany in the Italian press and broadcasts; the immobilization of as large British and French forces as possible; and, as a great favour, Italian manpower for industrial and agricultural work.

In the hope of saving face, Mussolini revived his suggestion of a conference and at the last moment, on the evening of 31 August, offered to act as mediator. Hitler thanked Mussolini for his trouble but declined to be drawn. There was no open breach between the Axis partners, but Hitler refused to alter his plans in order to save Mussolini's reputation. The Pact of Steel had failed to provide the support he had expected; he was not going to let it act as a brake.

Apart from a letter to Daladier, in which he repeated with considerable skill the argument that Danzig and Poland did not represent a sufficient issue to justify war between France and Germany, Hitler paid little attention to France in the final stages of the crisis. Hitler judged, correctly, that it was on

London that everything, including the French decision, would depend.

On the evening of the 26th Dahlerus returned from London, bringing with him a letter which he had persuaded Lord Halifax to write to Göring expressing Britain's desire for peace and wish to come to an understanding with Germany. Halifax's letter was couched in the most general and non-committal terms, but Göring declared it to be of enormous importance and drove to Berlin with Dahlerus to see Hitler. When they reached the Chancellery at midnight, Hitler was in bed, but Göring insisted on waking him up.

Entirely ignoring the letter Dahlerus had brought from London, Hitler began with a twenty-minute lecture justifying German policy and criticizing the British. After that he spent half an hour eagerly questioning Dahlerus, a Swedish friend of Göring's, about the years he had passed in England. Only then did he return to the current crisis, becoming more and more excited, pacing up and down and boasting of the armed power he had created, a power unequalled in German history. Suddenly, Dahlerus writes, Hitler 'stopped in the middle of the room and stood there staring. His voice was blurred, and his behaviour that of a completely abnormal person. He spoke in staccato phrases: "If there should be war, then I shall build U-boats, build U-boats, U-boats, U-boats, U-boats." His voice became more indistinct and finally one could not follow him at all. Then he pulled himself together, raised his voice as though addressing a large audience and shrieked: "I shall build aeroplanes, build aeroplanes, aeroplanes, aeroplanes, and I shall annihilate my enemies." He seemed more like a phantom from a story-book than a real person. I stared at him in amazement and turned to see how Göring was reacting, but he did not turn a hair.'

The upshot of the meeting was that Dahlerus

agreed to return to London with a new offer from Hitler to the British Government.

The British Government now had two considerably different sets of proposals before them, the first the offer Hitler had made to Henderson on the 25th, to which no reply had yet been sent, and the second, that now brought by Dahlerus. In the first Hitler had offered to guarantee the British Empire, but only after he had settled with Poland. Now the offer brought by Dahlerus suggested that Hitler was prepared to negotiate through the British for the return of Danzig and the Corridor, and the German colonies, guarantee Poland's new frontiers, and then pledge herself to defend the British Empire. The circumstances in which the proposals had been thrown together, however, made the British Government sceptical as to their value. It was finally agreed that Dahlerus should return to Berlin with a reply to the second offer, and report on its reception by Hitler, before the official reply to the first offer was drafted and sent over by Henderson on the 28th.

So, once again, Dahlerus flew back to Berlin, and on Sunday night delivered the British Government's message to Göring. In principle the British were willing to come to an agreement with Germany, but they stood by their guarantee to Poland; they recommended direct negotiations between Germany and Poland to settle questions of frontiers and minorities, stipulating that the results would have to be guaranteed by all the European Powers, not simply by Germany; they rejected the return of colonies at this time, under threat of war, though not indefinitely; and they emphatically declined the offer to defend the British Empire. Göring at once went off to Hitler, this time alone. To Dahlerus's surprise, Hitler accepted the British terms. Dahlerus got the British Chargé d'Affaires out of bed at 2 a.m. and sent an account of his reception off to London.

Sir Nevile Henderson flew back to Berlin on the

evening of Monday the 28th, and was received in state. Despite further outbursts on Hitler's part against the Poles, the interview was conducted in a reasonable manner. The British politely declined Hitler's bribe. Britain would maintain her obligations to Poland, and must, in any case, insist that any settlement arrived at should be guaranteed by the other European Powers as well as by Germany. The British Government had secured the agreement of the Polish Government to discussions: they now asked the German Government if they too were prepared to negotiate.

Having got the British and the Poles' offer to start negotiations, Hitler characteristically put up his demands. The German reply, handed to Henderson the next day, began with a lengthy indictment of the Poles, their refusal of the German demands, the provocation and threats they had since offered to Germany, and their persecution of the German minority—a state of affairs intolerable to a Great Power. All this was skilfully used to heighten the effect of the concession Hitler was ready to make in order to win Britain over, a declaration that the German Government was prepared to enter into negotiations with Poland.

The catch was Hitler's demand, that the Polish emissary was to leave at once and reach Berlin the following day—and was to come provided with full powers. If the Poles accepted, it meant capitulation. To send a plenipotentiary to Berlin, with full powers to commit the Polish Government, was to invite a repetition of what had happened to the Austrian Chancellor, Schuschnigg, and to President Hacha. But if the Poles refused, Hitler hoped, the British and French might come to regard them in the same light as Benes and the Czechs the year before, as the sole obstacles to a peaceful settlement, which Germany was only too eager to sign. After all, was Danzig worth a war?

This time the British Government did not fall into the trap. Although they continued to work for negotiations between Germany and Poland and to urge this course in Warsaw, they declined to put pressure on the Poles to comply with Hitler's demand for a plenipotentiary within twenty-four hours, a condition which Halifax described as wholly unreasonable. A last attempt by Göring to influence the British by sending Dahlerus to London on the 30th failed to alter the situation.

During the course of Wednesday the 30th a precise statement of the German claims against Poland was for the first time drawn up under sixteen heads. It included the return of Danzig; a plebiscite, under international control, on the Corridor; extra-territorial communications between Germany and East Prussia, and between Poland and Gdynia; an exchange of populations; and the guarantee of minority rights. Hitler himself later said, 'I needed an alibi, especially with the German people, to show them that I had done everything to maintain peace. That explains my generous offer about the settlement of Danzig and the corridor.' The sixteen points might also tempt the British to put pressure on the Poles.

When Henderson arrived to present the British reply to Ribbentrop at midnight on 30–31 August, he was greeted with a violent display of temper. Although Ribbentrop read out the sixteen points, he refused to let Henderson see the text on the grounds that the time limit for negotiations with the Poles had elapsed.

Henderson had already been sufficiently impressed by the 'generosity' of the German terms to get Lipski, the Polish Ambassador, out of bed at 2 a.m. and urge him to take steps to start negotiations. But the Poles were not prepared to accept German dictation, and Lipski's instructions were simply to inform the Germans that the British suggestion of direct negotiations was being favourably considered by the Polish

Government, and that they would send a reply in a few hours.

Immediately on receiving this message from Warsaw, Lipski requested an interview with Ribbentrop, but the Germans had already broken the Polish diplomatic cipher and knew that Lipski was not empowered to act as a plenipotentiary and agree to Hitler's demands on the spot. The Foreign Minister brusquely informed Lipski there was no further point in talking.

There was not much longer to wait. The High Command of the Army was pressing Hitler for a decision one way or the other. On Thursday 31 August, Hitler signed 'Directive No. 1 for the Conduct of the War'.

Every preparation was complete, even down to the necessary 'incidents'. Since 10 August one of Heydrich's S.S. men, Naujocks, had been waiting at Gleiwitz, near the Polish frontier, in order to stage a faked Polish attack on the German radio station there. Condemned criminals dressed in Polish uniforms were to be given fatal injections, then gunshot wounds, and left dead on the ground by Müller, the head of the Gestapo. At 8 p.m. on 31 August Naujocks picked up one of these men near the Gleiwitz radio station, seized the station as he had been ordered, broadcast a short proclamation and fired a few pistol shots. The 'attack' on Gleiwitz was one of the Polish infringements of German territory cited by the Germans as the justification for their attack the next day.

At nine o'clock on the night of the 31st, the Berlin radio broadcast the sixteen points of the German demands as proof of the moderation and patience of the Führer in face of intolerable provocation. The Poles were represented as stubbornly refusing to undertake negotiations.

Away to the east of Berlin, tanks, guns, lorries, and division after division of troops were moving up the roads towards the Polish frontier all through the night. At dawn on 1 September, the precise date fixed in the Führer's directive at the beginning of April, the guns opened fire. Hitler's war had begun. Not for five and a half years, until he was dead, were they to be silenced.

On 1 September 1939 there were no scenes of enthusiasm, no cheering crowds in Berlin like those in Munich in which Hitler had heard the news of the declaration of war twenty-five years before.

Hitler's speech to the Reichstag at 10 a.m. was on a characteristic note of truculent self-defence. Not only was the whole blame for failure to reach a peaceful settlement thrown on the Poles, but they were actually accused of launching an offensive against Germany, which compelled the Germans to counter-attack.

It was not one of Hitler's best speeches. He made a great deal of the Pact with Russia, but was uncertain in his attitude towards the Western Powers, disclaiming any quarrel with France or Britain and insisting on his desire for a final settlement with both. He was also clearly embarrassed by the need to refer to Italy. Towards the end he announced that, if anything should happen to himself, Göring would be his successor, and after him, Hess.

The nomination of Göring was as much a purely personal decision on Hitler's part as the decision to attack Poland. No Cabinet had now met for two years past, and anything that could be called a German Government had ceased to exist. By assuming the right to name his own successor Hitler demonstrated the arbitrary character of his rule.

Hitler was still not convinced that Britain and France would intervene and the delay in their declaration of war strengthened his belief. Not until more than forty-eight hours after the invasion did an ally of Poland enter the war. At nine o'clock on Sunday morning, the 3rd, Sir Nevile Henderson delivered the ultimatum of the British Government: unless satisfactory assurances of German action to call off the attack on Poland were received by 11 a.m. a state of war would exist between Britain and Germany from that hour.

According to interpreter Paul Schmidt, when Hitler received the message he 'sat immobile, gazing before him.' Goring contented himself with the remark: 'If we lose this war, then God help us!' Goebbels stood apart, lost in his own thoughts.

Hitler's embarrassment was soon relieved by the remarkable progress of the German armies in Poland. This seems to have been almost the only campaign of the Second World War in which the German generals did not have to submit to Hitler's direct interference. Hitler's interest in what was happening, however, was so great that he at once left Berlin for the Eastern Front. He established his headquarters in his special train near Gogolin, and every morning he drove to the front line.

Despite the utmost bravery the Polish forces were overwhelmed by the speed and impetus of the German armoured and motorized divisions, supported by an air force which had swept all opposition out of the skies in the first two or three days. By the end of the second week the Polish Army had virtually ceased to exist as an organized force. Warsaw and Modlin alone held out, much to Hitler's anger.

On 18 September, Hitler moved to the luxurious Casino Hotel at Zoppot, on the Baltic, and from there made his triumphal entry into Danzig the following day. He was certainly in no mood to hold out any olive-branches. After the inevitable justification of his actions, Hitler turned angrily upon the British warmongers, who, he declared, had manoeuvred the

Poles into provoking and attacking Germany. The real British motive was hatred of Germany. However long the war went on, Germany would never capitulate. 'We know too well what would be in store for us: a second Versailles, only worse.'

The authentic Hitler touch appears in his references to the Poles. Poland, like Czechoslovakia, was dismissed as an artificial creation. Hitler's contempt for the Poles as a people of inferior culture and inferior rights was not concealed.

The speed of the German advance in Poland took the Soviet Government by surprise. They had hastily to prepare for an immediate occupation of the territory allotted to them by the August Agreement. The German and Russian armies met at Brest-Litovsk, and the Russian advance brought Soviet troops to the frontiers of Hungary, news which raised the utmost alarm throughout the Balkans and Rome.

In meetings at the Kremlin on 27 and 28 September, so skilfully did Stalin play his hand that the final result of the German campaign in Poland was to strengthen the Russian, even more than the German, position in Eastern Europe, and to hand over to the Soviet Union half Poland and the three Baltic States. The Germans had even to withdraw their troops from the oil region of Borislav-Drohobycz, which they were eager to acquire, but which Stalin insisted on retaining in the Soviet half of Poland. The only concession offered in return was a Russian promise to export to Germany a quantity of oil equal to the annual production of the area.

Hitler can scarcely have enjoyed the price he had to pay for agreement with Russia, especially in the Baltic States, traditional outposts of German civilization in the East which had now to be abandoned to the Slavs. But he saw the advantages of shelving the problems of Eastern Europe, at least for a time, and leaving himself free to concentrate all his attention and forces on dealing with the West.

Ostensibly, Hitler still held the view that, after the defeat and partition of Poland, no further cause for war existed between Germany and the two Western Powers. There is every indication that the German people and the Army leaders would have warmly welcomed peace after the successes in the east.

The talk of a peace offensive, which was widespread in Berlin after Hitler's return from Poland in the last week of September, found a fervent echo in Rome. The elimination of Catholic Poland, a country for which Italians felt a traditional friendship; the advance of Russia to the threshold of the Balkans; Hitler's neglect of Italy for his new-found Russian friends; the feeling of being left out and no longer informed of what was afoot, greatly increased the mixed emotions of chagrin, envy, and resentment with which the Duce had watched Hitler's success in the past month. Mussolini was eager for peace, if only to save his own face. He was also anxious to discover Hitler's intentions, and an invitation to Ciano to visit Berlin was at once accepted.

Ciano's record of his conversation on 1 October leaves no doubt that, while still nominally committed to a peace offer, Hitler was already thinking in terms of its failure and the deliberate extension of the war to the west. The inactivity of the Western Powers during the Polish campaign was evidence of their feebleness, an invitation to rid himself of their interference for good. The Pact with Russia, which had given him a free hand in Poland, now guaranteed his freedom of action in the west, without the need to worry about his rear. The fact that Hitler no more trusted the Russians than they trusted him was an additional argument for settling with France and Britain as soon as possible.

Thus the much-advertised peace offer of 6 October,

to which so much importance had been attached in advance, was largely discounted by Hitler before it was made. Hitler's main purpose in making the speech seems to have been to carry German opinion with him, and convince the German people that, if the war continued, it was through no fault of his. The unpopularity of the war and the longing for peace impressed everyone who was in Berlin in the autumn of 1939. It was, almost certainly, with this undercurrent of disaffection in view that the German Press had been ordered to build up Hitler's peace offer in advance, and that Hitler now presented himself in his most plausible mood.

Hitler opened his speech to the Reichstag with an exultant description of the triumph of German arms, followed by a long and venomous attack upon the Polish nation and its leaders, and ending with a grossly distorted account of atrocities, represented as rising to new heights of infamy after the British guarantee.

He underlined the importance of Germany's new relationship with Russia. Where the League of Nations had totally failed to provide the much-promised revision of the Peace Treaties, Germany and Russia had carried out a resettlement which had removed part at least of the material for a European conflict. The new settlement with Poland was the culminating achievement in his policy of ridding Germany of the fetters fastened on her by the Treaty of Versailles. This last revision of the Treaty, too, could have been brought about in the same peaceful way as in the other cases, but for the malignant opposition of the warmongers abroad. Hitler then proceeded to recite all the efforts he had made to improve relations and live in peace with Germany's neighbours.

Hitler made it quite clear that the reorganization of Central Europe was a subject on which he would not permit 'any attempt to criticize, judge or reject my actions from the rostrum of international presumption'. But the future security and peace of Europe was a matter which must be settled by international conference and agreement, and 'it would be more sensible to tackle the solution before millions of men are first uselessly sent to their death.'

Every paper in Germany at once broke into headlines: 'Hitler's Peace Offer. No war aims against France and Britain. Reduction of armaments. Proposal of a conference.' As propaganda it was a well-turned trick; as a serious offer of peace it was worthless. Daladier's and Chamberlain's replies left no doubt that they were not prepared to consider peace on terms which, as Chamberlain put it, began with the absolution of the aggressor. On the 13th an official German statement announced that Chamberlain had rejected the hand of peace and deliberately chosen war. Once again Hitler had established his alibi.

It was scarcely more than that. On 27 September Hitler had already informed his three Commanders-in-Chief and General Keitel that he intended to attack in the west.

Three developments during September persuaded Hitler to embark on a European war: the speed and ease with which the German Army had eliminated Poland; the passivity of the French and British during the Polish campaign; and the German-Russian Agreement, which eliminated worry about the eastern frontiers.

Neither agreement with Russia nor the decision to attack in the west represented any change in Hitler's ultimate intention to carve out Germany's *Lebensraum* in the east. The elimination of French and British opposition was a prerequisite, not a substitute, for his eastern ambitions. But he was in no hurry to attack Russia; that remained in the future and he was prepared to draw every advantage he could from the new relationship with Moscow. Although he was prepared to change course and adapt himself to any new circumstance that appeared, the drift of his ideas is clear enough: deal with Poland before turning west; deal with France and Britain, before turning back to the east.

To a man of Hitler's hesitations the decision to resort to war represented a severe psychological test. He had supported himself with the assurance that it was only a localized war on which he was embarking. But now that the ordeal was past, now that military success had proved both so easy and so gratifying, his confidence bounded up and he let his ideas expand. Instead of waiting to see what the Western Powers were going to do, he would take the initiative himself. Released from the hesitation and anxieties of the time before he had risked war, he was already on the way to that assumption of his own infallibility which marked the deterioration of his judgement.

BOOK III

WAR-LORD
1939–45

CHAPTER TEN

THE INCONCLUSIVE VICTORY
1939–40

With the beginning of the war Hitler became more and more immersed in the political and strategical calculations by which he hoped to win it. If it is difficult at times in the intricate history which follows to detect the figure of the man, this is not due solely to the character of the records which have survived—minutes of conferences, diplomatic exchanges, and military directives: it corresponds to the actual conditions of his life during these years. The human being disappears, absorbed into the historic figure of the Führer. Only in the last two years of his life, as the magic begins to fail, is it possible to discover again the mortal and fallible creature beneath. The greater part of this section must therefore necessarily be concerned with Hitler as the war-lord of Nazi Germany, with the situations that confronted him and with the decisions that he took.

In the autumn of 1939 Hitler was well aware that the professional soldiers were opposed to extending the war by an attack in the west, particularly at the time he proposed, in the closing months of the year. Before he could develop his plans he had to master this opposition, and to that end he drafted a memorandum setting out his views and read it to the three Commanders-in-Chief, Halder, and Keitel at a conference held on 10 October.

The memorandum, a well-constructed piece of work, began with the defensive argument that Germany must strike in the west to prevent the occupation of Belgium and Holland by the French, and sought to prove that time was on the side of the enemy, not least because of the uncertainty of Russia's intentions.

In his eagerness to launch the attack during the autumn Hitler insisted that the process of refitting and reinforcing formations used in Poland must be carried out with a speed which might leave much to be desired. If necessary, the attacking forces must be prepared to go on fighting right into the depths of the winter—and they could do this, he argued, so long as they kept the fighting open and did not let it become a war of positions. The German Army was to sweep across Holland, Belgium, and Luxembourg, and destroy the opposing forces before they could form a coherent defensive front.

Hitler's arguments did not convert the opposition in the Army. They did not share his confidence in the superiority of the German Army over the French; and they were sceptical of the claims which Hitler made for their advantage in armour and in the air. If the gamble failed to come off, and no quick victory was obtained, Germany would find herself involved in a second world war, for which, they felt, her resources were inadequate.

While the Army leaders tried to dissuade Hitler by playing up the practical difficulties, in the background the same group of men who had urged Brauchitsch and Halder to remove Hitler by force in 1938 was again active—among them General Beck, the former Chief of Staff; Goerdeler; Hassell, the former Ambassador in Rome; General Oster, of the Counter-Intelligence (the Abwehr); and General Thomas,

Head of the War Economy and Armaments Office.

A lull followed, but at the end of October Hitler announced that the attack would begin on 12 November, and Brauchitsch, as Commander-in-Chief of the Army, was faced with the choice between giving orders for an offensive which he believed was bound to end disastrously for Germany, and organizing a putsch against the man who was the Commander-in-Chief of the German Armed Forces and the Head of the State.

The one institution in Germany which possessed the authority and the forces to carry out a *coup d'état* was the Army. As a result the history of the active German Opposition is a history of successive attempts to persuade one or other of the military leaders to use armed force against the Nazi régime.

For a few days at the beginning of November the conspirators were hopeful. Their argument that, if Hitler were removed, it would be possible to reach a settlement with the Western Powers and save Germany from another disaster appeared to be making an impression on the Army High Command. Discussions were held with the Army Chief of Staff, General Halder, on 2 and 3 November, and General Oster was assured that preparations for a military putsch had been made, if Hitler should insist on giving the final order for the attack. A meeting between General Brauchitsch and Hitler was fixed for Sunday, 5 November, after which the Army leaders promised a final decision.

The interview between Hitler and Brauchitsch did not last long. Hitler listened quietly enough as Brauchitsch set out his anxieties over the proposed attack, but when the Commander-in-Chief remarked that the spirit of the German infantry in Poland had fallen far short of that of the First World War, Hitler flew into a rage, shouting abuse at Brauchitsch and forbidding him to continue with his report. Under this direct attack the Commander-in-Chief crumpled up. Furious at the defeatism of the High Command, Hitler peremptorily ordered the preparations to continue and the attack to begin at dawn on the day fixed, 12 November.

After the dressing-down he received from Hitler, Brauchitsch (and Halder) hastily disavowed all interest in the conspiracy. Chance offered them a way out. On 7 November the attack had to be postponed, owing to an unfavourable weather forecast, and the High Command was able to make use of the same excuse to secure further postponements throughout the winter.

While the conspirators of the Opposition group around Beck and Oster had been trying, with diminishing hopes, to stir the Army High Command to action, they were startled by the news that a bomb explosion had wrecked the Bürgerbräukeller in Munich a short time after Hitler had finished speaking there on the anniversary of the 1923 putsch. It seems likely that the attempt on Hitler's life was organized by the Gestapo as a means of raising the Führer's popularity in the country. Goebbels made the utmost use of the incident to stir up resentment against those who were lukewarm towards the war, and to portray Hitler as an inspired leader whose intuition alone had preserved him from death. The timing had been a little too perfect, however, and the German people remained stolidly sceptical of their Führer's providential escape.

On 20 November Hitler ordered the state of alert to be maintained, so that immediate advantage could be taken of any improvement in the weather, and on 23 November he summoned the principal commanding officers of the three services to the Chancellery for another conference.

Hitler laid great stress on the fact that for the first time since the foundation of the German Empire by Bismarck, Germany had no need to fear a war on two fronts. The Pact with Russia brought no security for the future: 'Everything is determined by the fact that the moment is favourable now: in six months it may not be so any more. As the last factor I must in all modesty name my own person: irreplaceable. . . . My decision is unchangeable. I shall attack France and England at the most favourable and quickest moment. Breach of the neutrality of Belgium and Holland is meaningless. No one will question that when we have won.'

Although Hitler had designed the whole occasion in order to stamp the impression of an inspired leadership upon his commanders, it is difficult to believe that he was still only acting a part; the megalomania of the later years is already evident. He failed to convince the senior generals, but it was certain that the vacillating doubts of the High Command would prove insufficient to halt Hitler in a mood which had been hardened into reckless determination. All that the generals could do was to make use of the continued bad weather to delay the start of the offensive until well into 1940.

Meanwhile the impression which Hitler had formed of the lack of enthusiasm for war among his senior commanders strengthened the feelings of distrust with which he was coming to regard the professional soldiers. The clash between Hitler and his generals in the winter of 1939–40 bore fruit in his refusal ever again to let himself be influenced by their advice. When the attack in the west was followed by the most startling victories of the war, Hitler was encouraged to believe that his judgement was as superior to theirs in strategy, and even tactics, as he had always known it to be in politics—with disastrous results for both Hitler and the Army.

Throughout the autumn and winter of 1939–40 Hitler's mind was filled with the prospects of the offensive which he meant to launch in the west. Despite the successive postponements, he still thought of the attack as no more than a few days distant, and not until the New Year did he reluctantly agree to defer it to the spring or early summer. In the meantime, awkward problems were raised by German relations with Italy and with Russia.

The brief honeymoon period of the Pact of Steel was long since over. The alliance was popular in neither country, and Hitler admitted at the meeting with his commanders on 23 November that Italy's reliability depended solely upon Mussolini's continuation in power. The Duce's own attitude was far from stable. On 26 December, hoping for a German defeat, Mussolini told Ciano to let the Dutch and Belgian Governments know surreptitiously that their countries were threatened with an imminent German invasion. At other times Mussolini swung round and talked of intervening on Germany's side. But the ill-concealed contempt which many Germans felt for their 'non-belligerent' ally kept Mussolini's resentment alive, while German policy on a number of important issues was strongly criticized in Rome. The situation was not eased by Russia's invasion of Finland in early December. In January Mussolini sent Hitler a letter which represents the high-water mark of his independence towards his brother dictator.

The main burden of Mussolini's letter was the unfortunate consequences of Hitler's Pact with Russia —discontent in Spain and Italy; the sacrifice of Finland in a war for which Italian volunteers had come forward in thousands, as well as the failure to preserve an independent Polish State. Opposed to an extension of the war in the west, Mussolini urged Hitler to turn back and seek Germany's *Lebensraum* in the east, in Russia. He concluded: 'I have a duty to perform in adding that one step further in your relations with Moscow would have catastrophic results in Italy.'

Mussolini's letter was not only evidence of the

troubled state of the relations between the two allies; its arguments touched on a side of his own policy—the Nazi–Soviet Pact—about which Hitler was never altogether at ease.

The major problem in Germany's relations with Russia was no longer represented by Poland. After the partition, while part of the western half of Poland was annexed to the Reich, the rest was formed into the Government-General, under Hans Frank. On 7 October, the day after his so-called 'peace speech,' Hitler appointed Himmler as Head of a new organization, the R.K.F.D.V., the Reich Commissariat for the Strengthening of German Folkdom. Its first task was to carry out the deportation of Poles and Jews from the provinces annexed to Germany.

In the Government-General to which they were deported, Frank, besides recruiting forced labour and stripping the country of food and supplies, undertook the responsibility for the 'Extraordinary Pacification Action', the liquidation of the Polish educated class. The treatment of the Jews was handed over to Himmler and the S.S. 'The final solution of the Jewish problem', that sinister and terrible phrase, was the cover name used to disguise S.S. plans for the extermination of all men, women, and children of Jewish blood in Europe: its first phase was put into operation in the Polish Government-General where, at Auschwitz, close on a million human beings perished before the end of the war. Poland, in fact, became the working model of the Nazi New Order based upon the elimination of the Jews and the complete subjugation of inferior races, like the Slavs, to the Aryan master-race represented by the S.S.

Whatever fears, therefore, the Russians may have had that Hitler would use the Poles living under German rule to stir up trouble among those living under Soviet rule were soon removed by the brutal way in which the Germans treated all Poles. The test of German–Russian relations in the winter of 1939–

40 was not Poland, but Finland. Hitler, who had accepted the absorption of the three Baltic States into the Russian sphere of influence, now had to sit by silently while the Russians used force to coerce the Finns, a people who had close ties with Germany and whose brave resistance to the Russians inevitably roused admiration.

Hitler can scarcely have been blind to the fact that the measures taken by Russia to strengthen her position in the Baltic were obviously aimed at defending herself against an attack by Germany. Yet he still judged the price of co-operation with the Soviet Union to be worth paying for its economic, political, and strategic advantages.

So important were the Russian supplies of raw materials for Germany that, on 30 March, Hitler ordered the delivery of German equipment to the U.S.S.R. to be given priority over deliveries to Germany's own Armed Forces. In return the Soviet Union's shipments to Germany in the first year included a million tons of cereals, half a million tons of wheat and 900,000 tons of oil. In the autumn of 1939 Russian propaganda and the Communist parties abroad gave support to the German thesis that the responsibility for continuing the war rested upon the Western Powers—the so-called Peace Offensive. The two Governments cooperated in bringing pressure to bear on Turkey to prevent the Turks from abandoning their neutrality, while the division of North-eastern Europe into spheres of influence was put into effect without a hitch. But the supreme advantage to Hitler was strategic: the possibility of making his attack in the west without worrying about the defence of Germany's eastern frontiers.

Like Stalin, Hitler was taking advantage of a situation which neither side expected to last long. For the moment the balance of advantage appeared to be in Stalin's favour, but Hitler was undismayed. If he could use Russian neutrality to inflict a defeat upon the Western Powers, he would more than redress the balance.

On 10 January 1940 Hitler ordered the attack in the west to begin on the 17th at fifteen minutes before sunrise.

The very day of Hitler's decision—10 January—a German Air Force staff officer made a forced landing in Belgium while flying from Münster to Cologne. He had with him the complete operational plan for the opening of Hitler's offensive, and although he burnt part of it, enough fell into Belgian hands to alarm the Germans. This settled the matter, to the barely disguised relief of the High Command. Until the winter was over, and new plans could be prepared, Hitler at last agreed that there was no further point in postponing the start of the operation from day to day.

Hitler was partly reconciled to this decision by the interest he began to feel in a new project, the initiative for which came from the Naval High Command. The Commander-in-Chief of the German Navy, Admiral Raeder, searching for a means of increasing the Navy's power of attack against Britain's sea-routes, hit on the idea of securing bases in Norway. On 10 October he made the suggestion to Hitler, who at that time was absorbed in the plans for invading the Low Countries and France. The Norwegian project was not mentioned again until the middle of December, when the situation had been transformed by the Russian attack on Finland. The possibility of British and French troops being sent to aid Finland through Norway, even of an allied occupation of Norway, was taken seriously by the Germans. The most vulnerable link in Germany's war economy was her dependence on large supplies of iron ore from Sweden. By occupying Norway the Western Powers could not only interfere with this vital traffic, but could block the movement of German ships from the Baltic into the North Sea and the Atlantic.

Raeder suggested a means of securing control of Norway which at once attracted Hitler. Through Rosenberg, head of the Party's Foreign Policy Bureau, the Admiral had been put in touch with Quisling and Hagelin, the leaders of the small Norwegian Nazi Party. Quisling encouraged the belief that, with German support, he would be able to carry out a *coup d'état*. The German forces needed could thus be reduced to proportions which would not disturb the main concentration on the western frontier or endanger the plans already made to attack France.

Hitler was sufficiently impressed by Raeder's argument to see Quisling three times between 14 and 18 December, and to give orders for German support to be made available to him. Hitler preferred the method of the *coup d'état* because of its obvious economy, but, in case of its failure, he agreed to plans being prepared for an occupation of Norway by force. It was decided that the occupation of Denmark was to be carried out at the same time as the operation against Norway.

On 17 February the British destroyer *Cossack* intercepted the German prison-ship *Altmark* in Norwegian waters and rescued a number of British prisoners. This incident roused Hitler's anger and put an end to any hesitation. He appointed General Nikolaus von Falkenhorst, who had served in Finland in 1918, to take over the preparations for the occupation of Norway. The force employed was to be kept as small as possible, and Falkenhorst was given no more than five divisions. Quisling's part had dwindled into insignificance, and the Germans were preparing to carry out a military occupation, relying on surprise to supplement the small forces.

The risks involved in such an operation were great. As Raeder pointed out to Hitler, it was contrary to all the principles of naval warfare to attempt such an undertaking without command of the sea. Reports of British and French preparations for the occupation of Norwegian ports as part of the plan to come to Finland's aid kept the German High Command in a state of constant alarm lest they should be forestalled by the Allies.

Their anxiety was relieved when the Finns were driven to ask the Russians for an armistice and the danger of an Anglo-French landing receded. But, having gone so far, Hitler refused to turn back. 'Exercise Weser' (Norway) was to come before 'Case Yellow' (the attack in the west), and the preparations were to go forward whether the Allies proceeded with their plans or not. Hitler confirmed the order for operations to begin on 9 April.

Evidence of the German concentration of troops and naval forces along the Baltic coast in March and early April was not taken seriously by the Norwegian and Danish governments. The Germans, therefore, enjoyed the advantage of complete surprise.

Hitler signed the preliminary directive for 'Exercise Weser' on 1 March, and received Sumner Welles, the U.S. Under-Secretary of State, on the 2nd. The purpose of Welles's visit was to sound out the possibilities for the re-establishment of peace before the conflict began in earnest, and in particular to strengthen the reluctance of the Italians to be drawn into the war. Neither purpose commended itself to Hitler, and Sumner Welles's reception in Berlin was cool.

The American envoy's visit to Berlin, as he soon recognized, was a waste of time, so Welles returned to Rome after visits to Paris and London. The possibility that in Rome he would find a more sympathetic audience than in Berlin caused Hitler some anxiety. The Duce's letter of January had been left unanswered for two months: now, quite unexpectedly, the German Ambassador informed Ciano on

8 March that Ribbentrop would arrive in Rome in two days' time, before Welles returned, bringing with him Hitler's long-delayed reply.

The letter was couched in the most cordial terms, and Hitler played skilfully on the conflict in Mussolini's mind between fear and the desire to play a historic role. 'Sooner or later', he wrote, 'fate will force us after all to fight side by side. . . .'

Mussolini accepted the argument without dispute and replied that at 'the given moment' Italy would enter the war, but Ribbentrop could not pin him down to a specific date. Ribbentrop had, however, succeeded in reinforcing the Axis by his visit. Sumner Welles, who saw Mussolini again on 16 March, was impressed by the change in the Duce: 'He seemed to have thrown off some great weight.'

As soon as Ribbentrop returned to Berlin, he telephoned to Rome to ask Mussolini to meet Hitler on the Brenner on the 18th, and the Duce agreed. Ciano wrote that Mussolini still hoped to persuade Hitler to desist from his attack in the west, but he added gloomily: 'It cannot be denied that the Duce is fascinated by Hitler, a fascination which involves something deeply rooted in his make-up. The Führer will get more out of the Duce than Ribbentrop was able to.'

However much he might try to bolster up his resolution, Mussolini could not overcome the sense of inferiority he felt in face of Hitler. The sole successes which Mussolini valued were military successes. Hitler had dared to risk war, and he had not. Hitler had only to play on this feeling of humiliation to stimulate the Duce's longing to assert himself and revive his flagging belligerency. When they met on the Brenner, Hitler overwhelmed Mussolini with a flood of talk. Mussolini used the few minutes that were left him to re-affirm his intention of coming into the war.

Back in Rome the Duce might grumble at the

way in which Hitler talked all the time, but face to face with him he was unable to conceal an anxious deference. Hitler handled him with skill. The impression of German strength which he created and the confidence with which he spoke stirred Mussolini's old fear of being left out at the division of the spoils. Three months later Italy entered the war.

Hitler said not a word to Mussolini of his intention of attacking Norway. From the beginning of April, however, all his attention was directed to the Baltic and the north.

One day after Hitler confirmed 9 April as the date for 'Exercise Weser', the British Cabinet at last authorized the Royal Navy to mine Norwegian waters, an operation fixed for 8 April. In case of German counter-action, British and French forces were embarked to occupy the very same Norwegian ports selected by the German Navy as its own objectives. Thus, between 7 and 9 April, two naval forces were converging on Norway, and the scraps of news which reached Berlin of British preparations heightened the atmosphere of tension in Hitler's headquarters. Raeder was staking virtually the whole German fleet on the Norwegian gamble, and if it had the ill-luck to encounter the British fleet in any force, disaster could follow within a few hours.

The tactics of surprise proved brilliantly successful. Oslo, Bergen, Trondheim, Stavanger, and Narvik were captured at a blow. Quisling's *coup* was a miserable failure; the Norwegian King and Government escaped, and six weeks' hard fighting lay ahead before the Allied troops, now hurriedly landed, were driven out. None the less, the British had been 'completely outwitted' (the phrase is Churchill's) on their own native element of the sea. The British Navy inflicted considerable losses on the German forces, but they were a small price to pay for safeguarding the iron-ore supplies, securing the Baltic, and breaking out into the Atlantic, with bases along the whole of the Norwegian coast at the disposal of the German Navy and the German Air Force.

In the Norwegian expedition, Falkenhorst was directly responsible to Hitler, whose own command organization (the O.K.W., the Supreme Command of the Armed Forces) replaced the Army High Command (O.K.H.) for the planning and direction of operations. This led to considerable friction and departmental jealousy, and it marks the beginning of that continuous personal intervention in the daily conduct of operations which was more and more to absorb Hitler's attention and drive his generals to distraction. Hitler's temperament was singularly ill-fitted for the position of a commander-in-chief. He easily became excited, talked far too much and was apt to blame others for his own mistakes, or for adverse circumstances out of their control.

At the end of April Hitler felt sufficient confidence in the outcome of the Norwegian operations to fix a provisional date for the opening of the western campaign in the first week of May. After a slight delay due to bad weather, at dawn on 10 May 1940 the battle in the west was joined at last.

Little though Hitler may have realized it then, he owed his success in the Battle of France more than anything else to the long delay in opening the attack which had so much irked him at the time.

The original plan for the attack had assigned the chief role to the most northerly of the three German Army Groups in the west, Army Group B under von Bock. This was to carry out a wide sweeping movement through the Low Countries, supported by Army Group A (Rundstedt), which held the centre of the German line opposite the Ardennes, and by Army Group C (Leeb), which held the left wing facing the Maginot Line. For this purpose virtually the whole of the German panzer forces were assigned to Bock on the right wing. However, this was a repetition of the German advance in 1914 and therefore unlikely to take the Western Allies by surprise. It meant sending the tank forces into country broken by innumerable canals and small rivers, where they would collide head-on with the pick of the British and French armies advancing into Belgium.

An alternative plan had already been worked out by Rundstedt's Chief of Staff, General von Manstein, who argued that the decisive thrust should be made in the centre, through the Ardennes, aiming at Sedan and the Channel coast. Such a move would take the French completely by surprise, for they had written off the Ardennes as unsuitable for tank operations, and this part of the French Front was more weakly defended than almost any other. If the German plan proved successful, it would destroy the hinge upon which the British and French advance into Belgium depended, severing their lines of communication, cutting them off from France and forcing them into a trap with their backs to the Belgian coast.

Manstein's suggestion was frowned on by the Army High Command, but, thanks to the delays, he succeeded in getting his scheme brought to Hitler's attention. The proposals had precisely those qualities of surprise and risk to which Hitler attached so much importance. In the course of February he ordered the whole plan of attack to be recast along Manstein's lines, transferring the all-important panzer forces from the right wing to the centre under Rundstedt. By March Manstein's plan had become Hitler's own—he seems to have believed that he had thought of it himself—and by May the new orders were ready to be put into operation.

The German Army which invaded the Low Countries and France on the morning of 10 May consisted of eighty-nine divisions, with a further forty-seven held in reserve. It included the formidable weapon of ten panzer divisions, with three thousand armoured vehicles, a thousand of which at least were heavy tanks. The first sensational success was the overrunning of the Dutch and Belgian defence systems. The key to this was the use of small forces of highly trained parachute and glider troops which captured the vital bridges before they could be destroyed. The famous fortress of Eben Emael on the Albert Canal was taken according to a plan conceived by Hitler. A detachment of less than a hundred parachute engineers equipped with a powerful new explosive landed on the roof. The success was built up by German propaganda as a demonstration of the power of Germany's secret weapons.

But the crux of the operation, still unsuspected by the Western Powers, was the thrust through the Ardennes. Rundstedt's Army Group, ranged along the frontier from Aachen to the Moselle, disposed of forty-four divisions, including three panzer corps under the command of General von Kleist. The armoured column was over a hundred miles long, stretching back fifty miles the other side of the Rhine. The German armour quickly traversed the Ardennes, passed the French frontier on 12 May and were over the Meuse on the 13th. The High Command indeed became alarmed at the ease with which Kleist was advancing.

Hitler shared this anxiety. Preoccupied with the possibility of a French counter-attack from the south, he personally intervened to halt the advance of General Guderian's leading panzer divisions, which had reached the Oise on the night of the 16th. But on the evening of the 18th he was persuaded to allow the tanks to resume their advance. Backed by an irresistible superiority in the air, the German armoured thrust broke right through the French front and threw the Allied plans into confusion, trapping the British Expeditionary Force and the First French Army between the converging forces of Bock's Army Group on the north and Rundstedt's on the south.

The German plan of encirclement was defeated only by the brilliant improvisation of the Dunkirk evacuation. Between 27 May and 4 June a total of 338,000 British and French troops were got away by sea from the beaches and harbour of Dunkirk. Yet the possibility of such an evacuation might well have been denied to the British if Guderian's tanks had not been ordered by Hitler to halt a few miles south of Dunkirk on 24 May, before the British Army had fought its way back to the coast. Hitler wanted at all costs to preserve his armoured force for the next phase of the offensive which would decide the battle for Paris and France. Forty-eight hours later he reversed his decision, and on 27 May the panzer troops were allowed to resume their advance. But by then it was too late. This first of Hitler's military mistakes was to have momentous consequences for the future of the war.

At the time, however, the failure at Dunkirk appeared slight beside the continuing news of German successes. So Mussolini, as well as Hitler, judged. Fear of the consequences of going to war gave place in the Duce's mind to fear of arriving too late. Gauging the mood of his brother dictator with skill, Hitler found time in the midst of his preoccupations to write a series of letters to Mussolini in which he poured scorn on the feebleness of the British and French. Mussolini's replies were each more enthusiastic than the last, culminating in the decision to declare war on 10 June.

Hitler was delighted at Mussolini's decision. The German people, however, showed no more enthusiasm for the alliance than the Italians. Mussolini's timing, indeed, was so bad that his declaration of war in June 1940 made their new allies appear even more contemptible in German eyes than the Italian failure to fight in September 1939.

Meanwhile, on 5 June, the German Army renewed the attack by driving south across the Somme. The

French had lost nearly a third of their Army in Belgium and only two out of the fourteen British divisions now remained in France. In eleven days the battle was over. On 14 June Paris was occupied by the Germans and the panzer divisions were racing for the Rhône Valley, the Mediterranean, and the Spanish frontier. On the evening of 16 June M. Reynaud resigned, and the same night Marshal Pétain formed a new French Government, whose sole aim was to negotiate an armistice. Less than six weeks after the opening of the campaign Hitler was on his way to Munich to discuss with Mussolini the terms to be imposed on France.

Before the war Hitler had scored a series of political triumphs, culminating in the Nazi–Soviet Pact, which could challenge comparison with the diplomacy of Bismarck. Now he had led the German Army to a series of military triumphs which challenged comparison with the victories of Frederick the Great and even Napoleon. Hitler, the outsider who had never been to a university or a staff college, had beaten the Foreign Office and the General Staff at their own game.

It is customary to decry this achievement, to point, for instance, to the luck Hitler had in encountering such weakness and incompetence on the other side, to his good fortune in finding a Manstein to construct his plan of campaign for him and men like Guderian to put it into operation. But if there was weakness and incompetence on the other side, it was Hitler who divined it. If Manstein designed the plan of campaign it was Hitler who took it up. If Guderian was the man who showed what the German panzer divisions could do when used with imagination it was Hitler who grasped the importance of armour, and provided the new German Army with ten such divisions at a time when there was still strong opposition inside the Army itself to such ideas. If Hitler, therefore, is justly to be made responsible for the later disasters of the German Army, he is entitled to the major share of the credit for the victories of 1940: the German generals cannot have it both ways.

But what use did Hitler intend to make of his victory? The conflict with Britain and France arose not from any demands he had to make on the Western Powers themselves, but from their refusal to agree to a free hand for Germany in Central and Eastern Europe. The British had now, he felt, lost any reason for continuing to adhere to their former policy. Their last ally on the Continent had gone, their Army had been driven into the sea, they must now surely accept the impossibility of preventing a German hegemony in Europe and, like sensible people, come to terms. For his part, he was perfectly ready to conclude an alliance with Great Britain and to recognize the continued existence of the British Empire. England would have to return the German colonies and recognize Germany's dominant position in Europe, but that was all.

A final settlement with France could be allowed to wait until the end of the war, but the character of the armistice terms offered to France now might have considerable influence on the British. In particular, a French decision to continue the fight from North Africa or the departure of the French Fleet to join the Royal Navy would strengthen the British determination to go on fighting themselves. On the other hand, French acceptance of the German armistice terms might well make the British think twice.

Mussolini was in a very different mood. He was eager to become the heir of the French Empire in North Africa and to secure the mastery of the Mediterranean. As guarantees, the Duce wanted to occupy the whole of French territory and to enforce the surrender of the French fleet. But, as he confessed with some bitterness to Ciano on the train to Munich, Hitler had won the war and Hitler would have the last word.

The German reception of the Italians was cordial, but Ribbentrop made it quite plain that Hitler would not agree to make demands of the French which might drive them to continue the war from North Africa or England. For this reason, no less than for the effect on British opinion, Mussolini's annexationist ambitions would have to be deferred. Hitler proposed to occupy only three-fifths of France, to allow a French Government in Unoccupied France, to promise not to make use of the French fleet during the war and to leave the French colonies untouched.

These were heavy blows for Mussolini, but he was in no position to argue. Hither agreed not to conclude an armistice with France until she had come to terms with Italy as well, but he declined Mussolini's suggestion of joint German-Italian negotiations with the French. He had no intention of sharing his triumph.

For the signing of the armistice Hitler had appointed the exact place in the Forest of Compiègne, north-east of Paris, where Foch had dictated the terms of capitulation to the German delegation on 11 November 1918. The old restaurant car in which the negotiations had taken place was brought out from its Paris museum and set up on the identical spot it had occupied in 1918.

It was a hot June afternoon when Hitler led his small but impressive retinue—Göring, Keitel, Brauchitsch, Raeder, Ribbentrop, and Hess—through the elms and pines up to the block of granite on which the French inscription read: 'Here on 11 November 1918 succumbed the criminal pride of the German Reich . . . vanquished by the free peoples which it tried to enslave.'

Hitler received the French delegates in silence. He stayed to listen to the reading of the preamble, then rose, gave a stiff salute with his outstretched arm, and, accompanied by his retinue, left the railway car.

As he strode back down the avenue of trees to the waiting cars the German band played *Deutschland über Alles*, and the Nazi Horst Wessel Song. The one-time agitator, who told the Munich crowds in 1920 that he would never rest until he had torn up the Treaty of Versailles, had reached the peak of his career. He had kept his promise: the humiliation of 1918 was avenged.

With France Hitler had every reason to be satisfied. The French Government had been relieved by the comparative moderation of the German demands, and the armistice was signed without further difficulty. But the news which Hitler had been waiting for since the middle of June, a sign for London that the British were willing to consider peace negotiations, still failed to come. Tentative soundings through the neutral capitals produced no result. On 18 June Churchill, speaking in the House of Commons, declared the Government's determination to fight on, whatever the odds, 'so that, if the British Empire and its Commonwealth last for a thousand years, men will still say: "This was their finest hour".'

Hitler at last summoned the Reichstag for 19 July, more than a month after the collapse of France. The month's silence left little doubt that the British, to Hitler's genuine astonishment and even regret, were resolved to continue the war. After waiting in vain for a move on the part of the British Government Hitler decided to make a last gesture. 'In this hour, I feel it to be my duty before my own conscience to appeal once more to reason and common sense in Great Britain as much as elsewhere. I consider myself in a position to make this appeal since I am not the vanquished begging favours, but the victor speaking in the name of reason. I can see no reason why this war must go on.'

Hitler had already issued the directive for the invasion of Britain three days before his speech. If Britain would not come to terms she must be forced to submit. So, throughout the rest of the summer of 1940 and well into the autumn, the preparations for a direct assault on the British Isles continued, and all the world waited for the news that Hitler had launched his invasion armada across the Channel.

How then did it come about that, five months after his speech to the Reichstag, Hitler signed the order for the invasion, not of Britain, but of the Soviet Union? Why did he make the mistake of attacking Russia before he had finished with Britain, thereby deliberately incurring the dangers of a war on two fronts?

Until the summer of 1940 Hitler had never seriously considered how a war against Britain could be fought and won. Once the French had been defeated and the British thrust out of continental Europe, he had no further interest in continuing the war in the west. Although, as early as November 1939, Raeder had set his naval staff to work on the problems involved in crossing the English Channel, until the end of June 1940 Hitler was not interested. He still assumed the British would come to terms.

In July Hitler rather reluctantly directed that preparations for the invasion of England be completed by the middle of August. Clearly, the problems to be overcome were formidable. Hitler spoke to Raeder in gloomy terms of the difficulties of the operation.

Raeder needed no convincing. Not only was he powerfully impressed by Germany's naval inferiority, but he saw no possibility at all of providing sufficient transports for the forty divisions which the Army planned to land. An acrimonious debate followed between the Army and the Navy on the number of divisions to be put across the Channel and the width of the front on which landings were to be made.

In mid-August Hitler finally accepted the Navy's

arguments against attempting a landing on the scale originally planned. In effect, this meant reducing invasion to a subordinate role dependent upon the ability of the Luftwaffe to strike a decisive blow first at the British power and will to defend themselves. Göring had no doubts that it could and, on 13 August, 1,500 German aircraft took part in the opening attack of 'Operation Eagle' planned to eliminate the R.A.F.

On 1 September the movement of shipping to the German embarkation posts began. But when Raeder saw Hitler on the 6th, he found him inclined to regard 'Operation Sea-Lion' as unnecessary: 'he is finally convinced that Britain's defeat will be achieved, even without the landing'.

On 7 September, the German Air Force made the first of its mass raids on London, with 625 bombers escorted by 648 fighters, the heaviest attack ever delivered on a city. Further night raids followed and on Sunday, the 15th, the Luftwaffe set out to deliver a final assault on the battered capital. This was the climax of the Battle of Britain. The switch from attacks on the R.A.F. to the bombing of London proved to be a major mistake in tactics. The week's comparative respite gave the R.A.F. Fighter Command the chance to recover when it had been on the verge of exhaustion: on the 15th it broke up the waves of German bombers and drove them off with heavy losses to the attackers. Thus the conditions for a successful landing were still lacking. 'Operation Sea-Lion' was postponed indefinitely, and was quietly cancelled in January 1942.

Already, before the issue was decided by the failure of the Luftwaffe, Hitler was beginning to ask himself whether the final defeat of the British, if it was not to be quickly obtained, need hold up the further plans which he was revolving in his mind. Before July was out, even before the Luftwaffe had begun its all-out offensive against the British, he gave orders to start preliminary planning for an attack on Russia.

Hitler was strengthened in his resolve not to delay a settlement with Russia too long by the advantage which the Russians were taking of his preoccupation with the west. In June 1940 while the Battle of France was still in progress, the Soviet Government annexed the Baltic States without informing the Germans in advance, and followed this by heavy pressure on Rumania to make a further cession of territory. Always sensitive to any change in the east, Hitler's suspicions of the Russians mounted in proportion to the treachery of his own intentions.

The Luftwaffe maintained its attacks on London and other big cities well into the winter of 1940–1 and Hitler was perfectly willing to see whether the heavy German air-raids might not shake the British resolution to continue the war. In the meantime, however, the planning of Operation Barbarossa (the invasion of Russia) went steadily forward independently of a decision on the operation against Britain. On 29 July General Jodl told the head of the O.K.W., General Warlimont, that Hitler had made up his mind to prepare for war against Russia.

Warlimont's staff at once set to work, and on 9 August got out the first directive to start work in the deployment areas in the east for the reception of the large masses of troops which would be needed.

At roughly the same time Hitler also gave orders to the Army General Staff under Halder, quite independently of Warlimont's team, to prepare a plan of campaign for operations against the Soviet Union. The plan was presented on 5 December, and Hitler approved it, stressing that the primary aim was to prevent Russian armies withdrawing into the depths of the country and to destroy them in the first encounter. The number of divisions to be committed was fixed at 130 to 140 for the entire operation.

The movement of troops to the east began in the

summer of 1940. In November the Economic Section of the O.K.W. set to work. A special department for Russia was established and among its tasks was a survey of the whole of Russian industry (especially the arms industry) and of the sources of raw material supplies (especially of petroleum).

Thus, when Hitler came finally to express his intention of invading Russia in the directive of 18 December, it was no hastily improvised decision, but one which already had behind it several months of hard work by the planning departments of the O.K.W. and the O.K.H.

There was a third possibility, an alternative to the invasion of either Britain or Russia, which Admiral Raeder persistently urged Hitler to consider—the Mediterranean, and the adjacent territories of North Africa and the Middle East. Here, Raeder argued, was the most vulnerable point in Britain's imperial position, the weak link against which Germany ought to concentrate all her strength.

As he developed his case in two discussions with Hitler, on 6 and 26 September, Raeder brought forward additional arguments, the economic importance of this area for supplying the raw materials Germany so badly needed, and the dangers of a British, or even American, landing in French West Africa by way of the Spanish and Portuguese islands in the Atlantic. Raeder proposed that Gibraltar and the Canary Islands should be secured, and the protection of north-west Africa strengthened in cooperation with Vichy France. At the same time, the Germans, in cooperation with Italy, should launch a major offensive against Suez, and from there advance northwards through Palestine and Syria to Turkey.

There was much to be said for his proposals. They kept Britain in the centre of the picture as the chief enemy; they made full use of Germany's alliances with Italy and Spain, and the projected operations were much more within the compass of German strength than the conquest of Russia proved to be. If Hitler had appreciated the importance of sea-power he would have recognized the force of the Admiral's arguments.

The Führer showed interest in Raeder's suggestions, and in the last four months of 1940 devoted considerable time and energy to plans for operations in the western Mediterranean, but his mind was already firmly made up on the invasion of Russia, and his interest in the Mediterranean was governed by very different assumptions from those of the German Naval Staff.

Hitler's aims in the Mediterranean and African theatres were, first, to add to Britain's difficulties by closing the Mediterranean to her shipping and, second, to safeguard North-west Africa and the Atlantic islands—the Cape Verde group, the Azores, the Canaries, and Madeira—against possible Allied landings. The Atlantic islands could be used as a stepping-stone to West Africa, and Africa was the back door to Europe. In August French Equatorial Africa had declared for General de Gaulle, and to secure that area was a common-sense precaution on Hitler's part.

From beginning to end, however, Hitler looked to Franco's Spain to undertake the main responsibility in the western Mediterranean, and, less hopefully, to Vichy France for the defence of North-west Africa, just as he insisted that the burden of operations in the eastern Mediterranean must fall on Italy. At no time did he contemplate the use of German forces in anything more than a supporting role. There is no evidence at all that he ever seriously considered Raeder's proposals as an alternative to his own plan for attacking Russia. What he did was to single out, against Raeder's advice, those parts which fitted in with his own very different view of the future of the war. At the same time, he did all he could to secure the entry of Spain, and if possible France as well, into the war in order to safeguard the west and keep Britain fully occupied while he turned to the east.

Even these limited objectives proved to be beyond Hitler's power to achieve, largely because of the superior skill of the Spanish dictator in avoiding the trap, into which Mussolini had fallen, of identifying his régime too closely with a German alliance.

General Franco had first expressed his willingness to enter the war in June 1940 at a time when it seemed likely that the division of the spoils was about to begin. As the summer wore on, however, and the capitulation or invasion of Britain failed to take place, Franco's enthusiasm cooled and he began to lay stress on the conditions which were a prerequisite of Spanish intervention. These included territorial claims, large-scale economic assistance in grain and petroleum, and the provision of military equipment. In September 1940 Franco sent his future Foreign Minister, Serrano Suñer, to Berlin, partly in order to allay the growing German irritation with Spain and partly in order to spy out the land.

Ribbentrop pressed Suñer hard for a definite date by which Spain would enter the war, but he described the Spanish demands for aid as excessive, and was evasive on the question of Spain's territorial claims. In his turn, Suñer avoided any definite commitments and continued to insist that Spain's demands must be met before she could risk intervention.

Suñer took a violent dislike to Ribbentrop's preposterous vanity and overbearing methods in trying to get his own way. Hitler made a different impression on him. The Führer had assumed the role of the 'world-historical' genius for the occasion of Suñer's visit, exhibiting the calm confidence of the master of Europe and leaning over the maps to demonstrate with assured gestures the ease with which he could take Gibraltar. Unlike Ribbentrop, he took care neither to utter any complaints nor to exert pressure on his visitor.

For the rest of Suñer's visit everything possible was done to impress him with the power and efficiency of the Third Reich. At the end, exhausted and oppressed, the Spanish Minister escaped with relief. But before he left, Suñer had a second interview with Hitler, in which the Führer, dropping the impressive part he had played at their first meeting, displayed a childlike pleasure in the gift Franco had sent him and astonished the Spaniard by his lack of dignity and by the unaffected behaviour of a German *petit bourgeois*. This contrast between the grandiose pretensions of the régime and the underlying vulgarity and childishness of its rulers was the most permanent impression which Suñer carried away from his visit to Berlin.

From the point of view of the war nothing had been decided. The Spaniards refused to commit themselves to a precise date, even if they were to receive all they asked for.

A week later, on 4 October, Hitler and Mussolini met on the Brenner to review the situation. This time Hitler gave a double reason for refusing to cede French Morocco to Spain: his plan for a large German empire in Central Africa, for which part of the Moroccan coast would be needed as an intermediate base; and his fear that such a step would lead the French colonies in North and West Africa to join de Gaulle. Hitler developed the theme of collaboration with France at some length, much to the irritation of Mussolini, who feared that Vichy might be ingratiating itself with Hitler, and so depriving him of his anticipated reward, the major share of the French colonial empire. The idea of drawing France more fully into the Axis camp, however, continued to intrigue Hitler, despite the Duce's protests, and towards the end of October he resolved to clear up his difficulties with Spain and France by a personal visit to Franco on the Spanish frontier and a meeting with the leaders of Vichy France on the way.

The meeting with Franco proved to be one of the few occasions since he had become the Dictator of Germany on which Hitler found himself worsted in a personal encounter. The memory of his failure never ceased to vex him. For Hitler went out of his way to flatter the Spanish leader. He put himself to the trouble of a long journey across France and he offered Spain immediate aid in the capture of Gibraltar. Admittedly, Hitler was unwilling to agree to the cession of French Morocco. Nevertheless, he was confident that with vague promises for the future and with the impression which his dynamic personality was bound to leave on Franco he would succeed in persuading the Spaniards to agree to enter the war in the near future and play the part for which he had cast them.

The two dictators met at the Spanish frontier town of Hendaye on 23 October. Hitler began with an impressive account of the strength of Germany's position and the hopelessness of England's. He then proposed the immediate conclusion of a treaty, by which Spain would come into the war in January 1941. Gibraltar would be taken on 10 January and would at once become Spanish.

To Hitler's mounting irritation, however, Franco appeared unimpressed and began to insist on Spain's need of economic and military assistance, and to ask awkward questions about Germany's ability to give either in the quantities required. He even ventured to suggest that, if England were conquered, the British Government and Fleet would continue the war from Canada with American support. Barely able to control himself, Hitler at one point stood up and said there was no point in continuing the talks, only to sit down and renew his efforts to win Franco over. Franco refused to commit himself to anything beyond vague generalities. The Führer for once had to admit defeat, after a conversation lasting nine hours.

By contrast, Hitler's interview with Pétain at Montoire on the following day appeared to go well. The aged Marshal of France was one of the few men who ever impressed Hitler, perhaps because he had played a prominent part in the defeat of Germany in 1918. Hitler made no attempt to disguise the difficulty of France's position. 'Once this struggle is ended,' he told the Marshal, 'it is evident that either France or England will have to bear the territorial and material costs of the conflict.' When, however, Pétain expressed himself ready to accept the principle of collaboration, it was agreed that the Axis Powers and France had an identical interest in seeing Britain defeated as soon as possible and that France should support the measures they might take to this end. In return, Hitler agreed that France should receive compensation for territorial losses in Africa from Britain and be left with an empire equivalent to the one she still possessed.

A meeting with Mussolini took place almost immediately after Hitler had seen Franco and Pétain. Its occasion was an item of news which took Hitler by surprise and was to prove the turning-point in his plans for the Mediterranean and Africa.

On 28 October the Italians attacked Greece, not merely without Hitler's agreement, but in flat contradiction of his wishes. This affected the whole future of the war, but here we are concerned only with its effects on Hitler's interest in the southern theatre of operations.

These did not at first appear to be great. True the Italian setbacks in Greece faced Hitler with the prospect of having to come to Mussolini's aid in the Balkans. This, and the improvement in Britain's position as a result of her occupation of Crete and a number of the Aegean Islands certainly put an end to any idea of German offensive operations in the eastern Mediterranean. Even German aid for the Italian drive on Suez, which was the most to which Hitler had ever agreed, was now held back; it was only to be given, if at all, after the Italians had reached Mersa Matruh. But plans for action in the western Mediterranean were confirmed. 'Operation Felix' was to be carried out with German troops supporting the Spaniards in the assault on Gibraltar.

During November Hitler pressed Franco hard, but the Spaniards warily refused to give any definite commitment. On 7 December Admiral Canaris presented the Spanish dictator with a proposal from Hitler to send German troops across the frontier on 10 January and begin operations against the British on that date. This time Franco gave Hitler a blunt refusal. He had little confidence in the plans suggested, he replied, nor was Spain in an economic condition to enter the war yet. The sudden opening of the first British Desert Offensive and the victory of Sidi Barrani in the second week of December confirmed all Franco's doubts. The Italian Army was soon in full retreat across the desert. Forced to recognize his failure with Franco, on 11 December Hitler issued the brief notice: '"Operation Felix" will not be carried out as the political conditions no longer obtain.'

For a moment Hitler even feared that the British successes might lead to the break-away of the French colonial empire under General Weygand, and on 10 December he ordered preparations to be made for an emergency operation, 'Attila', which would secure, if necessary, the occupation of the whole of France and the capture of the French Fleet and Air Force. On 13 December Laval, the advocate of collaboration, was dismissed from his office in the Vichy Government and placed under arrest. The Germans soon secured his release, but not his return to office, and Marshal Pétain stubbornly refused to go to Paris to receive the ashes of the Duke of Reichstadt, Napoleon's son, which Hitler had ordered to be sent from Vienna as a symbolic gesture to the French. Hitler had to abandon the hopes he entertained in the autumn of 1940 of winning the French over to active cooperation against the British.

In the New Year Hitler made one final effort to persuade Franco to come into the war, writing personally to the Caudillo on 6 February and invoking Mussolini's intervention as well. Neither approach had any effect. Mussolini met Franco at Bordighera on 12 February 1941, but, far from changing the Spanish point of view, only echoed Franco's complaints of Hitler's illusions about France. Hitler's own letter to Franco was strongly worded. 'Spain will never get other friends than those given her in the Germany and Italy of today. . . . Caudillo, I believe that we three men, the Duce, you, and I, are bound together by the most rigorous compulsion of history, and that thus we in this historical analysis ought to obey as the supreme commandment the realization that, in such difficult times, not so much an apparently wise caution as a bold heart can save nations.'

Three weeks later Hitler received Franco's reply. The Caudillo was profuse in his protestations of loyalty, but he maintained that attitude of polite evasion which, throughout the negotiations, had baffled Hitler's clumsy efforts to pin him down. The Spanish dictator was to appear high on the list of those who disappointed the Führer by their failure to fulfil the historic role for which he had cast them. It was a stigma which that wily politician knew how to bear with fortitude and eventually to turn to profit.

CHAPTER ELEVEN

'THE WORLD WILL HOLD ITS BREATH'

1940–1

In the summer of 1940 the Russians, alarmed by the extent of the German victories in the west, had hurriedly taken advantage of Hitler's preoccupation to occupy the whole of their sphere of influence under the 1939 Agreement. By August, Estonia, Latvia, and Lithuania were incorporated in the Soviet Union. A Russian ultimatum to Rumania at the end of June demanded the cession of the provinces of Bessarabia and the Northern Bukovina. Hitler could only advise the Rumanians to comply, but henceforward he was determined to prevent any further Russian move towards the west.

His immediate object was to avoid the development of a situation in the Balkans which would provide the Soviet Government with an excuse for intervention. The danger came from Rumania's neighbours, whose territorial ambitions had been aroused by the Russian acquisition of Bessarabia. The Bulgarian claim to the South Dobrudja was soon settled, but the Hungarian demand for Transylvania was more than Rumanian national pride would accept, and relations between the two States rapidly deteriorated to a point where war was a possibility.

Hitler could not afford to see Russian troops occupying the Rumanian oilfields in the event of the Rumanian State disintegrating. Behind the scenes, therefore, he used his influence to bring the Hungarians and the Rumanians into a more reasonable frame of mind, and, when advice proved ineffectual, Ribbentrop summoned both parties to Vienna at the end of August to accept a settlement dictated by the two Axis Powers. Ribbentrop only obtained the Hungarians' consent by shouting at them in a threatening manner, while the Rumanian Foreign Minister, Manoilescu, fell across the table in a faint when he saw the line of partition which Ribbentrop had drawn.

The crisis ended very much to Hitler's advantage. A few days after the Vienna Award, King Carol abdicated in favour of his son, and General Antonescu, an admirer of the Führer, became Rumanian Prime Minister. Before the end of September Antonescu, who was soon to become one of Hitler's favourites, set up a dictatorship, adhered to the Axis Pact (23 September) and 'requested' the dispatch of German troops to help guarantee the defence of Rumania against Russia. A secret order from the Führer's H.Q. on 20 September directed that the troops should be sent. 'To the world their tasks will be to guide friendly Rumania in organizing and instructing her forces.' However, the directive said, their real tasks—to be kept secret from both the Rumanians and the troops themselves—included protecting the oil district and preparing for possible war with Russia.

The reorganization of the Rumanian Army on German lines began in the autumn of 1940, and the German Military Mission was followed by German troops, including A.A. regiments to protect the oilfields, and the 13th Panzer Division which was transferred for training. Hitler had soon established a hold over Rumania as a satellite State which was not to be shaken until the end of the war.

These German moves were far from welcome in Moscow. On 1 September Molotov summoned the German Ambassador and described the Vienna Award as a breach of the Nazi–Soviet Pact, which provided for previous consultation. Molotov pointed out that the German guarantee to Rumania was directed against the U.S.S.R.

These protests, however, Hitler could afford to discount, although he offered a sop in the form of Russian membership of the new Danubian Commission, from which the Soviet Union had originally been excluded. Much more serious was the resentment with which the establishment of a German protectorate over Rumania was received in Rome.

Mussolini had for long entertained ambitions to extend Italian influence in the Balkans and along the Danube. It was fear of German expansion towards the south-east which had led him originally to oppose the *Anschluss,* and all Hitler's fair words had failed to eradicate the suspicion with which he watched any German move in the direction of the Danube or the Adriatic. Hitler was well aware of Mussolini's ambitions in the Balkans and also alive to the possibility of Mussolini taking action there to forestall him. At the interview he had with Ciano on 7 July 1940 Hitler was at pains to impress on the Italian Minister the need to delay any such action in the case of Yugoslavia, a country long marked down by the Duce as an object of his imperial designs. This warning was renewed in the succeeding weeks and extended to Greece, the other possible objective of an Italian move.

These German hints were not much to the Duce's liking. But in a letter of 27 August he assured Hitler that the measures he had taken on the Greek and Yugoslav frontiers were purely defensive: all the Italian resources would be devoted to the attack on Egypt. Hitler took care to associate the Italians with him in the settlement imposed on Hungary and Rumania, and when Ribbentrop visited Rome in the middle of September he repeated that 'Yugoslavia and Greece are two zones of Italian interest in which Italy can adopt whatever policy she sees fit with Germany's full support.'

German-Italian cooperation at this time appeared to be closer than usual, and Mussolini readily fell in with Ribbentrop's proposal of a new Tripartite Pact to be signed by Germany, Italy, and Japan. At the end of September Ciano travelled to Berlin for the signature of the Pact. Every effort was made to impress its importance on the minds of a people who were becoming more sceptical as the war entered its second year.

A week later another meeting between the Führer and the Duce on the Brenner Pass appeared to confirm the solidarity of the Axis partnership. What Hitler did not mention, however, was the steps he was already taking to secure German control over Rumania.

When the movement of German troops became known during the next week, Mussolini's anger at Hitler's duplicity showed how fragile were the bonds of confidence between the two régimes. Once again the Italian dictator felt that Hitler had stolen a march on him. Belated attempts to send an Italian contingent as well were unsuccessful, and the indignant Duce burst out to Ciano: 'Hitler always faces me with a *fait accompli*. This time I am going to pay him back in his own coin. He will find out from the newspapers that I have occupied Greece.'

This time the consequences of Mussolini's pique were more serious than Italy's occupation of Albania as a tit-for-tat after Hitler's march into Prague. The role which Hitler had assigned the Italian forces in his strategic plan was the invasion of Egypt, and on 13 September the Italian Army under Graziani had crossed the Egyptian frontier and begun a slow advance eastwards. Even against the scanty British forces opposing them this soon proved to be a task demanding all the resources Mussolini could command. The Duce's sorely bruised vanity demanded a bold *coup* to restore Fascist prestige, and early on 28 October Italian troops began the invasion of Greece from Albania.

At the last minute Mussolini informed Hitler of

his intention in a long letter written on 19 October. It did not reach Berlin until the 24th and was only communicated to Hitler personally late that night after his interview with Pétain at Montoire. At precisely the moment when he had succeeded in pacifying the Balkans by the virtual occupation of Rumania, the Italians were about to set the whole peninsula in turmoil again by their ill-timed attack. Bulgaria and Yugoslavia, both with claims on Greece, were bound to be aroused; Russia would be provided with a further pretext for intervention, while the British would almost certainly land in Greece and acquire bases on the European shores of the Mediterranean. On top of his unsatisfactory interview with Franco on the 23rd the news from Rome strained Hitler's temper to the limit. Yet the manner in which Mussolini had acted was a clear enough indication of the resentment he felt at high-handed behaviour by the Germans, and Hitler, quick to see the danger of alienating his one reliable ally, for once hesitated to intervene too forcefully. In the hope that a personal appeal to the Duce before the attack began might persuade him to change his mind, a meeting was hurriedly arranged at Florence. Two hours before he got there, Hitler was informed that Italian troops had begun the assault that morning, and Mussolini, smirking with self-satisfaction, could not wait to leave the station platform before announcing his first successes.

It is an interesting sidelight on Hitler's character that, in such provocative circumstances, he controlled himself without difficulty and throughout the talks showed no trace of his real feelings. On the contrary, he began by offering the Duce Germany's full support in the new campaign and placed German parachute troops at his disposal if they should be required for the occupation of Crete. He followed this with a long report to his Italian partner on his negotiations with Spain and Vichy France—clever tactics in view of Mussolini's suspicion of France—and ended with a belated but reassuring account of his relations with Rumania.

Appearances had been preserved. But Hitler's actions on returning to Germany show that he had no illusions about the problems with which the Italians' blunder confronted him. New orders were issued in the directive of 12 November. Although the dispatch of German forces to the support of the Italian drive on Suez was to be considered only after the Italians had reached Mersa Matruh, provision was made for the rapid transfer of a German armoured division to North Africa if necessary. Meanwhile the German forces in Rumania were to be reinforced, and an Army Group of ten divisions assembled to march into Greek Thrace if the need should arise. Hitler still hoped to be able to carry 'Operation Felix' out against Gibraltar, but it is evident that he anticipated trouble in the Balkans, either from British air attacks on Rumania or from an Italian failure in Greece, and was already making preparations to meet it in advance.

Hitler, moreover, had to look still further ahead. The plans for an attack on Russia, on which the Army General Staff had been engaged since August, were now taking shape. While his immediate anxiety, therefore, was the possibility of a British landing in Greece and an Italian collapse, Hitler was bound to view any action he might be obliged to take in the context of his larger design.

Russian suspicions had already been aroused, not only by the guarantee given to Rumania, but also by an agreement between the German and Finnish Governments for the movement of German troops through Finland to the outlying garrisons in Northern Norway. When the German Chargé d'Affaires in Moscow called on Molotov on 26 September he was pressed by the Soviet Foreign Minister to provide a copy of the agreement recently concluded between Germany and Finland, 'including its secret portions'. The German Foreign Office at once complied with the Russian request, but Molotov was not satisfied, asking for more information about the agreement and about the dispatch of a German Military Mission to Rumania. Anxious to allay Russian fears, Ribbentrop suggested that Molotov should visit Berlin. After a week's consideration Stalin agreed that Molotov should go in the first half of November. The extension of the war to the Balkans in the meantime, and the possibility of German intervention in Greece —which would necessitate the passage of German troops through Bulgaria—added to the importance of the Soviet Foreign Minister's visit.

Molotov arrived in the German capital on 12 November. The Führer at once placed the discussion on the most lofty plane: he believed that 'an attempt had to be made to fix the development of nations . . . so that friction could be avoided and the elements of conflict precluded so far as was humanly possible. This was particularly in order when two nations such as the German and Russian nations had at their helm men who possessed sufficient authority to commit their countries to a development in a definite direction. . . .'

Molotov, a cold and stubborn negotiator, precise to the point of pedantry, waited for Hitler to finish his characteristic long and high-flown monologue; then, in equally characteristic fashion, he asked a series of pointed questions about German-Russian cooperation in the present. What were the Germans doing in Finland, which had been assigned to the Russian sphere of influence in their earlier agreement? What was the significance of the Tripartite Pact? 'There were also issues to be clarified regarding Russia's Balkan and Black Sea interests with respect to Bulgaria, Rumania and Turkey.' On all these points, Molotov said, he would like to have explanations.

Hitler was so taken aback that he made the excuse of a possible air raid and broke off the discussion until the following day.

When they met again next morning, Hitler made an effort to forestall Molotov's remarks by admitting that the necessities of war had obliged Germany to intervene in areas where she had no permanent interests, such as the Balkans and Finland. Germany had lived up to her side of the Agreement, he said, while Russia had occupied the Northern Bukovina and part of Lithuania, neither of which had been mentioned in the Agreement at all. He made a determined effort to keep his temper and bring the discussion round to 'more important problems', the partition of the British Empire after its impending defeat.

But as soon as Hitler finished, Molotov resumed where he had left off: the next question was the Balkans and the German guarantee to Rumania, 'aimed against the interests of Soviet Russia, if one might express oneself so bluntly'. With mounting impatience, Hitler went over the familiar ground again: Germany had no permanent interests in the Balkans, wartime needs alone had taken her there, the guarantee was not directed against Russia, and so on.

The talks left Hitler in a state of violent irritation. Franco had only angered him by evasion; Molotov had answered back and argued with him. That night he was unexpectedly absent from the banquet which Molotov gave to his hosts in the Russian Embassy.

Half-way through the banquet a British air-raid drove the two foreign ministers to take shelter below ground. Ribbentrop with characteristic maladroitness seized the occasion to confront Molotov with the draft of a new agreement which would have brought the Soviet Union into the Tripartite Pact.

The core of the treaty was Article II, an undertaking to respect each other's natural spheres of influence. The significance of this was made clear by two accompanying protocols, both of which were to remain secret. The first defined the Four Powers' spheres of influence. Apart from revisions in Europe, Germany's territorial aspirations centred in Central Africa, and Italy's in northern and northeastern Africa. Japan's domain was in the area of eastern Asia to the south of the Island Empire of Japan, and the Soviet Union's area was south of the U.S.S.R. in the direction of the Indian Ocean.

If he could persuade Molotov and Stalin to accept such a settlement, Hitler believed that he would be able to divert Russia from her historic expansion towards Europe, the Balkans and the Mediterranean—areas in which she was bound to clash with Germany and Italy—southwards to areas such as the Persian Gulf and the Indian Ocean, where Germany had no interest and where Russia would at once become embroiled with the British. It was a bold but transparent proposal which cut right across both the traditions and the interests of Russia. Hitler and Ribbentrop hoped, however, to make it more attractive by the second protocol, which promised German and Italian cooperation in detaching Turkey from her commitments to the West and winning her over to collaboration with the new bloc of Powers. In addition, the German Foreign Minister spoke in vague but tempting terms of German help in securing for Russia a Non-Aggression Pact with Japan, as a result of which Japan might be persuaded to recognize the Soviet spheres of influence in Outer Mongolia and Sinkiang and to do a deal over the island of Sakhalin, with its valuable coal and oil resources.

Molotov, who had had no chance to examine the draft in advance, was the last man to let himself be carried away by Ribbentrop's barnstorming diplomacy. His reply made it unmistakably clear that

Russia was not prepared to disinterest herself in Europe. Turkey, Bulgaria, Rumania and Hungary were also of concern to the Soviet Union. 'It would further interest the Soviet Government to learn what the Axis contemplated with regard to Yugoslavia and Greece, and likewise what Germany intended with regard to Poland. . . . The Soviet Government was also interested in the question of Swedish neutrality . . . and the question of the passages out of the Baltic Sea.'

Ribbentrop, complaining that he had been 'queried too closely' by his Russian colleague, made one last effort to pull the conversation back to the agenda which he had proposed. 'He could only repeat again and again that the decisive question was whether the Soviet Union was prepared and in a position to cooperate with us in the great liquidation of the British Empire.'

But Ribbentrop's last exasperated plea met with no more response than Hitler's. To Ribbentrop's repeated assurances that Britain was finished, Molotov replied: 'If that is so, why are we in this shelter and whose are these bombs which fall?'

On 25 November, less than a fortnight after Molotov's visit to Berlin, the Soviet Government sent an official reply accepting Ribbentrop's suggested Four-Power Pact, on condition that the Germans agreed to a number of additional demands. These included the immediate withdrawal of German troops from Finland; a mutual assistance pact between Russia and Bulgaria, including the grant of a base for Russian land and naval forces within range of the Straits; a further Russian base to be granted by Turkey on the Bosphorus and Dardanelles; and Japan's renunciation of her rights to coal and oil concessions in northern Sakhalin. Provided these claims were accepted, Russia was prepared to sign the Pact, rewriting the definition of her own sphere of expansion to make its centre the area south of Baku and Batum in the general direction of the Persian Gulf.

No reply was ever sent to the Soviet counter-proposals, despite German assurances that the Russian Note was being studied. Hitler's offer had been designed to divert Russia away from Europe. Once it became clear from Stalin's reply that Russia insisted on regarding Eastern Europe as within her sphere of influence, Hitler lost interest in any further negotiations. If he had still entertained any doubts about giving the order to prepare for 'Operation Barbarossa' before Molotov's visit he had none left after it.

Hitler now reinforced his determination to invade Russia and thus secure Germany's future *Lebensraum* in the east by the argument, of which he soon convinced himself, that Russia was preparing to attack Germany. Russian objections to German intervention in Finland and the Balkans were twisted into evidence of a Russian intention to cut off German iron-ore supplies from Sweden and oil supplies from Rumania. From this it was only a step to postulating the existence of an agreement between Russia and Great Britain. Thus Germany was once more threatened with encirclement, and Hitler was able to adopt the indignant attitude of the innocent man driven to defend himself. The captured German papers reveal this for the lie it was, and document, step by step, the systematic preparation of an act of aggression against a people whose government to the last day was anxious to avoid giving any pretext for war to the German dictator.

Meanwhile, Italy's attack on the Greeks had run into heavy trouble, and in a letter to Mussolini on 20 November 1940, Hitler told the Duce frankly that the consequences of the Italian action were likely to prove grave. The reluctance of Bulgaria, Yugoslavia, Turkey, and Vichy France to commit themselves had been fortified; Russian alarm about the Balkans and the

Straits had been increased, while Britain had been given the opportunity to secure bases in Greece from which to bomb Rumania and southern Italy.

The measures with which Hitler proposed to meet these difficulties were comprehensive. Spain must come into the war at once, seize Gibraltar and guarantee north-west Africa. Russia must be turned away from the Balkans, Turkey persuaded to stop any threats against Bulgaria, Yugoslavia induced to collaborate with the Axis against Greece, and Rumania pressed to accept German reinforcements. To these political tasks, Hitler added increased air attacks on the British Navy and its bases in the eastern Mediterranean, in which the German Air Force would assist the Italians. The German squadrons must, however, be sent back by 1 May at the latest, and land operations against Egypt would have to be abandoned for the time being. The principal military effort would go into a German attack by March 1941 to clear the British out of Thrace.

Mussolini's comment on reading Hitler's letter was brief: 'He has really smacked my fingers.' However, he accepted Hitler's proposals.

In the meantime Hitler was engaged in securing the political prerequisites for his intervention in Greece. A succession of Balkan rulers was imperiously summoned to Germany. On 5 December Hitler wrote again to Mussolini. Yugoslavia and Bulgaria were proving difficult—the latter under Russian pressure—but he had hopes of bringing them over, and Mussolini was much relieved at the more confident tone of the letter.

Unfortunately for Mussolini, the degree of Fascist incompetence had not yet been fully revealed. On 7 December the Italian Ambassador begged Ribbentrop for immediate help to relieve the situation on the other side of the Adriatic, where the Italians were in danger of a complete rout. When Hitler received the Ambassador the next day and asked for an early meeting with the Duce, Mussolini refused to face him. To add to the Duce's troubles, the Battle of Sidi Barrani, which began on 9 December, led to the collapse of the Italian threat to Egypt and the headlong retreat of Graziani's forces back across Libya.

In this crisis Hitler kept his head. He refused to be diverted from his main objectives. Between 10 December and 19 December he issued a series of orders which were designed not only to prop up his failing Italian ally, but to carry out his long-range plans.

On 10 December Hitler ordered formations of the German Air Force to be moved to the south of Italy, from where they were to attack Alexandria, the Suez Canal, and the Straits between Sicily.

On 13 December a directive for the invasion of Greece ('Operation Marita') was issued. A German task force was to be formed in Rumania ready to thrust across Bulgaria as soon as favourable weather came, and to occupy the Thracian coast of Greece. The first objective was to deny the British air bases in Thrace, from which they could bomb Rumania and Italy, but if necessary the operation was to be extended to the occupation of the whole of the Greek mainland.

On 18 December Hitler signed Directive No. 21 for 'Barbarossa': 'The German Armed Forces must be prepared to crush Soviet Russia in a quick campaign even before the end of the war against England. . . .' The active cooperation of Finland, Hungary, and Rumania was counted on, and in December both the Chief of the Finnish General Staff and the Hungarian Minister of War visited Germany. General Antonescu, who had already seen Hitler in November, came to Berchtesgaden a second time in January 1941.

Finally, on 19 December, Hitler saw the Italian Ambassador and promised increased economic aid for Italy, on condition that German experts should go to Italy and advise on its use. In return more Italian workmen were to be sent to Germany. This was one more step in the reduction of Italy to the status of a German satellite.

With these measures put in train, Hitler was confident that he could master the crisis and still be ready for the attack on Russia by 15 May, the date fixed for the completion of preparations. He said nothing of such a possibility to Mussolini, but in the letter which he wrote to the Duce on the last day of 1940 he was cordial in tone and did his best to encourage Mussolini and to assure him of his own unshaken confidence in the future.

Early in the New Year the chiefs of the three Services were summoned to the Berghof, where a war council lasting two days was held, on 8–9 January 1941. Hitler reviewed what could be done for Italy. His general mood was still one of confidence. Britain could not win; she went on fighting only because of the hopes she entertained of American and Russian intervention. Hitler described Stalin as 'a cold-blooded blackmailer who would, if expedient, repudiate any written treaty at any time.'

Ten days later Mussolini reluctantly visited the Berghof. Smarting under the humiliations of Libya and Greece, he looked forward without relish to the Germans' patronizing condolences. To his surprise Hitler behaved with tact and cordiality.

Mussolini found Hitler in a very anti-Russian mood, and Ribbentrop called a sharp halt to ill-timed Italian attempts to improve their relations with Moscow. On the second day of the visit, Monday 20 January, Hitler made a speech of two hours on his coming intervention in Greece which much impressed the Italian military men by the grasp of technical matters it displayed. Demonstrating his points with expressive gestures on the map, he impressed upon his audience the picture of a master of strategy who had foreseen every possibility and who was in complete command of the situation.

Hitler did not, however, reveal his intention of attacking Russia. It could be argued that Italy was in no position in 1941 to give any help at all in the operations about to begin in Eastern Europe. But this argument could certainly not be applied to the second of Hitler's partners in the Tripartite Pact, Japan, whose relations with the U.S.S.R. had balanced precariously on the edge of war since the Japanese invasion of Manchuria in 1931. For ten years Japan and Russia had eyed each other with mutual suspicion and hostility. Yet Hitler made no effort to bring Japan into the war he was proposing to wage against Russia; on the contrary, he did everything he could to divert her away from Russia's Far Eastern territories towards the south.

Hitler wanted Japan to enter the war at the earliest possible moment, but it was against England, not against Russia, that he sought her cooperation. The war in Europe, Hitler and Ribbentrop assured Japanese Foreign Minister Matsuoka when he visited Berlin in March 1941, was virtually over; it was only a question of time before Britain was forced to admit defeat. An attack by Japan upon Singapore would not only have a decisive effect in convincing Britain that there was no further point in continuing the war, it would also provide the key to the realization of Japanese ambitions in Eastern Asia at a time when circumstances formed a unique combination in her favour.

Hitler admitted that there were risks, but he dismissed them as slight. England was in no position to defend her possessions in Asia. America was not yet ready, and an attack on Singapore would strengthen the tendency towards non-intervention in the United States. If, none the less, America should attack. Japan could rely on German support, and Germany would at once attack Russia if she moved against Japan.

In all their conversations Hitler and Ribbentrop

persistently urged on Matsuoka the importance of an attack on Singapore at the earliest possible date. Ribbentrop asked the Japanese Foreign Minister for maps of the British base, 'so that the Führer, who must certainly be considered the greatest expert of modern times on military matters, could advise Japan as to the best method for the attack on Singapore.' Japan, in short, was to play in the Far East the role for which Hitler had cast Franco's Spain and Mussolini's Italy: the capture of Singapore was the Far Eastern version of the capture of Gibraltar and the drive on Suez.

Had Hitler succeeded in persuading his allies to fall in with his plans, Britain's strength would have been stretched to the limit. This time it was not his strategy but his diplomacy that was at fault. Between the defeat of France and the attack on Russia, Hitler conducted a considerable number of diplomatic negotiations: it is a striking fact that, in every case where he was unable to use the threat of force if his wishes were not met, these negotiations failed. Spain, Italy, Vichy France, and now Japan, all chose different paths from those the Führer had mapped out for them. It is not difficult to see why. Hitler's overbearing manner and his total inability to cooperate with anyone on equal terms; Ribbentrop's belief that the most effective method of diplomacy was to nag and, if possible, to threaten, produced in most of their visitors only a feeling of relief when the interview came to an end. It was too patent on every occasion why the Germans wanted what they were asking for, too obvious who was to benefit from it.

To clumsiness the Germans added falseness. Hitler and Ribbentrop deceived their allies, even when there was no need. Nothing so much angered Mussolini as the fact that his allies told him lies and then sprang surprises on him. Hitler showed surprising loyalty to Mussolini, but it never extended to trusting him. His golden rule in politics remained: Trust nobody.

It is not surprising, therefore, that Hitler should have concealed his purpose to attack Russia from the Japanese Foreign Minister. It is one more piece of evidence pointing to the confidence which Hitler felt in his ability to conquer Russia, as he had conquered France, in a single campaign and without the need of help from outside which, when victory had been won, might prove an embarrassment. When Germany attacked Russia three months later Matsuoka's failure to warn the Japanese Government led to his fall. Thereby Hitler lost his best ally in the Tokyo Cabinet, and the Japanese quickly made up their minds to follow their own plans and keep the Germans in ignorance. As in the case of the Italian attack on Greece, the Germans had little justification either for surprise or complaint.

His absorption in the war left Hitler with less time for public appearances on other than military occasions. In the whole of 1940 he made only seven speeches of any importance, even less in succeeding years, a fact of some importance when it is recalled how great a part Hitler's oratory played in the history of the years before the war.

It was on his military rather than his political gifts that Hitler now relied, and with the spring of 1941 he looked forward eagerly to the moment when he could once more give the order to advance on Russia, the greatest of all his schemes, as a necessary preliminary to which he had now come to accept sweeping the British out of Greece and the Balkans.

Between Germany and Greece lay four countries—Hungary, Rumania, Yugoslavia, and Bulgaria—whose compliance had to be secured before Hitler could reach the Greek frontier. Hungary and Rumania had already accepted the status of German satellites, and throughout the winter months German troop trains steadily moved across Hungary to Rumania. In Bulgaria a sharp tussle for influence took place between the Germans and the Russians. The Germans won, and on the night of 28 February German forces from Rumania crossed the Danube and began to occupy key positions throughout the country. The following day Bulgaria joined the Tripartite Pact.

Yugoslavia proved more difficult. Recognizing this, Hitler did not ask for the passage of German troops, but he put strong pressure on the Yugoslav Government to accede to the Tripartite Pact. Hitler's bribe was the offer of Salonika, and it was taken. On 25 March, the Pact was signed in Vienna. Given favourable weather, Hitler now told Ciano, the decision in Greece could be brought about in a few days.

Hitler's satisfaction, however, was premature. On the night of 26–7 March a group of Yugoslav officers, rebelling against their Government's adherence to the Axis cause, carried out a *coup d'état* in Belgrade in the name of the young King Peter II.

The insolence of a nation which ventured to cross him roused Hitler's fury. A hurried council of war summoned to the Chancellery learned of the Führer's decision to inflict exemplary punishment on Yugoslavia, while the Japanese Foreign Minister was kept waiting in another room. Hitler took the decision, then and there, to postpone the attack on Russia up to four weeks, so completely was he prepared to sacrifice everything to the satisfaction of his desire for revenge.

Never was the man's essential character more clearly illuminated. The brutal tone of the orders reflects this mood. Not content with taking steps to ward off any threat to his plans from Yugoslavia, he was bent upon the entire destruction of the state and its partition. The blow, he insisted, must be carried out with 'merciless harshness'.

The military preparations for this new and unexpected campaign had to be improvised, but Hitler issued his directive that very day, and again included the sentence: 'Yugoslavia, despite her protestations of loyalty, must be considered as an enemy and crushed as swiftly as possible.' Hitler wrote to Mussolini on the evening of the 27th, requesting him to cover the Yugoslav-Albanian frontier. At the same time imperious messages were sent to Hungary and Bulgaria, and General von Paulus hastily dispatched to Budapest to coordinate the military measures to be taken by the satellite forces against the isolated Yugoslavs. Hitler's political preparations contained provision not only for stirring up the hatred and greed of Yugoslavia's neighbours, but also for disrupting the Yugoslav State internally by appealing to the Croats, whose grievances against the Belgrade Government had long been fostered by Nazi agents.

By 5 April, ten days after he had received the news of the *coup d'état*, Hitler had completed his preparations, and at dawn on the 6th, while German forces pushed across the frontiers, squadrons of German bombers took off for Belgrade to carry out 'Operation Punishment'. Flying at rooftop height, the German pilots systematically bombed the city for three whole days, killing more than seventeen thousand people.

Simultaneously, other German divisions operating from Bulgaria began the invasion of Greece. Both operations, mounted with overwhelming force, were rapidly carried to success. On 17 April the Yugoslav Army was driven to capitulate; six days later the Greeks, after their six months' heroic resistance to the Italians, were forced to follow suit. On 22 April the British troops, who had landed in Greece less than two months before, began their evacuation. On the 27th the German tanks rolled into Athens, and on 4 May Hitler presented his report to a cheering Reichstag. The Balkan war, which Mussolini had begun in an attempt to assert his independence, had ended in a German triumph which completely eclipsed the Italian partner of the Axis, and which was by implication a public humiliation of the Duce, who had been driven by his failures to turn to Germany for help.

In his speech to the Reichstag Hitler did his best to disguise this unpalatable fact, but the real relationship between Berlin and Rome was revealed by the partition of Yugoslavia. Italy's claims to the divided state were treated on a level with those of the other satellites, and the Duce had perforce to accept Hitler's unilateral decisions.

Italian dependence upon Germany was further emphasized by the course of events in North Africa. Italian failure to check the British advance led to the British conquest of Cyrenaica. Hitler discounted the military danger in losing North Africa, but he was worried about the effect on Italy. Mussolini, he decided, must be given effective assistance. Recognizing that air support was no longer enough, he reluctantly agreed to the transfer of an armoured division from the Balkans and secured Mussolini's consent to the creation of a unified command of all mechanized and motorized forces in the desert under General Rommel.

Rommel took not only the British, but the German High Command, by surprise. Ordered to submit plans for consideration by 20 April, he actually began his attack on 31 March, and by 12 April had driven right across Cyrenaica and recaptured Bardia within a few miles of the Egyptian frontier.

Indeed, by the early summer of 1941, the situation in the eastern Mediterranean had been changed out of recognition. The British had been thrown out of Greece and pushed back to the Egyptian frontier. In Iraq the pro-German Prime Minister, Rashid Ali, led a revolt against the British garrison, and at the beginning of May appealed to Hitler for help for which Syria, under the authority of Vichy, provided a convenient base. Finally, between 20 May and 27 May, German parachute troops captured the island of Crete.

With the small British forces available stretched to the limit to hold Egypt, Palestine, and Iraq, it appeared to the German Naval Staff and to Rommel that it needed only a sharp push to destroy the whole edifice of Britain's Middle Eastern defence system. Accordingly, on 30 May, Raeder revived his demand for a 'decisive Egypt–Suez offensive for the autumn of 1941 which [he argued] would be more deadly to the British Empire than the capture of London'. The anxiety revealed in Churchill's and General Wavell's dispatches at this time lends retrospective support to Raeder's arguments. Even a quarter of the forces then being concentrated for the attack on Russia could, if diverted to the Mediterranean theatre of war in time, have dealt a fatal blow to British control of the Middle East.

But Hitler, his mind wholly set upon the invasion of Russia, declined to look at the Mediterranean as anything more than a sideshow which could be left to the Italians with a stiffening of German troops. In vain both Raeder and Rommel tried to arouse his interest in the possibilities open to him in the south.

Hitler's mind had been made up at the beginning of the year. On 15 February he announced that any large-scale operations in the Mediterranean must wait until the autumn of 1941, when the defeat of Russia would have been accomplished. Then Malta could be taken and the British expelled from the Mediterranean—but not before. The capture of Crete was the end of the operations in the Balkans, not a stepping-stone to Suez and the Middle East. On 25 May Hitler gave orders to support Rashid Ali's revolt in Iraq, but help was to be limited to a military mission, some assistance from the German Air Force and the supply of arms.

All this time the building up of the German forces in the east had steadily continued. On 3 February General Halder presented the Army's detailed estimate of the situation to the Führer. The huge forces to be engaged and the vast distances to be covered excited Hitler's imagination, and he was in no mood to listen to doubts. He insisted that the Russians must be prevented from falling back into the depths of their country. Everything depended upon the encirclement of the main Russian forces as near to the frontier as possible. The participation of Finland, Rumania, and Hungary in the attack was assured, but Hitler added that—with the exception of Rumania—agreements could be made only at the eleventh hour, in order to keep the secret well guarded. After examining the operational plans for each Army Group, Hitler expressed himself as satisfied. 'It must be remembered,' he declared, 'that the main aim is to gain possession of the Baltic States and Leningrad. . . . When "Barbarossa" begins, the world will hold its breath and make no comment.' By a double bluff, meanwhile, the concentration of German troops in the east was to be represented as a feint to disguise renewed German preparations for the invasion of England and the attack on Greece.

A month later, early in March, Hitler held another military conference, to which he summoned all the senior commanders who were to take part in the attack. Hitler presented the invasion as a step forced on him by Russia's imperialistic designs in the Baltic and the Balkans. A Russian attack on Germany was a certainty, he assured them, and must be forestalled. A secret agreement had even been arrived at between Russia and England, and this was the reason for the British refusal to accept German peace offers. 'The war against Russia', he continued, 'will be such that it cannot be conducted in a chivalrous fashion. This struggle is one of ideologies and racial differences and will have to be conducted with unprecedented, merciless and unrelenting harshness.' Breaches of international law by German soldiers were to be excused since Russia had not participated in the Hague Convention and had no rights under it. A number of the generals protested after the conference that such a way of waging war was intolerable. The most Brauchitsch felt able to do was surreptitiously to issue an order instructing officers to preserve strict discipline and to punish excesses.

The generals were even more disturbed at the proposals for the administration of the territories occupied in the east. A special directive issued on 13 March provided that 'in the area of operations, the Reichsführer S.S. [Himmler] is entrusted, on behalf of the Führer, with special tasks for the preparation of the political administration. . . . Within the limits of these tasks, the Reichsführer S.S. shall act independently and under his own responsibility'.

This could only mean that Himmler and the S.S. were to be given a free hand to stamp out all traces of the Soviet system. The directive also provided for handing over the areas occupied to the political administration of special commissioners appointed by Hitler himself, and for the immediate economic exploitation of the territory seized under the direction of Göring. Even the most unpolitical of German generals can have had little doubt what all this amounted to. Hitler was taking steps in advance to make sure that no scruples or conservatism on the part of the Army Commanders should stand in the way of the treatment of the occupied territories on thorough-going National Socialist lines.

On 20 April Hitler appointed Alfred Rosenberg, the half-forgotten figure who had played a great part in forming his views on German expansion in the east, as Commissioner for the East European Region. It was an unhappy choice. Himmler, as Reichsführer of the S.S. corps d'élite, already claimed the responsibility for laying the racial foundations of the New Order in the east, and Göring, as Plenipotentiary for the Four-Year Plan, claimed the right to organize the economic exploitation of the territories in the east so as to guarantee Germany's present and future needs in food and raw materials. Against two such powerful empire builders as Himmler and Göring, Rosenberg was quite unable to defend his own position, and there was thus from the beginning a conflict of authority in the east between the Army, Himmler, Göring, and the nominal Commissioner, Rosenberg, which only became worse as time went on.

The ruthlessness of the German treatment of the occupied territories in the east was not fortuitous; it was part of a methodical system of exploitation and resettlement planned in advance and entered upon with a full appreciation of its consequences. The overriding need, as defined in a directive of Göring's Economic Staff East dated 23 May 1941, was the use of the food-producing areas of the east to supplement Germany's and Europe's supplies both during and after the war. The directive, in discussing the consequences for Russia's industrial population, noted: 'Many tens of millions of people in the industrial areas will become redundant and will either die or have to emigrate to Siberia. Any attempt to save the population there from death by starvation by importing surpluses from the Black Soil Zone would be at the expense of supplies to Europe. It would reduce Germany's staying-power in the war, and would undermine Germany's and Europe's power to resist the blockade. This must be clearly and absolutely understood.'

This, it should be pointed out, is not Hitler talking late at night upon the Obersalzberg; this is the translation of those grim fantasies into the sober directives and office memoranda of a highly organized administration, methodically planning economic operations which must result in the starvation of millions. Not far away, in the offices of Himmler's S.S., equally methodical calculations were being made of how this process could be accelerated by the use of gas chambers (including mobile vans) for the elimination of the racially impure.

On 30 April with the Balkan operations completed, apart from the pendant of Crete, Hitler fixed 22 June as the new date for the opening of the attack in the east. By May the armoured divisions which had overrun Greece were on their way north to join the concentration of German troops in Poland, and an ominous lull settled over the battle fronts.

In order to camouflage his intentions Hitler ordered Russian orders for goods placed in Germany to be fulfilled and deliveries maintained till the last moment. The Russians continued to make a prompt dispatch of raw materials and food to Germany up to the day of the attack. Indeed, in the last three months before the attack, the Soviet Government, while building up its defences in the west, did everything it could to conciliate and appease the Germans.

Early in May Stalin took over the Chairmanship of the Council of People's Commissars, a step universally regarded as indicating the prospect of a crisis with which only Stalin himself could deal. Immediately afterwards, however, on 8 May, *Tass* denied reports of troop concentrations in the west; on 9 May the U.S.S.R. withdrew its recognition from the legations of the exiled Governments of Belgium, Norway, and Yugoslavia, and on 12 May established relations with the pro-Nazi Government of Rashid Ali in Iraq. All through this period the Soviet Press was kept under the strictest restraint in order to avoid provocation, and as late as 14 June *Tass* put out a statement categorically denying difficulties between Germany and Russia.

There is not a scrap of evidence to show that, in the summer of 1941, the Soviet Government had any intention of attacking Germany. Warnings from the British (who knew the date fixed for the invasion before the end of April) were dismissed by the Russians as trouble making. But Hitler was interested only in reports that could be used to support the pretext for his decision, a decision reached long before without regard to Russia's attitude or the threat which he now alleged of Russian preparations to strike westwards.

In May Antonescu paid his third visit to Hitler, this time at Munich, and agreed that Rumania should take part in the attack. At the end of the month the Finnish Chief of Staff spent a week in Germany to discuss detailed arrangements for cooperation between the two armies. Still Hitler said nothing to Mussolini. When they met at the Brenner on 2 June the most that Ribbentrop admitted was that Russo-German relations were not so good as they had been. Stalin, he told Ciano, was unlikely to commit the folly of attacking Germany, but if he did the Russian forces would be smashed to pieces.

A fortnight later Ribbentrop was more forthcoming, or more indiscreet, when he met Ciano at Venice. Ciano asked his colleague about the rumours of an impending German attack on Russia.

'Dear Ciano,' was Ribbentrop's expansive reply, 'I cannot tell you anything as yet because every decision is locked in the impenetrable bosom of the Führer. However, one thing is certain: if we attack, the Russia of Stalin will be erased from the map within eight weeks.'

From Venice Ribbentrop sent a telegram to Budapest warning the Hungarians to be ready. On 18 June a Non-Aggression Pact between Germany and Turkey was announced. On the 14th Hitler had summoned a conference of his commanders-in-chief in the Reich Chancellery. The generals showed none of the doubts they later claimed to have felt about the Russian 'adventure'. Now, one after another they explained their operational plans to the Führer, while Hitler nodded his approval. Satisfied with the preparations that had been made, in the following week Hitler left for his new headquarters, Wolfsschanze ('Wolf's Lair'), near Rastenburg in East Prussia.

There, on 21 June, the eve of the attack, he dictated a letter to Mussolini. It was the first official news Mussolini had been given of his intentions.

As so often before, Hitler proceeded to justify himself at length. Britain had lost the war, but held out in the hope of aid from Russia. The Russians, reverting to their old expansionist policy, prevented Germany from launching a large-scale attack in the west by a massive concentration of forces in the east. Until he had safeguarded his rear, Hitler declared, he dared not take the risk of attacking England.

Once again Mussolini was roused in the middle of the night with the usual urgent message from the Führer. While the Duce was still reading Hitler's letter the attack was already beginning. From the Arctic Circle to the Black Sea more than a hundred and fifty German, Finnish, and Rumanian divisions were pressing forward across the Russian frontiers. The German forces, divided into three Army Groups commanded by Leeb, Bock, and Rundstedt, included nineteen armoured divisions and twelve motorized, supported by over 2,700 aircraft.

In the 1920s Hitler, then an unsuccessful Bavarian politician, whose political following numbered no more than a few thousands, had written at the end of *Mein Kampf*: 'And so we National Socialist. . . . when we speak of new territory in Europe today we must think principally of Russia and her border vassal states. Destiny itself seems to wish to point out the way to us here. . . . This colossal Empire in the east is ripe for dissolution, and the end of the Jewish domination in Russia will also be the end of Russia as a state.'

At dawn on 22 June 1941, one year to the day since the French had signed the armistice at Compiègne,

Hitler believed that he was about to fulfil his own prophecy. He concluded his letter to Mussolini with these words: 'Since I struggled through to this decision, I again feel spiritually free. The partnership with the Soviet Union, in spite of the complete sincerity of my efforts to bring about a final conciliation, was nevertheless often very irksome to me, for in some way or other it seemed to me to be a break with my whole origin, my concepts and my former obligations. I am happy now to be delivered from this torment.'

It was to prove an irrevocable decision.

CHAPTER TWELVE

THE UNACHIEVED EMPIRE

1941–3

At the time Hitler gave two reasons for his decision to attack Russia: the first, that Russia was preparing to attack Germany in the summer of 1941; the second, that Britain's refusal to acknowledge defeat was due to her hopes of Russian and American intervention, and that Britain had actually entered into an alliance with Russia against Germany.

At most these arguments reinforced a decision already reached on other grounds. Hitler invaded Russia for the simple but sufficient reason that he had always meant to establish the foundations of his thousand-year Reich by the annexation of the territory lying between the Vistula and the Urals.

The novelty lay not so much in the decision to turn east as in the decision to drop the provision he had hitherto regarded as indispensable, a settlement with Britain first. Forced to recognize that the British were not going to be bluffed or bombed into capitulation, Hitler convinced himself that Britain was already virtually defeated. She was certainly not in a position, in the near future, to threaten his hold over the Continent.

Most important of all was the belief, a result partly of his conviction that the German Armed Forces were invincible, partly of an underestimate of Russian strength, that the Soviet armies could be defeated in a single campaign. Hitler knew that he was taking a risk in invading Russia, but he was convinced that the war in the east would be over in two months, or three at the most. He not only said this, but acted on it, refusing to make any preparations for a winter campaign. A series of sharp defeats, and he was certain that Stalin's Government would fall. 'We have only to kick in the door,' Hitler told Jodl, 'and the whole rotten structure will come crashing down.' Hitler was not blind to the numerical superiority of the Russians, but he was certain that the political weakness of the Soviet régime, together with the technical superiority of the Germans, would give him a quick victory.

The opening of the campaign seemed to justify Hitler's optimism. The German armoured divisions struck deep into Russian territory. By 5 July they had reached the Dnieper, by the 16th Smolensk, little more than two hundred miles from Moscow.

But, although the German troops rapidly gained ground, they did not succeed in destroying the Russian armies, and all the time the German Army was being drawn deeper in.

At this point a divergence began to appear between Hitler's and the Army High Command's views of the objectives to be gained. Hitler laid the greatest stress on clearing the Baltic States and capturing Leningrad; once the initial battles were over, the Centre Army Group was to support this northerly drive through the Baltic States and not push forward to Moscow. At the same time the Southern Army Group was to drive south-east towards Kiev and the Dnieper, in order to secure the agricultural and industrial resources of the Ukraine.

Brauchitsch and Halder believed that the best chance of catching and destroying the Russian forces was to press on to Moscow. They were in favour of concentrating, not dispersing, the German effort. This view was supported by Bock, the Commander-in-Chief of the Centre Army Group, and by his two panzer commanders, Guderian and Hoth, but it was rejected by Hitler, who insisted on ordering part of Bock's mobile forces to assist the northern army group's drive on Leningrad and the rest to wheel south and support the advance into the Ukraine. Brauchitsch temporized, the dispute rumbled on, and the Centre Army Group remained halted east of Smolensk.

By September Hitler was beginning to lose interest in Leningrad, and he agreed to launch a major offensive against Moscow, but he insisted that the battle of encirclement in the Ukraine must be put first. Reluctantly the General Staff were forced to assent, but General Halder has since argued that this was the turning-point of the campaign and that Hitler threw away the chance of inflicting a decisive defeat on the Russians for the sake of a prestige victory and the capture of the industrial region of the Ukraine.

For not only had this dispute seriously worsened the relations between Hitler and his generals, it also led to the waste of valuable time. The southern encirclement proved a great success and over 600,000 Russians were taken prisoner east of Kiev, but it was late in September before the battle was ended. The onset of the autumn rains, which turned the Russian countryside, with its poor roads, into a quagmire, promised ill for the attack on Moscow, which the Army High Command had wanted to launch in August. Beyond the autumn loomed the threat of the Russian winter. Hitler, however, elated by his success in the south, now pushed forward the attack on Moscow which he had held back for so long.

On 2 October the advance of the Centre Army Group was resumed, after a halt of two months. On 8 October Orel was captured, and the next day Otto Dietrich, the Reich Press Chief, caused a sensation with the announcement that the war in the east was over. Between Vyazma and Bryansk, another 600,000 Russians were trapped and taken prisoner. A week later the German spearheads reached Mozhaisk, only eighty miles from the Russian capital.

Yet even now Hitler could not make up his mind to concentrate on one objective. In the north Leeb was ordered at the same time to capture Leningrad, link up with the Finns and push on to cut the Murmansk railway. In the south Rundstedt was ordered to clear the Black Sea coast (including the Crimea) and strike beyond Rostov, eastwards to the Volga and south-eastwards to the Caucasus. 'We laughed aloud when we received these orders,' Rundstedt later declared, 'for winter had already come and we were almost seven hundred kilometres from these cities.'

Thus, with forces which were numerically inferior to the Russians, and fanned out across a thousand-mile front, Hitler had fallen into the trap against which he had warned his generals before the invasion began, that of allowing the Russians to retreat and draw the Germans farther and farther into the illimitable depths of their hinterland. When the dreaded winter broke over them, the German armies, despite their victories and advances, had still not captured Leningrad and Moscow, or destroyed the Russian capacity to continue the war.

Once the attack on Russia had been launched, the war on the Eastern Front absorbed all Hitler's thoughts and energies. Not content with fixing the strategic objectives of his armies in the east, he began to interfere in the detailed conduct of operation.

It was not only the military operations in the east which absorbed him: he saw himself about to realize his historical destiny by the foundation of a new German Empire in the lands conquered from the Russians. The prospect gripped and excited his imagination. From this period, the summer of 1941, date the records of his conversations taken under Bormann's supervision and subsequently published as his talk. They give a vivid impression of Hitler's mood at the peak of his fantastic career, the peer as he saw himself of Napoleon, Bismarck, and Frederick the Great, pursuing, to use his own words, 'the Cyclopean task which the building of an empire means for a single man'.

On the evening of 17 October, with the Russians (as he believed) already defeated, Hitler let his imagination ride as he talked of populating the Russian desert with two or three million people from Germany, Scandinavia, the Western countries, and America. 'We shan't settle in the Russian towns and we'll let them go to pieces without intervening. And above all, no remorse on this subject! We're absolutely without obligations as far as these people are concerned. To struggle against the hovels, chase away the fleas, provide German teachers, bring out newspapers—very little of that for us! We'll confine ourselves, perhaps, to setting up a radio transmitter, under our control. For the rest, let them know just enough to understand our highway signs, so that they won't get themselves run over by our vehicles.

'For them the word "liberty" means the right to wash on feast days . . . There's only one duty: to Germanize this country by the immigration of Germans and to look upon the natives as Redskins.'

Mussolini visited Hitler at his East Prussian headquarters towards the end of August 1941. Wolfsschanze was hidden in the heart of a thick forest, miles from any human habitation. The dim light of the forest produced a feeling of gloom in everyone who went there.

Two conversations took place between the Führer and the Duce on the 25th. The first meeting was taken up with an exposition of the military situation in the east, during which Mussolini was reduced to the role of admiring listener. Hitler, he noted, spoke with great confidence and precision, but admitted that faulty intelligence work had completely misled him as to the size and excellence of the Russian forces as well as the determination with which they fought. In their second talk, the same evening, the two dictators ranged over the rest of the world. Hitler spoke bitterly of Franco and was evasive on the subject of the French, who were, as always, the object of jealous complaints by Mussolini. He showed some embarrassment at the Duce's pressing offer of more Italian troops for the Eastern Front, but 'concluded by expressing the most lively desire to come to Italy—when the war is over—in order to pass some time in Florence, a city dear to him above all others for the harmony of its art and its natural beauty.'

Later in the week Hitler and Mussolini flew to Rundstedt's headquarters at Uman in the Ukraine. There Mussolini inspected an Italian division and lunched with the Führer in the open air, surrounded by a crowd of soldiers. At the end of the meal Hitler walked about among the crowd talking informally, while Mussolini, to his annoyance, was left with Rundstedt. Mussolini had his revenge, however, on the return flight, when he insisted on piloting the plane. Hitler's own pilot, Bauer, remained at the controls all the time, but Hitler never took his eyes off Mussolini and sat rigid in his seat. The Führer's congratulations were mingled with undisguised relief. Mussolini was childishly delighted and insisted on his performance being recorded in the communiqué.

The visit appears to have been organized more for propaganda purposes than to provide an occasion for serious discussions. Earlier in August, Churchill and Roosevelt had met off the coast of Newfoundland and from there issued the joint declaration of war aims known as the Atlantic Charter. The meeting of Hitler and Mussolini and the final communiqué, with the prominence which it gave to the slogan of the 'European New Order', was designed as a counter-demonstration. The dictators pledged themselves to remove the causes of war, eradicate the threat of Bolshevism, put an end to 'plutocratic exploitation', and establish close and peaceful collaboration among the peoples of Europe. This was an expansion of

Hitler's earlier idea of a 'Monroe doctrine for Europe' directed against the Anglo-Saxon powers.

Hitler was in Munich for the traditional celebration of the 8 November anniversary. In his speech he developed an argument which was to provide a companion theme to the European New Order in Nazi propaganda—Germany as the society in which class divisions and privileges had been abolished, the New Germany in the New Europe.

Confident that the Russian campaign would be finished before the snows, Hitler and his staff had made no provision for winter clothing to be issued to the troops. From early November the Germans were fighting in sub-zero temperatures, intensified by a bitter wind, the few hours of daylight and the long nights, and fighting in an unfamiliar land against an enemy inured to the conditions, warmly clothed and equipped for winter operations.

But Hitler insisted that the Russian resistance was on the verge of collapse. Warnings and appeals were of no avail. He categorically refused to admit that he had been wrong. Whatever the cost in men's lives, his armies must make good his boasts, and he drove them on relentlessly. On 2 December Kluge's Fourth Army made a last desperate effort to break through the Russian defences in the forests west of Moscow. A few parties of troops actually reached the outskirts of the capital, but they had to be pulled back.

At that moment, on 6 December, to the complete surprise of Hitler and the German High Command, the Russians launched a major counter-offensive along the whole Central Front with one hundred fresh divisions, and swept away the German threat to Moscow. The German troops, already driven to the limit of endurance, wavered; for a few days there was great confusion and the threat of a Russian breakthrough. Hitler was faced with the most serious military crisis of the war so far. Even if he surmounted it, one thing was already clear: the great gamble had failed and 1941 would end without the long-heralded victory in the east.

On 7 December, the day after the Russians opened their offensive to relieve Moscow, the Japanese took the American Fleet by surprise in Pearl Harbor. At the beginning of the month, Oshima, the Japanese ambassador in Berlin, had informed the German Government that Japanese–American relations had reached a crisis and that war might be imminent. In fact the Japanese task force had already sailed for Pearl Harbour on 25 November, but taking a leaf out of Hitler's book, the Japanese kept their own counsel and the news of the attack on Pearl Harbour came as a surprise to Hitler.

The one course which Hitler had never recommended to the Japanese had been to attack the U.S.A. It might have been expected therefore that the Führer would show some irritation at the independent course adopted by the Tokyo Government in face of his advice. On the contrary, he appears to have been delighted with the news of Pearl Harbour. He rapidly decided to follow the Japanese example by declaring war on the United States himself. When Ribbentrop pointed out that the Tripartite Pact only bound Germany to assist Japan in the event of an attack, and that to declare war on the U.S.A. would be to add to the number of Germany's opponents, Hitler dismissed these as unimportant considerations.

Hitherto, Hitler had shown considerable patience in face of the growing aid given by the U.S. Government to the British. But he was coming to the conclusion that a virtual state of war already existed with the U.S.A. and that there was no point in delaying the clash which he regarded as inevitable. Knowing nothing of the United States, Hitler disastrously underestimated American strength. The mixture of races in its population, as well as the freedom and lack of authoritarian discipline in its life, predisposed him to regard it as another decadent bourgeois democracy, incapable of any sustained military effort. The ease with which the Japanese struck their blow at Pearl Harbor confirmed these prejudices.

Another factor in Hitler's decision is more difficult to assess. The prospect of a war embracing the whole world excited his imagination with its taste for the grandiose and stimulated that sense of historic destiny which was the drug on which he fed. Elated by the feeling that his decisions would affect the lives of millions of human beings, he declared in the speech of 11 December, in which he announced Germany's declaration of war on America: 'I can only be grateful to Providence that it entrusted me with the leadership in this historic struggle. . . . A historical revision on a unique scale has been imposed on us by the Creator.'

Most of Hitler's speech on 11 December was devoted to abuse of the America of President Roosevelt, whom he depicted as the creature of the Jews. He drew a comparison between the success of National Socialism in rescuing Germany from the Depression and what he described as the catastrophic failure of the American New Deal: it was the desire to cover up this failure which led Roosevelt to divert American attention by a provocative foreign policy.

At the end of his speech Hitler announced that a new agreement had been concluded between Germany, Italy, and Japan, binding them not to conclude a separate armistice or peace with the U.S.A. or with England, without mutual consent.

It was with Russia, however, far more than with the United States or Great Britain, that Hitler was still concerned in the winter of 1941–2. The Russian counter-offensive, launched on 6 December, faced him with a crisis, which, if mishandled, might well have turned to disaster.

Hitler rose to the occasion. In face of the professional advice of his generals and in total disregard of the cost to the troops, he ordered the German armies to stand and fight where they were, categorically refusing all requests to withdraw. This order was enforced in the most ruthless fashion. By this remarkable display of determination he succeeded in holding the German lines.

The toll taken by the Russians, and even more by the terrible winter, was high. Thousands of German soldiers died of the cold, and in certain places Hitler had reluctantly to accept the withdrawal of decimated German divisions. But the Russians did not break through, and when the spring came, the German Army still stood on a line deep in the interior of Russia. More than this, by drawing on his own country and his allies, Hitler brought up the forces on the Eastern Front to sufficient strength to enable him to propose a resumption of the offensive in 1942.

The winter crisis marks a decisive stage in the development of Hitler's relations with the Army which was to have considerable consequences for the future.

After the invasion of Russia there was no longer a High Command or General Staff in Germany. Hitler ordered the C.-in-C. of the Army and his Staff (O.K.H.) to confine themselves to the conduct of the war in the east (excluding Finland). The other fronts were to be left to his own Supreme Command of the Armed Forces (O.K.W.). But the O.K.W. was in turn excluded from the Eastern Front, and in any case lacked the independent authority which the High Command of the Army traditionally possessed in Germany. The responsibility for the conduct of operations was thus divided, and the strategic picture of the war as a whole remained the concern of Hitler alone.

Hitler was far from being a fool in military matters. He had read widely in military literature and he took an eager interest in such technical matters as the design of weapons. His gifts as a politician gave him notable advantages in war as well. He was a master of the psychological side, quick to see the value of surprise, bold in the risks he was prepared to take and receptive of unorthodox ideas.

His faults as a military leader were equally obvious. He had too little respect for facts, he was obstinate and opinionated. His experience in the First World War, to which he attached undue importance, had been extremely limited. He had never commanded troops in the field or learned how to handle armies as a staff officer. He lacked the training to translate his grandiose conceptions into concrete terms of operations. The interest which he took in technical details, instead of compensating for these deficiencies, only made them clearer. Moreover, he allowed himself to become intoxicated with figures, with the crude numbers of men or of armaments production, which he delighted to repeat from memory without any attempt to criticize or analyse them.

These were precisely the faults which the professional training of the generals qualified them to correct. A combination of Hitler's often brilliant intuition with the orthodox and methodical planning of the General Staff could have been highly effective. But this was ruled out by Hitler's distrust of the generals.

Well aware of the Army's unrivalled prestige as the embodiment of the national tradition, he was quick to suspect its leaders of a lack of enthusiasm, if not active disloyalty, towards the new régime. The German generals, Hitler complained, had no faith in the National Socialist idea. The German Officer Corps was the last stronghold of the old conservative tradition, and Hitler never forgot this. His class-resentment was never far below the surface; he knew perfectly well that the Officer Corps despised him as an upstart, as 'the Bohemian corporal', and he responded with a barely concealed contempt for the 'gentlemen' who wrote 'von' before their names and had never served as privates in the trenches.

To political distrust and social resentment was added Hitler's inveterate suspicion of the expert. Nothing so infuriated him as the 'objectivity' of the trained mind which refused to accept his own instinct for seeing all problems in the simplest possible terms. It required great tact to get him to accept a view which differed from his own, and this was a quality which few of the German generals possessed.

So long as the German Army was successful the underlying lack of confidence between Hitler and his generals could be papered over. But the moment Hitler found himself faced with a situation like that on the Eastern Front in the winter of 1941–2 he made it only too clear that he had no faith at all in the High Command's ability to deal with it. Brauchitsch, feeling that he was placed in an impossible position, offered his resignation on 7 December. Ten days later Hitler accepted his offer, and on the 19th announced that he would himself take over the command-in-chief of the German Army in the field. This step was the logical conclusion to the policy of concentrating all power in his own hands which Hitler had steadily pursued since 1933.

To the German people Brauchitsch was made to appear as the man responsible for having, as Goebbels wrote in his diary, 'completely spoiled the entire plan for the eastern campaign as it was designed with crystal clarity by the Führer.' The Führer could do no wrong. If the promise of victory by the autumn had proved illusory, it was because the High Command had failed.

Hitler did not stop to reflect that, in his new position, it would be less easy to find scapegoats in the future. For even when he had been most calculating in his exploitation of the image of the inspired

Führer, Hitler had never lacked belief in the truth of the picture he was projecting. As success followed upon success the element of calculation was completely overshadowed by the conviction that he was what he had so long claimed to be, a man marked out by Providence and endowed with more than ordinary gifts. The image he had himself created took possession of him until he became the last victim of his own propaganda.

The success of his intervention in checking the Russian counter-offensive exalted his sense of mission and his confidence in his military genius. After the winter of 1942 he was less prepared than ever to listen to advice—or even information—which ran contrary to his own wishes. This was the reverse side of the strength which he derived from his belief in himself—and it was the weakness which was to bring him down, for in the end it destroyed all power of self-criticism and cut him off from all contact with reality.

Now that he had taken over the direction of operations himself, Hitler believed that 1942 would infallibly produce the knock-out blow which had eluded him the previous year, and he began to draw up his plans for the new campaigning season. When Halder told him that the Army Intelligence Service had information that six or seven hundred tanks a month were coming out of the Russian factories, Hitler thumped the table and said it was impossible—the Russians were 'dead'.

The German Army had come through the winter without a major disaster, but at a heavy price. Casualties numbered 1,168,000, not counting the sick; out of the 162 divisions on the Eastern Front, only eight were ready for offensive operations at the end of March. Despite these losses, Hitler issued emphatic orders to prepare for a resumption of the offensive, with the south as the principal theatre of operations and the oil of the Caucasus, the industries of the Donbass and Stalingrad on the Volga as the objectives.

The Home Front, no less than the Army, needed its faith in the Führer's leadership restored, and in the first four months of 1942 Hitler found time to make three big speeches.

In his speech of 26 April, Hitler, with the winter now behind him, gave the fullest expression to his renewed faith in Germany's eventual triumph. This time he made no attempt to conceal how near the German Army had been to disaster. He deliberately exaggerated the seriousness of the situation on the Eastern Front in order to throw into more effective contrast his own decision to assume personal responsibility and the news that the crisis had been mastered. Then, picking up the allusion to Napoleon's Retreat from Moscow, so often invoked during the winter, he added: 'We have mastered a destiny which broke another man a hundred and thirty years ago.'

Hitler's picture of the conditions under which the Army had fought in the east during the past few months was a prelude to a demand for still greater powers to be vested in himself, the counterpart on the Home Front to his decision to take over the personal conduct of operations on the Eastern Front.

The law, duly passed by the Reichstag without discussion, proclaimed: 'The Führer must have all the rights demanded by him to achieve victory. Therefore—without being bound by existing legal regulations . . . Führer must be in a position to force, with all the means at his disposal, every German, if necessary . . . to fulfil his duties. In case of violation of these duties the Führer is entitled, regardless of rights, to mete out punishment and remove the offender from his post, rank, and position without introducing prescribed procedures.'

Hitler's request for a confirmation of the arbitrary

power which he already possessed is at first sight puzzling. The explanation of the decree of 26 April 1942 is to be found in Goebbels's diaries, in which the Minister of Propaganda continually complains of the shortcomings of the state and party administration, and of the failure to organize German economy and civilian life to meet the demands of 'total' war. During his visit to headquarters in March, Goebbels pressed Hitler to adopt much more drastic measures to control war-profiteering and the black market, to increase production, reduce the swollen staffs of overgrown ministries, and provide additional manpower.

Goebbels and Hitler laid the blame for these shortcomings on the conservatism of the German Civil Service and judiciary. But they were only paying the penalty for treating the administration of the State as 'spoils' for the Nazi Party once it had come to power. The Nazis remained what they had always been, gangsters, spivs, and bullies—only now in control of the resources of a great state. It is astonishing that they had not ruined Germany long before the end of the war with their corruption and inefficiency. The fact that they did not was due to the stolid virtues and organizing ability of the permanent officials of the civil service, of local government and industry, who, however much abused, continued to serve their new masters with an unquestioning docility.

Hitler was the last man to remedy this situation. Without administrative gifts, disliking systematic work and indifferent to corruption, Hitler was at the same time far too jealous of his authority to make any effective delegation of his powers.

In the 1930s Hitler spoke of the Party as 'a chosen Order of Leadership' whose task was 'to supply from its membership an unbroken succession of personalities fitted to undertake the supreme leadership of the State'. On closer inspection the new élite was far from impressive. Even amongst the Reich leaders of the Party there were few men of ability, integrity, or even education. One of the exceptions was Goebbels. Göring, too, undoubtedly displayed ability in 1933–4, but by 1942 this had long been overlaid by the habits of indolence and the corruption of power. Men like Ley, Ribbentrop, Funk, Darré, and Rosenberg were wholly unfitted to hold positions of responsibility.

In February 1942, however, Hitler had the luck to make one of the few good appointments he ever made. Albert Speer, whom he chose as Minister for Armaments and Munitions was a young architect. Disinterested as well as able, he soon showed himself to be an organizer of remarkable powers and was entrusted with one job after another until he became virtual dictator of the whole of German war production. Finding himself faced with great difficulties in the way of procuring manpower from the obstruction of the Gauleiters, Speer shrewdly suggested that one of them should be made responsible for increasing Germany's labour force. This led in March 1942, to the appointment as Plenipotentiary-General for Manpower of Fritz Sauckel, a former sailor, who was Gauleiter of Thuringia. The powers given to Speer and the use he made of them produced a sensational rise in German war-production in 1942 and 1943 without which Hitler could never have continued the war at all.

To make good the German losses in manpower, Hitler demanded more of the satellite states. The bulk of the new divisions were to come from Rumania and Hungary. But Hitler now began to ask for the Italian troops which he had scorned to accept the year before. He had not seen the Duce since August 1941 and, now that the winter crisis was over, he thought it desirable to remove any doubts in Mussolini's mind and to revive his flagging faith in an Axis victory. Accordingly the Duce and Ciano again set out for the north at the end of April 1942, and spent two days with Hitler at Salzburg.

Ciano reported that Hitler looked tired and grey, but he was even more impressed by his loquacity. The discussions followed familiar lines, and on the way back Mussolini complained that he could not see why Hitler had asked them to make the journey. Resentment at his own reduced role was beginning to be tinged with the uneasy fear that he, as well as the Germans, would have to pay for the mistakes of an overconfident Hitler.

Before the attack on Russia Hitler had evaded Raeder's proposals for intensifying the war in the Mediterranean, with the promise to take up these plans after Russia was defeated. Although Hitler was forced to send stronger forces to the Mediterranean theatre, throughout 1941 and the winter of 1941–2, the sole purpose behind these moves was defensive, to prevent an Italian collapse in North Africa. At the end of the winter, however, Raeder returned to the attack and succeeded in rousing Hitler's interest in the Mediterranean, largely because of the grandiose way in which the plan (known as the 'Great Plan') was dressed up as a drive through the Middle East to join the Japanese in a vast encirclement of Britain's Asian Empire. Hitler agreed to a two-fold operation for the summer of 1942 to serve as the prelude to the 'Great Plan': the renewal of the desert offensive against Egypt, Suez, and beyond to Persia; and the capture of Malta, the key to the security of Rommel's supply route.

The operations began well with Rommel's capture of Tobruk and the invasion of Egypt. By 30 June 1942, a month after the offensive had opened, the Afrika Korps reached the El Alamein line, only sixty-five miles from Alexandria. But Hitler showed a curious reluctance to undertake the second part of the plan, the assault on Malta. He proposed to starve and bomb Malta into submission, arguing against Raeder and Kesselring that its capture was no longer necessary with Rommel on the verge of occupying Egypt.

As the summer passed, the British had time to build up their forces in Egypt and to strengthen Malta; the losses on the Italy-North African run began to mount again. By the autumn the Afrika Korps was still at El Alamein and Malta still unsubdued.

At the beginning of September Hitler reassured Rommel: 'I mean to give Africa all the support needed. Never fear, we are going to get Alexandria all right.' But in fact Hitler's interest in the Mediterranean and North Africa, never more than fitful, was beginning to waver again. In 1942, while agreeing to the 'Great Plan', he never once displayed that energy and singleness of purpose in forcing it through which had held the Eastern Front firm in the winter. He never grasped the importance of North Africa in the total picture of the war.

In fact, for all his talk of a war between continents Hitler showed little understanding of sea-power, the element which bound together the alliance which opposed him. As long as he held the initiative he went on thinking in terms of land war. For years Admiral Raeder had tried to persuade Hitler that the one certain way of defeating Great Britain was by attacking her trade routes and blockading her ports. Even after the directive for 'Barbarossa' had been issued, Raeder argued in December 1940: 'What is being done for U-boat and naval-air construction is much too little. . . . Britain's ability to maintain her supply lines is the decisive factor for the outcome of the war.'

Hitler's reply was to promise Raeder that once Russia had been defeated he should have all he asked for. Meanwhile the Navy had to be content with what it could scratch together in face of the competition of the Army and the Air Force. Raeder was not allowed to establish a naval air force, nor was he able to secure the effective cooperation of the Luftwaffe in attacks on British shipping, harbours, and shipyards. Göring, who was on bad terms with the Commander-in-Chief of the Navy, was a law unto himself, and Hitler simply let the quarrel between the two Services drag on.

Up to February 1941, Raeder found it impossible to keep more than some six U-boats at sea at a time. By the end of 1941 this had been increased to sixty. With these limited forces the U-boats achieved such remarkable successes in 1942, sinking over nine hundred vessels, that Hitler was converted and began to talk of the U-boats as the factor which would decide the outcome of the war. When Raeder demanded that no workman engaged on U-boat construction or repair should be drafted for military service, Hitler at once agreed, and more than three hundred U-boats were completed during 1942.

But Hitler's interest in the possibilities of the U-boats came too late. Although the shipping losses between the beginning of 1942 and the spring of 1943 taxed the Allies to the limit, they now had at their disposal resources which Hitler, deeply committed in Russia, could not hope to equal. The Battle of the Atlantic, which might—as Raeder had so often argued—have been decisive, was destined to prove one of his greatest failures. It was a failure which sprang from his defective grasp of the war as a whole and which was confirmed by the decision to invade Russia, a campaign into which a disproportionate amount of Germany's resources in men and machines was drawn at the expense of every other front.

On the third front, the western seaboard of Europe and Northern Africa, Hitler was not blind to the threat of an Anglo-American landing, but the problem of how to defend so vast a coastline was one to which he never found a satisfactory answer. Moreover, from the autumn of 1941, he displayed a growing conviction that Britain and the U.S.A.—possibly in cooperation with Russia—were planning a large-scale assault on Norway.

There was little enough evidence to support such a view, but so impressed was Hitler by his intuition that virtually the whole of the German surface fleet was concentrated in Norwegian waters. Not until 1943 was Hitler prepared tacitly—but never openly—to admit that he had been wrong. By then the Allied armada had safely landed an army in north-west Africa, unmolested by the German naval forces a thousand miles away to the north, where they remained vainly keeping guard against an attack which never came.

Against a much more ominous threat from the west which began to develop in 1942 Hitler found himself without adequate defences. The German Air Force, already roughly handled in the Battle of Britain, never recovered from the demands made on it in Russia. The first thousand-bomber raid by the R.A.F., on Cologne, took place on the night of 30–31 May 1942, and it was a portent for the future. The war was beginning to come home to Germany.

Hitler had chosen the south as the main theatre for his operations on the East Front in 1942, and powerful German forces drove fast down the corridor between the Don and Donetz rivers. While one wing pushed east towards the Volga at Stalingrad, the other drove past Rostov and, covering another four hundred and fifty miles, reached the Caucasus and the more westerly oilfield round Maikop in the first half of August.

Hitler moved his H.Q. to Winniza, in the Ukraine, during July, and from here he followed the progress of his armies with mounting excitement. Now, he declared, his faith and determination in the winter had been justified: Russia was on the verge of defeat.

Hitler, however, made exactly the same mistake he had made the year before. Overestimating the German strength, he did not limit himself to his original objective, to reach the Volga and capture Stalingrad, but tried to break into the Caucasus with its valuable oilfields as well, thus dividing his forces and ending by gaining neither Stalingrad nor the oil. At the end of July, when the 4th Panzer Army could probably have taken Stalingrad without much difficulty, it was diverted south to support Kleist's drive for the Caucasus. When it was freed to return north the Russians had gathered sufficient strength to hold Stalingrad. By September the battle for Stalingrad was beginning to assume proportions which made Halder doubt whether its capture was worth the effort or the risks. For the city's name and its historical association with Stalin during the Civil War made the Russians as eager to defend it as Hitler was to take it. As the Germans fought their way forward they exposed their long-drawn-out northern flank to grave danger from a Russian counterattack across the Don.

Halder's attempts to point out the dangers of the situation led to a repetition of the scenes of the previous autumn and winter. Hitler accused the General Staff of cowardice, ridiculing the Intelligence reports of growing Russian strength in preparation for a massive counter-attack.

Even a child could see the use Stalin was likely to make of the armies he was building up behind the front, but Hitler refused categorically to admit that such forces existed. When Halder recommended the breaking off of the attack at the end of September, Hitler dismissed him and replaced him as Chief of the Army General Staff with General Kurt Zeitzler, a younger man without Halder's experience or authority.

Meanwhile, the thrust into the Caucasus had been halted short of the main oilfields by stiffening Russian resistance. Hitler, beside himself with impatience, sent Jodl to investigate. When Jodl, on his return, ventured to defend the Commander-in-Chief in the Caucasus, Field-Marshal List, Hitler flew into another of his fits of fury. What particularly angered him was Jodl's citation of his own earlier directives to prove that List had only been obeying orders.

From that day on Hitler refused to eat any more with his staff officers at the common table. For several months he declined to shake hands with Jodl, and on 30 January 1943 he sent word that he was to be replaced. By a rare stroke of irony, Paulus, the man Hitler chose as Jodl's successor, the next day surrendered to the Russians at Stalingrad.

By the autumn of 1942 all Hitler's urging could not alter the fact that the German advance at Stalingrad, as well as in the Caucasus and North Africa, had been brought to a standstill. For the first time since he had proclaimed Germany's rearmament in 1935, more than seven years before, the initiative passed out of Hitler's hands, never to return.

On the night of 23 October 1942 the British 8th Army under General Montgomery attacked the German lines at El Alamein and after twelve day's heavy fighting broke out into the desert beyond.

On the night of 7–8 November British and American troops landed along the coast of Morocco and Algeria, and within a few days occupied the whole of French North Africa as far as the Tunisian frontier.

On 19 and 20 November three Russian Army Groups under the command of Generals Vatutin, Rokossovsky, and Eremenko attacked on a huge front north and south of Stalingrad and within five days succeeded in encircling twenty-two German divisions between the Volga and the Don.

Taken together, these three operations mark the turning-point in the war and the seizure of the initiative by the Allies. Henceforward Hitler was forced to stand upon the defensive.

Hitler's very success in halting a German retreat in the winter of 1941–2 now proved a fatal legacy. His one idea was to stand firm at all costs. When Rommel flew back to Germany at the end of November and told Hitler that Africa was lost, and the only course was to get the Afrika Korps out to fight in Italy, Hitler shouted at him that he was a defeatist and his troops cowards. Despite his categorical orders, however, neither Rommel nor anyone else could halt the Allies' advance.

The Allied landings in French North Africa in November took Hitler completely by surprise. He immediately summoned Laval and Ciano for a meeting at Munich. Hitler bluntly informed Laval that the Germans would occupy Tunisia at once, together with the rest of Unoccupied France. At last Mussolini had his way over France, but he was no longer in a position to derive much satisfaction from it: Italy was too obviously the Allies' next target after North Africa.

After his long neglect of the Mediterranean Hitler began to pour troops and supplies into Tunisia in order at all costs to hold a bridgehead covering Tunis and Bizerta. In something of a panic Mussolini urged Hitler to come to terms with Russia, or at least shorten his lines on the Eastern Front, so that the greatest number of divisions could be moved to the Mediterranean and the west. These suggestions were ignored. Hitler was determined to hold Tunisia, but he was equally determined not to give up anything elsewhere.

Meanwhile the Russians methodically tightened the net round the German Sixth Army at Stalingrad. The formula which had proved successful the previous winter was monotonously repeated: Stand and fight to the last man. Hitler refused to let Paulus attempt to break the ring from the inside. No conceivable military purpose was served by holding the German troops in their positions, but Hitler's personal prestige as a leader was now engaged, and in comparison with that the lives of the 330,000 men of the 6th Army were nothing.

Towards the end of January 1943 Paulus reported that the suffering of the troops, through cold, hunger, and epidemics, was no longer bearable, and that to continue fighting in such conditions was beyond human strength. Hitler was unmoved. For answer he sent Paulus the message: 'Capitulation is impossible. The 6th Army will do its historic duty at Stalingrad until the last man, in order to make possible the reconstruction of the Eastern Front.'

Hitler did not hesitate to stoop to bribes: at the last moment he promoted Paulus to the rank of Field-Marshal in order to buy the loyalty of the commander whose troops he had deliberately condemned to death. 'There is no record in military history,' he remarked to Jodl, 'of a German Field-Marshal being taken prisoner.'

The outcome was a far worse blow to Hitler's prestige than any order to withdraw could ever have been. On the night of 31 January, the Russians announced that they had completed the capture or annihilation of the remainder of the 6th Army and the 4th Panzer Army, adding that among the officers who had surrendered was Field-Marshal von Paulus himself.

At noon on 1 February, the day after the Russian communiqué, Hitler held his usual military conference. Totally oblivious of his own responsibility for what had happened, the Führer spared no thought for the men he had driven to death or captivity. He could think only of the commanders who had capitulated: such ingratitude and disloyalty, he declared, were beyond his comprehension. In the sufferings and defeat of a nation he saw only his own betrayal by a people unworthy of their Führer.

Amongst the reasons for the German failure one is of particular interest: Hitler's neglect of the political possibilities of weakening Russian resistance. At an earlier stage—for instance, in preparing to attack France—Hitler had shown a brilliant understanding of how war could be waged with other than military weapons. But, although he repeatedly described the war with Russia as an ideological conflict and counted on the overthrow of the Soviet Government by the Russian people, the harsh policy he adopted in the east worked in the opposite direction.

There is evidence to show that when the German armies entered the Ukraine and the Baltic States they were looked upon as liberators. The treatment the local population received from the civil administration and the S.S. who moved in behind the armies rapidly destroyed these illusions. Ignoring all that might have been done to drive a wedge between the people and the Soviet Government, especially in the Ukraine, Hitler preferred to treat the inhabitants of Eastern Europe indiscriminately as Slav *Untermenschen*, fit only for slave labour.

Goebbels was quick-witted enough to see the opportunities that were being lost. In September 1941 he dictated a lengthy memorandum on political plans for handling the Russian peoples, and the next year tried to get Hitler to issue a proclamation promising the Russians greater freedom and some relief from the oppressive exactions of the Soviet Government. But Hitler's policy in Eastern Europe was no hasty improvisation: it was the calculated expression of a mind which could conceive of politics only in terms of domination and could understand the exercise of power solely in terms of the whip.

The first need was to exploit the occupied eastern

territories for the strengthening and relief of the German war economy. This had been foreseen in the economic directives drawn up before the invasion, and was continually reaffirmed by Göring and others. In a conference on 6 August 1942 Göring, as Plenipotentiary for the Four-Year Plan, told the commissioners for the occupied regions: 'It used to be called plundering. But today things have become more genteel. In spite of that, I intend to plunder and to do it thoroughly.'

As the bombing of German industry and the losses of manpower and equipment began to exert a greater strain on the German economy, so the demands on the eastern territories mounted. These demands were not limited to raw materials, food, and machinery, but extended to manpower as well. Russia, like Poland and the other occupied countries in the west, France, Belgium, and Holland, was turned into a vast labour camp to provide the human material which German industry and agriculture needed. The organization of this new slave traffic was in the hands of Sauckel, and the brutality of the methods by which men, women, and children were rounded up, shipped to Germany and forced to work, often under unspeakable conditions, beggars description.

But even the five million Russian, Polish, French, Dutch, Yugoslav, and Italian workers did not satisfy Hitler. He constantly increased his demands on Sauckel, who in turn urged the local authorities to apply the most ruthless measures to secure more manpower.

Hitler's policy in the eastern occupied territories, however, was only in part determined by Germany's immediate economic needs. Under cover of the occupation he was determined to lay the basis of a German settlement in the lands between the Vistula and the Urals. Colonies of settlers from Germany and from the German minorities in other countries were to be established in Poland and European Russia, each settlement being linked by a network of military roads and protected by S.S. garrisons set up at key points, whose task was not only to guarantee the new frontiers after the war but to keep the native population in permanent subjection and extract from it slave labour for the industries and agriculture of the new German Empire.

The tasks of Himmler, as Reich Commissioner for the Strengthening of German Folkdom, were defined as the elimination of such alien groups as represented a danger to the Reich and the German Folk Community, and the formation of new German settlements from returning German citizens and racial Germans abroad. To carry out these duties Himmler set up special departments of the S.S. and outlined his programme in a number of speeches to his S.S. commanders which give an authoritative picture of Hitler's plans for the future.

The most interesting of these speeches of which we have a record is one which Himmler made to his S.S. Obergruppenführers at Posen on 4 October 1943. He began by insisting on the need for ruthlessness. 'One basic principle must be the absolute rule for the S.S. men: we must be honest, decent, loyal, and comradely to members of our own blood and nobody else. . . . Whether nations live in prosperity or starve to death interests me only in so far as we need them as slaves for our *Kultur:* otherwise it is of no interest to me. Whether ten thousand Russian females fall down from exhaustion while digging an anti-tank ditch interests me only in so far as the anti-tank ditch for Germany is finished. We shall never be rough and heartless when it is not necessary, that is clear. We Germans, who are the only people in the world who have a decent attitude towards animals, will also assume a decent attitude towards these human animals. But it is a crime against our own blood to worry about them and give them ideals. . . . Our concern, our duty, is our people and our blood. We can be indifferent to everything else. I wish the S.S. to adopt this attitude to the problem of all foreign, non-Germanic peoples, especially Russians.'

In passing, Himmler mentioned the extermination of the Jews: 'Most of *you* know what it means when a hundred corpses are lying side by side, or five hundred or one thousand. To have stuck it out, and at the same time—apart from exceptions caused by human weakness—to have remained decent fellows, that is what has made us hard.'

Among the particular duties of the S.S. was that of organizing the concentration camps. In 1942, Himmler, with Hitler's agreement, began to use these as a source of labour for armaments work, and the S.S. established its own factories. Certain categories of prisoners were agreed upon as suitable 'to be worked to death'. Among other uses to which concentration-camp prisoners were put was to serve as the raw material for macabre medical experiments by S.S. doctors.

The work of guarding the concentration camps and carrying out the brutal sentences of flogging, torture, and execution which were everyday occurrences was alloted to the S.S. Death's Head Units (Totenkopfverbände). In a speech which he delivered to S.S. leaders at Metz in April 1941, Himmler described such work as 'fighting the sub-humanity. This will not be a boring guard duty, but, if the officers handle it right, it will be the best indoctrination on inferior beings and the sub-human races.'

More terrible even than the concentration camps were the extermination camps. The largest of these was Auschwitz in Poland, where the four large gas chambers and crematoria were capable of a rate of extermination far above that of the others like Treblinka. Rudolf Höss, a member of the S.S. Totenkopfverbände since 1934, served in concentration camps for eleven years. From May 1940 to December 1943 he was at Auschwitz. In his affidavit he says:

'I was ordered to establish extermination facilities at Auschwitz in June 1941. . . . I visited Treblinka to find out how they carried out their extermination. The Camp Commandant told me that he had liquidated eighty thousand in the course of one half year. He was principally concerned with liquidating all the Jews from the Warsaw ghetto. He used monoxide gas and I did not think that his methods were very efficient. So at Auschwitz I used Cyclon B, which was a crystallized prussic acid dropped into the death chamber. It took from three to fifteen minutes to kill the people in the chamber, according to climatic conditions. We knew when the people were dead because their screaming stopped. We usually waited about half an hour before we opened the doors and removed the bodies. After the bodies were removed, our special commandos took off the rings and extracted the gold from the teeth of the corpses. Another improvement that we made over Treblinka was that we built our gas-chambers to accommodate two thousand people at one time. . . .'

Even the facilities at Auschwitz could not meet the demands made on them in 1944. In forty-six days during the summer of that year, between 250,000 and 300,000 Hungarian Jews alone were put to death at the camp and the S.S. resorted to mass shootings to relieve the pressure on the gas chambers.

When the invasion of Russia began Hitler and Himmler recruited four special units known as Einsatzkommandos to carry out the extermination of the Jewish population and also of Communist functionaries. Otto Ohlendorff, the Chief of the Security Police (SD), who commanded Einsatzgruppe D in southern Russia for a year, estimated that 90,000 men, women, and children were liquidated by his formation during that period. At first the victims were made to dig mass trenches into which they were thrown after execution by shooting. In the spring of 1942, however, the efficient Main

Office in Berlin began to supply gas vans for mobile extermination. Another formation, Einsatzgruppe A, in northern Russia, killed 135,000 Jews and Communists in its first four months of operations.

The subsequent capture and trial by an Israeli court of Eichmann, the head of the Jewish Office of the Gestapo, has provided further evidence of the sufferings inflicted on the Jewish people by the S.S. How many Jews perished in the extermination camps and at the hands of the Einsatzkommandos will never be known precisely. The best calculation that can be made of the cost of 'the Final Solution' in human lives puts the figure between 4,200,000 and 4,600,000.

It has been widely denied in Germany since the war that any but a handful of Germans at the head of the S.S. knew of the scope or savagery of these measures against the Jews. One man certainly knew. For one man they were the logical realization of views which he had held since his twenties, the necessary preliminary to the plans he had formed for the resettlement of Europe on solid racial foundations. That man was Adolf Hitler.

Himmler organized the extermination of the Jews, but the man in whose mind so grotesque a plan had been conceived was Hitler. As long ago as 1932–3, Hitler had spoken of the extermination of the Jews as the first step in the establishment of the imperial rule of the *Herrenvolk* over the whole Continent.

It is all too easy to dismiss such a conception as the fantasy of a diseased brain: it is well to remember, however, that in the sinister sites of Auschwitz and Buchenwald, and the well-kept records of the S.S., there are the proofs of how near the fantastic came to being realized.

CHAPTER THIRTEEN

TWO JULYS

1943–4

The immediate consequences of the Stalingrad disaster were not so great as might have been expected, and did not lead to a collapse of the German front in the east. The Russian attempt to cut off the army in the Caucasus was defeated by a skilfully conducted retreat. Moscow was relieved, but Leningrad remained under German shell-fire. When the winter fighting came to an end the German line, although withdrawn in the centre and the south, was still deep in Russian territory. It was not until the late summer of 1943 that the Russians renewed their attacks. By that time Hitler was faced with an even graver situation in Italy.

Hitler's rapid decision to seize Tunisia in November 1942 proved effective in balking the Allies of victory before the end of that year. The news from Italy, however, made him anxious: the Duce was ill, dislike of the Germans was widespread, and the one ambition of the Italian people was to get out of the war as soon as possible.

Something must clearly be done to stiffen his failing ally, and at the end of February 1943 Hitler sent Ribbentrop on a visit to Rome with a long personal letter to the Duce, following it with an agreement to allow the Italian workmen in Germany to return home—a considerable concession at a time when Sauckel was mobilizing the labour resources of the rest of Europe to work for Germany. But Mussolini, ageing, sick, and disillusioned, was fast losing control of the situation. Mass strikes in Turin and Milan, with the slogans 'Peace and Liberty', were a pointer to the impending collapse of the régime. All that Mussolini could think of was a renewed appeal to Hitler to make a separate peace with Russia. Hitler's reply was to press Mussolini to come to Salzburg, where they met in the middle of April.

Mussolini promised his lieutenants that this time he would stand up to Hitler: he was determined to urge peace with Russia, and the withdrawal of the Italian armies from abroad to defend their homeland. But, face to face with the dynamic Führer, he succumbed and sat silent while Hitler talked, and he returned to Rome a defeated man, no longer able to convince himself of the part he had to play. His despair was soon justified. On 7 May Tunis and Bizerta were captured by the Allies, and within a week the entire Axis forces in Africa, which Hitler, against Rommel's advice, had built up to more than 250,000 men, were taken prisoner with all their equipment.

It was obvious that the Allies would attempt a landing on the northern shores of the Mediterranean —and equally obvious that, with the loss of the troops in Tunisia, Hitler and Mussolini would be hard pressed to prevent them. The most difficult problem was whether the Italians could be relied on to fight. Cooperation between the Italian and German Armed Forces was increasingly strained. Hitler well knew the danger of the situation, but he feared to take drastic action lest this should drive the Italians into open revolt.

In this uneasy state of mind, foreseeing what might happen but unable to prevent it, Hitler waited for the Allied attack. It came on 10 July, in Sicily, and the Allies at once made good their landings.

Nine days later Hitler summoned Mussolini to meet him at Feltre, in Northern Italy. In a last effort to put new life into the alliance, Hitler talked for three hours on end before lunch. There was one course open to them, he declared, to fight and go on fighting, on all fronts—in Russia as well as Italy—and with a fanatical will to conquer.

After lunch Hitler summoned up his energies for a second performance. Once again, as Ciano had so often noted, he talked, talked, talked; and once again the Duce sat silent to the end. He even failed to get a promise of reinforcements from the Germans.

Immediately after the Feltre meeting Italian discontent with the German alliance and with the Duce as its representative came to a head. The Fascist Grand Council met on the night of 24–25 July, and Mussolini had to listen to violent criticism of his conduct of the war. The following evening the Duce was dismissed by the King and placed under arrest. The veteran Marshal Badoglio formed a non-Fascist government; the party itself was dissolved, and Fascist officials expelled from their posts. The basis of the new Government's authority was the Crown and the Army.

As soon as the news reached his headquarters Hitler summoned an immediate conference of all the Nazi leaders, together with Rommel, Dönitz, and other military figures. The fact that what he feared had at last materialized relieved rather than depressed Hitler, although it was an embarrassing situation involving his own prestige. Despite the strain, intensified by heavy fighting on the Eastern Front, he kept his head, showing not only determination and energy in dealing with the crisis, but considerable skill as well. This, combined with the slowness of the Allies in taking advantage of the situation, enabled him to make a brilliant recovery.

Hitler did not wait for his lieutenants to arrive before taking a number of key decisions on the spur of the moment. The first and most important was that the new Italian Government under Badoglio, however much it might protest its loyalty to the Axis, was only playing for time in order to make a deal with the Allies and must be treated accordingly. The second was to move in every man he could find in order to seize control and hold Italy when the time came.

By the time of his conference on the 26th, Hitler had prepared four sets of plans, and the forces to carry them out were steadily being collected. The first was a plan for the rescue of Mussolini, to which Hitler attached great importance; the second provided for the occupation of Rome and the restoration of the Fascist régime; the third covered the military occupation of Italy; and the fourth dealt with measures for the capture or destruction of the Italian fleet.

The practical question was one of timing. Hitler, Göring, and Goebbels wanted to act at once: the King, Crown Prince, and Badoglio's Government should be seized and brought to Germany, while Mussolini was restored to power in Rome. Rommel (whom Hitler had appointed Commander-in-Chief in Italy) and the other soldiers wanted to wait until the situation became clearer. They feared that precipitate action would drive Badoglio, whom they hoped to keep on their side, into the arms of the Allies; they were highly sceptical about the authority of Mussolini or the popularity of a revived Fascist régime, and they were impressed by the risks involved at a time when the German forces in Italy were still weak.

Hitler continued to defer final orders from day to day, much to his advantage. He was right in supposing that Badoglio would at once begin negotiations for a separate peace, but until he could reach agreement with the Allies he had to keep up the pretence of cooperation with the Germans. Hitler, realizing the game that was being played, made the most of the Allies' long delay to strengthen his forces in Italy before the showdown came. At the end of the six weeks which the Allies allowed to elapse between Mussolini's fall (25 July) and the publication of the armistice with the Badoglio Government (8 September) Hitler was in a very much stronger position to put his plans into effect.

The announcement of the Italian armistice again took Hitler by surprise, for the Italians kept up appearances to the last moment and succeeded completely in deceiving the Germans. The code-word for action, however, was immediately sent to Kesselring, who was in command in southern Italy.

By 10 September the German forces, some sixteen divisions, had disarmed the much more numerous Italian formations and seized the key positions, including control of Rome, without meeting any serious resistance. The King and Badoglio fled from the capital, and within a matter of hours the Germans were masters of the greater part of the country.

Simultaneously with the announcement of the armistice, the Allied Fifth and Eighth Armies had landed on the Italian mainland and begun to fight their way north. To Kesselring's relief, however, the Allies landed much farther south than he had dared to hope, at Salerno, to the south of Naples. Hitler had already written off the south of Italy, and the German defence plans were based on positions well to the north of Rome, but when Kesselring succeeded in holding up the Allied advance—even by the end of the year they had advanced no more than seventy miles from Salerno—Hitler agreed to allow Kesselring to fight on the Winter Line drawn across the peninsula not far to the north of Naples. This kept more than two-thirds of Italy, including the industrial north, in German hands, and it was not until June 1944 that the Allies succeeded in reaching Rome.

After his fall Mussolini was moved by the Badoglio Government from one place to another until he was finally taken to a small hotel at the Gran Sasso, high up in the Abruzzi Mountains. Hitler took a close personal interest in Mussolini's movements, and once he had been located a spectacular rescue from the air was planned. On 12 September this was carried out with success by an S.S. detachment under the command of Otto Skorzeny, and Mussolini was brought to the Führer's Headquarters at Rastenburg.

The first meeting between the two men was cordial, but a rapid disenchantment followed. Hitler's plan was to re-establish the Fascist régime in Italy, and to be successful, it must have Mussolini at its head. The Duce, however, was now a shrunken figure, an ageing man without political ambition, whose real wish was to be allowed to go home to the Romagna. Under Hitler's urging—and scarcely veiled threats—he agreed to play the part for which he had been cast, but it was without enthusiasm and, as it soon appeared, only with the help of vigorous prompting from the producer.

On 15 September Mussolini's restoration to the leadership of Fascism was proclaimed, and the new Italian Social Republic came into being. Its 'Government' followed a squalid and undistinguished career until the end of the war in Italy. Even when Mussolini returned to Italy, his villa was surrounded by S.S. Guards, ostensibly as a bodyguard. The new régime possessed neither independence nor authority: it was despised by the Germans and hated by the Italians.

The last phase of Mussolini's life was the most degrading of all. He was reduced to the rank of a puppet dictator, and Hitler exacted the full humiliation. In January 1944 Ciano was shot by a Fascist firing squad acting under the nominal authority of his father-in-law. The fascination which Hitler had once exerted over Mussolini was turned to hatred. But the Duce could not escape from the destiny he had forged for himself in making his pact with Hitler and, when the end came, his body was hung up on a gibbet in that same city of Milan in which he had proclaimed the Axis on 1 November 1936.

Hitler's conviction that Fascism could be revived in Italy proved as insubstantial as his belief in Mussolini as a brother Superman superior to the blows of Fate. But the Italian Social Republic served his purpose: it enabled appearances to be preserved, at least for a time. Taken with the German success in occupying the greater part of Italy and holding the Allies well south of Rome, the restoration of Mussolini could be presented as a triumphant ending to the crisis which had threatened in the summer to leave the southern frontiers of the Reich directly exposed to Allied attack.

Moreover, the Germans also took over the Italian zones of occupation in the Balkans, in Yugoslavia, Albania, and Greece, where Hitler had for some time been apprehensive of a British landing. Considering the course of events in the Mediterranean theatre since El Alamein and the landings in north-west Africa fourteen months before, Hitler might well congratulate himself at the end of 1943 on the effective way in which, by energy, determination, and luck, he had retrieved a disastrous situation.

Even so, the position in southern Italy was weakening in 1943. However slow might be the Allied advance, all that Kesselring could do was to fight a skilful rearguard action. The prospect elsewhere was still darker.

In the east, after throwing back the Russians in March 1943, in July the Germans launched a new offensive against their lines round Kursk. Half a million men, the finest troops left in the German Army, including seventeen panzer divisions equipped with the new heavy Tiger tanks, were used to carry it out. After heavy and costly fighting the Russians not only succeeded in bringing the German attack to a halt, but on 12 July themselves opened an offensive farther north. Gradually their attacks spread along the whole front. On 4 August they retook Orel, and on 23 August Kharkov. On 23 September they recaptured Poltava, and on the 25th Smolensk.

The sole result of Hitler's inflexible orders to stand and fight, without giving a yard, was to double the German losses and deprive his commanders of any chance of using their skill in defence. As the year ended the Red Army was steadily pushing the Germans back to the Polish and Rumanian frontiers.

The Russian advances, especially in the south, had political as well as military repercussions. As the Red Army drew nearer to their frontiers, fear began to spread among the satellite states, Rumania, Hungary, and Slovakia, whose loyalty to the Axis had been badly shaken by events in Italy. Hitler, already worried about the Balkans and the possibility of landings there, watched Turkey too with anxiety.

It was for these political as much as for military reasons that he rejected any suggestion of withdrawal on the southern sector of the Eastern Front, and obstinately refused to give up the Crimea at the cost of losing well over 100,000 men. When he was urged to evacuate the German garrisons in the Aegean and on Crete, he replied: 'To abandon the islands would create the most unfavourable impression. To avoid such a blow to our prestige, we may even have to accept the eventual loss of the troops and material.'

In the west, although the Allies had not yet attempted an invasion, 1943 saw two heavy blows to Hitler's hopes, the defeat of the U-boats and the intensification of the air war against Germany. In January 1943, angry at German naval losses, Hitler accused Raeder of lacking the will to fight and run risks. He appointed the Navy's U-boat specialist, Dönitz, as Commander-in-Chief of the Navy in Raeder's place, but as the Allies strengthened their defences against submarine attack, the figures for

U-boat losses began to rise, and at the end of May Dönitz was driven to withdraw all his vessels from the North Atlantic.

Hitler was no longer blind to the importance of the Battle of the Atlantic, but he lacked the resources to support it. Increased production of U-boats could only be at the expense of other equally urgent needs. By the end of 1943 the Battle of the Atlantic was lost.

During 1943 the American day-bombers joined the R.A.F. in keeping up an almost continuous offensive against targets in Germany and Western Europe. The scale of the raids began to rise too.

Hitler, worried most about the effect on German war production, was beside himself with fury at the failure of Göring and the Luftwaffe to fend off the attacks or to satisfy his demand for reprisals on Britain.

In 1938 and 1939 Hitler had been warned by Schacht and others that Germany had not the economic resources to wage another war. By 1943 the accuracy of these warnings was obvious. There was an increasing shortage of everything—manpower, raw materials, transport, oil, food, steel, armaments and planes. Even if the German people could withstand the strain of the air war, the effect on war production was such that it must in the end place Germany in a position of permanent inferiority.

Thus, if the last months of 1942 mark the turning-point of the war, 1943 may be taken as the year of Germany's defeat.

Hitler was now in his fifty-fifth year. The strain imposed on him by the war, particularly since the winter of 1941–2, had begun to leave its mark. During the course of 1943 he began to suffer from a trembling of his left arm and left leg, which became steadily more pronounced and refused to yield to any treatment. In an effort to control this tremor, Hitler would brace his foot against some object and hold his left hand with his right. At the same time he began to drag his left foot, as though he were lame.

To meet the demands which he made upon himself between 1930 and 1943 Hitler must have had an iron constitution. He was inclined to fuss about his health, believing that he had a weak heart and complaining of pains in his stomach and occasional bouts of giddiness. But his doctors found nothing wrong with his heart or his stomach, and until 1943 he actually suffered very little from ill-health.

Under the stress of war, however, Hitler began to take increasing quantities of drugs to stimulate his flagging energies. Since 1936 he had kept as his personal physician in constant attendance on him a Professor Morell, a quack doctor who had once practised as a specialist in venereal disease in Berlin. Morell won Hitler's confidence by curing him of eczema of the leg, and used his position to make a fortune by manufacturing patent medicines under the Führer's patronage. He is described by H. R. Trevor-Roper after the war as 'a gross but deflated old man, of cringing manners, inarticulate speech, and the hygienic habits of a pig'. Hitler himself never trusted Morell, trying constantly to trip him up and threatening him with ejection or worse, but his dependence upon him was incontestable. At every meal Hitler took a considerable number of tablets prepared by Morell and had frequent injections as well every day during the last two years of his life.

When Dr. Giesing examined Hitler in July 1944, he found that, to relieve the pains in his stomach, Morell had been giving him for two years at least a drug known as Dr. Koester's Antigas Pills which was compounded of strychnine and belladonna. Giesing believed that Hitler was being slowly poisoned by these pills and that this accounted both for the intensification of the pains and for the progressive discoloration of Hitler's skin. The only result, however, of telling Hitler was the dismissal of his other doctors, who supported Giesing, and the end of Dr. Giesing's own visits to the Führer's Headquarters. During the last two years of the Third Reich, not only Hitler but practically all the other members of his entourage kept themselves going on the drugs obligingly dispensed by Dr Morell.

To the strain of responsibility and the evil effects of Morell's ministrations must be added the effects of the life Hitler was now leading. From the summer of 1941 Hitler made his permanent headquarters at Wolfsschanze, in East Prussia. Under the threat of air-raids he soon moved to one of the massive concrete bunkers embedded in the ground, and made his home in a suite of two or three small rooms with bare, undecorated concrete walls and the simplest wooden furniture.

The austerity of Hitler's life at his headquarters matched the bleakness of the surroundings. General Jodl, who spent much time there, described it as 'a mixture of cloister and concentration camp. There were numerous wire fences and much barbed-wire. . . . Apart from reports on the military situation, very little news from the outer world penetrated this holy of holies.'

The main event of each day was the Führer's Conference at noon. To describe these as conferences is actually to misrepresent their character: they were a series of reports on the military situation, in which decisions were taken solely by the Führer. A certain number of officers were nearly always present, and other commanders or ministers would attend intermittently. Each of these officers was accompanied by his adjutants, who carried the maps to be spread out on the big centre table, or the memoranda and diagrams to be presented. As each report was made—Eastern Front, Italy, air war, and so on—Hitler would announce his decision, and the officers concerned would leave the room to send off the necessary instructions. There was no general discussion of the situation as a whole: only the Führer was allowed to concern himself with the over-all picture. Another more restricted military conference sometimes followed late in the evening, and there were frequent private meetings between Hitler and his chief lieutenants, Himmler, Bormann, the powerful head of the Party Chancery, or Goebbels on a flying visit from Berlin.

Hitler's day was almost entirely taken up with meetings of this kind. He rose late, breakfasted alone, and after the noon conference took lunch at any time between 2 and 5 p.m. Usually he rested in the late afternoon, resuming his talks at six or seven o'clock. Dinner might be served at any time between 8 p.m. and midnight. There followed further discussions and his day ended with tea in the company of his secretaries at four o'clock in the morning.

Apart from a short walk with his Alsatian bitch, Blondi, which had been given him by Bormann to raise his spirits after Stalingrad and to which he became very attached, Hitler took no exercise and enjoyed no form of relaxation. As the war went on he dropped the habit of seeing films after dinner, apart from newsreels. Up to the time of Stalingrad, he sometimes spent an evening listening to gramophone records, Beethoven, Wagner, or Wolf's *Lieder*. After Stalingrad, however, he would hear no more music, and his sole occupation as they drank their tea in the early hours of the morning was to recall the past, his youth in Vienna and the years of struggle. This was interspersed with reflections on history, on the destiny of man, religion, and other large subjects. Soon, his secretary complains, his remarks became as familiar as the records; they knew exactly what he would say and kept awake only with the greatest difficulty. On no account was the war or anything connected with it permitted as a subject of discussion.

The dominant impression derived from accounts of life at the Führer's Headquarters in 1943 and 1944 is one of intense boredom, punctuated by the excitement of crises like that caused by Mussolini's fall and by Hitler's unpredictable outbursts of rage, usually directed against the generals. Hitler saw little even of Eva Braun, who remained on the Obersalzberg. So long as Hitler remained at his headquarters, Goebbels thought that Blondi was closer to him than any human being.

Hitler's ostensible reason for shutting himself up in this way was the demands made on him by the war. But there was a deeper psychological compulsion at work. Here he lived in a private world of his own, from which the ugly and awkward facts of Germany's situation were excluded. He refused to visit any of the bombed towns, just as he refused to read reports which contradicted the picture he wanted to form. The power of Martin Bormann, Hitler's personal secretary, was built up on the skill with which he pandered to this weakness, carefully keeping back unpleasant information and defeating the attempts of those who tried to make Hitler aware of the gravity of the situation.

In the last years of his life Hitler deliberately refused to exercise the extraordinary powers he had once displayed as a mass-orator. Goebbels did everything he could to overcome the Führer's reluctance. Hitler's excuse was always the same: he was waiting for a military success. But again one may suspect a deeper reason. Hitler's gifts as an orator had always depended on his flair for sensing what was in the minds of his audience. He no longer wanted to know what was in the minds of the German people; at all costs he must preserve his illusions. Until he could force events to conform to the pattern he sought to impose, he hid himself away in his headquarters.

As the Allied armies began to press in on Germany in the course of 1944, some of the Nazi leaders began to look around for ways to disappear or make private deals with the enemy. This, if inglorious, may be regarded as a normal human reaction to such a situation. Hitler's was wholly different. He was fighting for something more than his power or his skin; he was fighting to preserve intact that image he had created of himself as one of Hegel's 'world-historical individuals'. The unforgivable sin was to fail, as Mussolini had failed, to rise to the measure of events. Hitler's faith was crystallized in the belief that if only he could survive the buffetings of the waves which were breaking over him he would be saved by some miraculous intervention and still triumph over his enemies. Everything depended upon the will to hold out.

This belief in turn depended upon the fundamental belief which he never abandoned to the end of his life—that he was a man chosen by Providence to act as the agent of the 'world-historical process'. Every incident in his life was used to support the truth of this assertion: the number of times he had escaped attempts at assassination, or the fact that the Russians had not broken through in the winter of 1941–2. Anything, however trivial, which went right in the last two years of the war served Hitler as further evidence that he had only to trust in Providence and all would be well.

There were other, more material factors on which Hitler based his hopes of a dramatic reversal of the war in his favour. In his speech of 8 November 1942, he referred to the new secret weapons which Germany was building and promised the Allies an answer to their bombing raids 'which will strike them dumb'. The weapons he had in mind were the V1, the flying bomb, and the V2, the rocket. To these must be added the new jet fighter-planes, with which the Luftwaffe was to sweep the enemy from the skies, and new types of U-boats, with which the Navy was to cut the Allies' supply lines.

The secret weapons actually existed, and the V1 and V2 were to play some part in the final stages of the war, but the hopes which Hitler and Goebbels placed upon them were exaggerated. Ignoring the almost insuperable difficulties of mass-production under the Allied air-attacks, they expected the weapons miraculously to transform the strategic situation, and set at naught rational calculations of manpower, economic resources, and military strength. This hope to which Hitler clung until the very end, his unfailing answer to every objection, was built upon the slenderest foundation, and the secret weapons, too—at least as they figured in Hitler's mind—soon belonged more to the realm of fantasy than that of fact.

More substantial, it may now appear, was the parallel set of hopes which he built up of a split between the partners in the Grand Alliance. No one, looking back at German anti-Bolshevik propaganda from the era of the Cold War, can fail to be struck by the aptness of much of the argument.

Subsequent events have shown how precarious was the basis of the wartime alliance between the Western Powers and the U.S.S.R. German propaganda, constantly repeating the theme of the Bolshevik threat to European civilization, was quick to pick on any hint of friction between the Allies, and Goebbels as well as Ribbentrop urged Hitler to follow this up with diplomatic action to split the alliance. Until the last week of his life Hitler expressed the firm conviction that the Allies were certain to fall out. He built the most extravagant hopes on such a quarrel, hopes which he kept alive in those around him by spreading rumours that negotiations were about to begin through a third party, or had already begun.

But all these hopes—the secret weapons and the break-up of the Grand Alliance—were subsidiary to the central pillar of Hitler's faith, the belief in himself, in his destiny and consequent ability to master any crisis. It was from this belief alone that he derived the strength of will to continue the war long after it had been lost, and to persuade not only himself, but many of those around him, against the evidence and their own common sense, that all was not yet hopeless. To the 'historic' image of the Führer, Hitler was prepared to sacrifice the German Army, the German nation, and in the end himself. From this course he never deviated: the only question was whether the German Army and the German nation were prepared to let him.

Little in the way of dissuading Hitler, still less of opposing him, could be expected from the other Nazi leaders.

Göring, still Hitler's successor, Reichsmarshal, Commander-in-Chief of the Air Force, Minister for Air, Plenipotentiary for the Four-Year Plan, Chairman of the Council of Ministers for the Defence of the Reich, Minister President of Prussia, President of the Reichstag and holder of a score of other offices, had steadily lost authority since the beginning of the war. In 1933–4 he was unquestionably the second man in Germany; by 1942, sloth, vanity, and his love of luxury had undermined not only his political authority but his native ability. Hitler was tolerant of Göring's weaknesses, but he was not blind to what was happening, and the failure of the Air Force finally discredited the Reichsmarshal in his eyes. There were angry scenes between the two men, Hitler accusing the Luftwaffe of cowardice as well as incompetence, and blaming Göring for letting himself be taken in by the Air Force generals. Some personal feeling for Göring remained until the end, but Hitler had no longer any confidence in him, and Göring kept out of his way.

The last of the original leadership, Joseph Goebbels, was both able and tough. He was a genius as a propagandist, but his cynical intelligence and caustic tongue did not make him popular in the Party. In the early years of the war, Hitler became very cool towards Goebbels, partly because of the scandal caused by his love affairs, partly out of mistrust of his malicious wit, but the later years, which marked the eclipse of Göring, saw Goebbels steadily rise in favour.

As early as 1942 Goebbels began to campaign for a more drastic mobilization of Germany's resources. This was a clever line to adopt when many people in responsible positions were beginning to hedge. He won back Hitler's confidence, and became one of the few men with whom Hitler could still exchange ideas.

Goebbels saw well enough the disaster which threatened Germany, and in 1943 and 1944 tried to persuade Hitler to consider a compromise peace. When this came to nothing he was too intelligent to suppose that there was any future for himself apart from Hitler, and instead of turning against the Führer he began to out-Herod Herod in his demands for still more drastic measures. It was Goebbels who, in 1945, proposed that Germany should denounce the Geneva Convention and shoot captured airmen out of hand, and who persuaded Hitler not to leave Berlin. His passion for self-dramatization was aroused by the idea of fighting on to the end, however hopeless the position, and he was the one member of the original group who joined Hitler in suicide in the Berlin bunker.

Of those who became prominent after 1933 only three are worth more than cursory mention: Himmler, Bormann, and Speer. Himmler's rise dated from 1934; in the following ten years he acquired sole power over the whole complex structure of the police state. As Minister of the Interior, Himmler controlled the Secret Police, the Security Service (SD) and the Criminal Police. As Reichsführer S.S. he commanded the political corps d'élite of the régime and, in the Waffen (Armed) S.S., possessed a rival army to the Reichswehr, numbering half a million men by the summer of 1944. Through the concentration camps, which he also controlled, he organized his own labour corps, which was set to work in factories run by the S.S. In the east he was in charge of all plans for the resettlement of the conquered territories. Himmler's Reich Security Main Office, in effect, administered a state in miniature, the embryonic S.S. State of the future, jealously defending its prerogatives and ceaselessly intriguing to extend them.

In 1944 the functions of military counter-intelligence were turned over to Himmler, who established a unified Intelligence Service. He became Commander-in-Chief of the Home Army, took over all prisoner-of-war camps from the Armed Forces and, before the end of the year, assumed the active command of an army group at the front.

Here was a concentration of power which even Hitler could not ignore. At one time, in 1943, an approach was made to Himmler by two members of the anti-Hitler opposition, in the hope of persuading him to take independent action. This led nowhere: Himmler was the last man of whom any such action could be expected, for two very good reasons. A man of undistinguished personality and limited intelligence, Himmler lacked the initiative to strike out a line for himself, particularly if it meant any conflict with the Führer. Moreover, Himmler had an unquestioning and single-minded faith in the doctrines of Nazism. He was a racist crank, to whom the Nazi *Weltanschauung* was the literal, revealed truth, and his humourless pedantry bored and irritated Hitler. This was poor material out of which to make the leader of an opposition, and only in the last days of the collapse was Himmler brought, with the utmost difficulty, to admit the possibility of acting on his own initiative to end the war.

The last of the great feudatories of the Nazi Court to carve out his demesne was Martin Bormann. In May 1941 Rudolf Hess, Head of the Party Chancery, took off for Scotland in an unauthorized attempt to conduct negotiations for peace. Hess's unexpected flight gave Bormann his chance. Succeeding to Hess's position, as early as January 1942 he was able to secure a directive laying down that he alone was to handle the Party's share in all legislation, jobs for Party members in the State administration, and all contacts between the various ministries and the Party.

This could be made into a powerful position, and Bormann was indefatigable in working to enlarge his claims. His agents were the Gauleiters, who were directly responsible to him. In December 1942, when all Gaue became Reich Defence Districts, the Gauleiters, now Reich Defence Commissioners as well, gained an effective control over the whole of the civilian war effort. After Himmler became Minister of the Interior in 1943 a clash between the two empires of the S.S. and the Party was inevitable. Bormann not only held his own against the powerful Reichsführer S.S., but by the end of 1944 had gained a lead in the struggle for power.

While both men controlled powerful organizations, Bormann grasped the importance of making himself indispensable to Hitler. In constant attendance on him, he succeeded in drawing most of the threads of internal administration into his hands. Hitler, preoccupied with the war, was glad enough to be relieved of the burden of administration, and in April 1943 Bormann was officially recognized as Secretary to the Führer. It was Bormann who decided whom the Führer should and should not see, what he should or should not read, it was Bormann who was present at nearly every interview and drafted the Führer's instructions. The importance of this position can scarcely be overestimated, for, as Weizsäcker, the State Secretary in the Foreign Office, says: 'Ministerial skill in the Third Reich consisted in making the most of a favourable hour or minute when Hitler made a decision, this often taking the form of a remark thrown out casually, which then went its way as an "Order of the Führer".'

In this way Bormann, a brutal and much-hated man, acquired immense power. It was a power, however, which he exercised not in his own right, but solely in the name of Hitler. Like his great rival Himmler, once separated from Hitler, Bormann was a political cypher. For all these men the road to power lay through acquiring Hitler's favour, not in risking its loss through opposition, and Bormann's voice, like that of Goebbels, was always raised in advocacy of more extreme measures.

As for the others, Ribbentrop still occupied the post of Foreign Minister, but had ceased to be taken seriously by Hitler or anyone else. Ley, when he was sober, ran anxiously from one group to another, trying to curry favour. The rest were minor figures, gratified by a nod of recognition from the Führer and wholly excluded from any knowledge of major decisions of policy.

Until the last few days of his life when Himmler and Göring made their last-minute attempts to negotiate with the Allies—and were promptly expelled from the Party—Hitler's hold over his Party remained intact. Its leaders were his creatures: had it not been for Hitler not one of them—with the possible exception of Goebbels and Göring—would ever have risen from the obscurity which was their natural environment. Their power was derivative, their light reflected. To turn against Hitler, to question his decisions, would have been to destroy the thread of hope to which they still clung. If Hitler failed, they would fall with him. If nothing else, the common crimes in which they had shared bound them together. But there was something more, according to Albert Speer: 'They were all under his spell, blindly obedient to him, and with no will of their own—whatever the medical term for this phenomenon may be. I noticed during my activities as architect, that to be in his presence for any length of time made me tired, exhausted and void. Capacity for independent work was paralysed.'

Speer is perhaps the most interesting case of them all, precisely because it is so different from that of the others. He only came into prominence in the spring of 1942, when Hitler suddenly nominated him as Minister for Armaments Production, but his rise in the next two years was rapid. By August 1944 he was responsible for the whole of German war economy. It was Speer who, by a remarkable feat of organization, patched up the bombed communications and factories, and somehow or other maintained the bare minimum of transport and production without which the war on the German side would have come to a standstill. Speer was not unaffected by the spell Hitler was still able to cast over those near him, but he stood apart from the contest for power; he was interested far more in the job than in the power it brought him.

A long illness kept him away from the Führer's Headquarters from February to June 1944, but on his return he became disquieted at the price which Germany was being made to pay for the prolongation of the war and—more disquieting still—realized that Hitler was determined to destroy Germany rather than admit defeat.

Speer systematically set about frustrating Hitler's design, and eventually, early in 1945, planned an attempt to kill Hitler and the men around him by introducing poison-gas into the ventilation system of his underground bunker. The plan had to be abandoned for technical reasons. Thereupon Speer continued his efforts to thwart Hitler's orders and to salvage something for the future. Yet he never again attempted to remove the man who was the author of the policy he opposed. The reason is interesting. Speer did not lack the physical courage to make a second attempt, but, as he admitted later, in the conflict of loyalties which divided his mind, he could not rid himself of the belief that Hitler was, as he claimed to be, the only leader who could hold the German people together, that he was, in Brauchitsch's phrase at the Nuremberg Trial, Germany's destiny, and that Germany could not escape her destiny.

Here, in the self-confessed failure of the one man among the Nazi leaders who retained the intellectual independence to see clearly the course on which Hitler was set and the integrity to reject it, is the clearest possible illustration of the hold which Hitler kept until the end over the régime he had established and the Party he had created.

If little had ever been expected of the Party by those Germans who saw in Hitler the evil genius of their country, much had been hoped of the Army. For a moment in the autumn of 1938 it had seemed possible that the Army High Command might lead a revolt against Hitler to avoid war, but the conspiracy came to nothing, and thereafter, however great their misgivings (at least in retrospect), the generals obeyed his orders, fought his battles for him and accepted the titles, the decorations and the gifts he bestowed on them.

In the strained relations which developed between Hitler and the Army after the invasion of Russia it was Hitler, not the generals, who took the offensive. Again and again he reversed the decisions of his senior commanders, ignored their advice, upbraided them as cowards, forced them to carry out orders they believed to be impossible to execute, and dismissed them when they failed. The generals submitted to treatment such as no previous German ruler had ever dared to inflict on the Army.

Hitler's criticism of the German Officer Corps was directed against its conservatism and its 'negative' attitude towards the National Socialist revolution. In practise, the revolutionary spirit meant willingness to carry out Hitler's orders without hesitation and without regard for the cost. Although he could not continue the war without the generals, those who retained office or secured promotion were the compliant, the ambitious who concealed their doubts, or rough-and-ready soldiers who went up to the front, drove their men to the limit and did not worry their heads too much about the strategic situation.

As the war went on Hitler came to rely more and more on the Waffen S.S. divisions, who were provided with the best equipment, given priority in recruitment and reserved for the most spectacular operations. Towards the end of the war the number of these divisions had risen to more than thirty-five. The growth of this rival S.S. Army was a particular grievance with the Regular Army officers.

After the fall of Mussolini, Hitler congratulated himself upon having no monarchy in Germany which could be used to turn him out of office. The thorough process of Nazification to which he had subjected the institutions of Germany, from the Reichstag to the Law Courts, from the trade unions to the universities, had destroyed, he believed, the basis for an organized opposition. However, two institutions in Germany still retained some independence.

The first was the Churches. Among the most courageous demonstrations of opposition during the war were the sermons preached by the Catholic Bishop of Münster and the Protestant pastor, Dr. Niemöller. Neither the Catholic Church nor the Evangelical

Church, however, as institutions, felt it possible to take up an attitude of open opposition to the régime. Yet without the support of some institution, any Opposition appeared to be condemned to remain in the hopeless position of individuals pitting their strength against the organized power of the State. It was natural, therefore, that those few Germans who ventured to think of taking action against Hitler should continue to look with expectation to the Army, the only other institution in Germany which still possessed a measure of independent authority, if its leaders could be persuaded to assert it, and the only institution which commanded the armed force needed to overthrow the régime.

There is some danger in talking of the 'German Opposition' of giving altogether too sharp a picture of what was essentially a number of small, loosely connected groups, fluctuating in membership, with no common organization and no common purpose other than their hostility to the existing régime. Their motives for such hostility varied widely: in some it sprang from a deeply felt moral aversion to the whole régime, in others from patriotism and the conviction that, unless he were halted, Hitler would destroy Germany. To diversity of motives must be added considerable divergence of aims, about the steps to be taken in opposing Hitler as well as the future organization of Germany and Europe.

Among those who continued to meet and discuss the chances of action against the régime were General Ludwig Beck, Dr. Karl Goerdeler, Ulrich von Hassel, and Colonel Hans Oster, the chief assistant of the enigmatic Admiral Canaris in the Abwehr, the counter-intelligence department of the O.K.W. The Abwehr provided admirable cover and unique facilities for a conspiracy, and Oster gathered a small group of devoted men around him of whom the outstanding members were Hans von Dohnanyi and Justus Delbrück; two Berlin lawyers, Joseph Wirmer and Claus Bonhöffer, and the latter's brother Dietrich Bonhöffer, a Protestant pastor and professor of theology who had once been minister of the Lutheran Church in London.

One of the uses to which the conspirators put the facilities of the Abwehr was to try and make contact with the British and Americans in the hope of securing some assurances as to the kind of peace the Allies would be willing to make if Hitler's government was overthrown. The Allies, however, were sceptical about any German opposition, and the conspirators had to face the need to act on their own without any encouragement from outside.

The conspirators devoted much time and energy to discussing how Germany and Europe should be organized and governed after the overthrow of Hitler. Discussion of such questions was the purpose of the group which Count Helmuth von Moltke, a former Rhodes Scholar at Oxford and bearer of one of the most famous names in German military history, brought together on his estate at Kreisau in Silesia. The Kreisau Circle was drawn from a cross-section of German society: amongst its members were two Jesuit priests, two Lutheran pastors; conservatives, liberals, and socialists, landowners and former trade unionists. The discussions at Kreisau were concerned, not with planning the overthrow of Hitler, but with the economic, social, and spiritual foundations of the new society which would come into existence afterwards. Moltke, who died for his beliefs with great courage, was strongly opposed to any active steps to get rid of Hitler.

An analysis of the different groups and shades of opinion represented in the German Opposition lies outside the scope of this study. From the point of view of Hitler their activities were important only in so far as they led to action.

At first, Goerdeler and Beck pinned their hopes on persuading one or other of the commanders in the field—amongst them Field-Marshal von Kluge, the commander of Army Group Centre in the East—to arrest or get rid of Hitler. All such hopes proved illusory, and after Stalingrad the more active conspirators accepted the fact that they would have to assassinate Hitler first before anyone in authority would be willing to move.

The first attempt was made on 13 March 1943, when Hitler paid a visit to Kluge's headquarters at Smolensk. General Henning von Tresckow and Fabien von Schlabrendorff hid a time-bomb in a package of brandy to be delivered to a friend, and placed it on the plane which carried Hitler back to East Prussia. The bomb failed to explode. Schlabrendorff flew at once to the Führer's Headquarters, recovered the bomb before it had been discovered, and took it to pieces on the train to Berlin.

As many as six more attempts on Hitler's life were planned in the later months of 1943, but all for one reason or another came to nothing. In the meantime Himmler's police agents, although singularly inefficient in tracking down the conspiracy, were beginning to get uncomfortably close. In April 1943 they arrested Dietrich Bonhöffer, Joseph Muller, and Hans von Dohnanyi. Too many threads led back to the Abwehr, which the rival S.S. Intelligence Service was eager to suppress, and in December 1943 General Oster, the key figure in the Abwehr, was forced to resign.

Fortunately, just as the Abwehr circle was being broken up, a new recruit joined the conspiracy who promised to bring to it the qualities of decision and personality which the older leaders lacked.

Klaus Philip Schenk, Count von Stauffenberg, born in 1907, came of an old and distinguished South German family. A brilliant young man, he served with distinction as a staff officer in Poland, France, and Russia. In Russia he was initiated into the conspiracy by Tresckow and Schlabrendorff. After recovering from wounds in the Tunisian campaign, which cost him his left eye, right hand, and two fingers of the other hand, Stauffenberg secured appointment to Olbricht's staff in Berlin and threw himself into preparations for a renewed attempt at a *coup d'état.*

Stauffenberg used the pretext of the dangers of a revolt by the millions of foreign workers in Germany to prepare plans for the Home Army to take over emergency powers in Berlin and other German cities. 'Operation Valkyrie' was worked out in detail and a series of orders and appeals drawn up to be signed by Beck as the new head of state and Goerdeler as chancellor.

With the help of men on whom he could rely at the Führer's headquarters, in Berlin and in the German Army in the west, Stauffenberg hoped to push the reluctant Army leaders into action once Hitler had been killed. Stauffenberg allotted the task of assassination to himself despite the handicap of his injuries.

The conspirators were now working against time. Further arrests were made early in 1944, including that of Moltke. In February, the greater part of the Abwehr functions were transferred to a unified Intelligence Service under Himmler, who told Admiral Canaris that he knew very well a revolt was being planned and that he would strike when the moment came.

The news that the Allies had landed in Normandy added confusion to the conspirators' plans. In July there were more arrests, and a warrant was out for Goerdeler. The plot was now in danger of being wrecked.

At the end of June Stauffenberg was appointed Chief of Staff to the Commander-in-Chief of the Home Army, a position which gave him frequent access to Hitler. On 11 July he attended a conference at Berchtesgaden with a time-bomb concealed in his brief-case, but in the absence of Himmler and Göring he decided to wait until there was a better chance of killing all the leaders at one blow. A second chance came on 15 July when he was again summoned to a conference at the Führer's Headquarters in East Prussia. Hitler unexpectedly cut the meeting short, and Stauffenberg did not have time to set off the bomb. Four days later, on 20 July, he flew to East Prussia determined that his third chance should be decisive.

For Hitler the first six months of 1944 had brought nothing but an intensification of all the familiar problems. In January the Russians freed Leningrad from its German besiegers, and by the end of June large sections of the German front ceased to exist. The German divisions Hitler insisted on holding in the Baltic States were threatened with encirclement, while the Russians were already thrusting towards the province of East Prussia, the first German territory to be threatened with invasion.

During the same six months the Allied air forces continued to bomb German towns and communications with monotonous regularity and in March the Americans made their first day raid on Berlin. In Italy Kesselring was forced to retreat, and the Allies entered Rome on 4 June.

Two days later at dawn the British and Americans began the long-awaited assault from the west. In preparation for the invasion, Hitler had recalled Rundstedt to act as Commander-in-Chief in the west. Considerable effort had been expended in building defences along the western coastline of Europe, but the Atlantic Wall was much weaker than German propaganda represented it to be. In June 1944 sixty German divisions were available to hold a front which extended from Holland to the south of France; few were of first-rate quality, and only eleven of them were armoured formations.

German Intelligence was badly at fault in forecasting the date, place, and strength of the invasion. Hitler rightly guessed that Normandy would be the part of the coast chosen by the Allies, but he also believed that a second landing would take place in the narrower part of the Channel, where the sites for the V1s were situated. The elaborate deception planned by the British to encourage this belief was accepted at face value, and powerful German forces were stationed north of the Seine and held there, on Hitler's orders, when their intervention in the fighting in Normandy might have had great effect.

The actual landing in the early hours of 6 June caught the Germans unawares, and the opportunities of the first few hours were missed. Once the bridgehead had been made good, Hitler refused to give his commanders a free hand, constantly intervened to dictate orders which were out of keeping with the situation at the front, and persisted in believing that the Allies could still be thrown back into the sea. Relations between Hitler and the generals on the spot rapidly became strained, and on 17 June he summoned both Rundstedt and Rommel for a conference at Margival, near Soissons.

The Führer was in a bitter mood. The fact that the Allied landings had succeeded he ascribed to the incompetence of the defence. When Rommel answered with an account of the difficulties of the situation, which were only increased by Hilter's rigid insistence on defending every foot of territory, Hitler went off into a monologue on the subject of the V-weapons which, he declared, would be decisive. Rommel's attempt to make him grasp the seriousness of the German position failed. Hitler talked of 'masses of jet-fighters' which would shatter the Allied air superiority, described the military situation in Italy and on the Russian front as stabilized, and lost himself in a cloud of words prophesying the imminent collapse of Britain under the V-bombs.

Further efforts to make Hitler realize that the

attempt to defeat the landings had already failed proved no more successful. At the end of June, the two Field-Marshals again tried to persuade Hitler to give them a free hand in the west and to end the war. Hitler responded by replacing Rundstedt with Field-Marshal von Kluge. But Kluge was no more able than Rundstedt or Rommel to stem the Allied advance, and by 20 July, although he still refused to recognize the fact, Hitler was confronted with as serious a military crisis in the west as in the east. For the first time he was being made to realize the meaning of 'war on two fronts'.

Hitler returned to his Headquarters in East Prussia in the middle of July. A conference was scheduled for 12.30 on the 20th, after which Mussolini was to visit the Führer.

Stauffenberg flew from Berlin during the morning and was expected to report on the creation of new Volksgrenadier divisions. He brought his papers with him in a brief-case in which he had concealed the bomb fitted with a device for exploding it ten minutes after the mechanism had been started. The conference was already proceeding when Stauffenberg arrived. Twenty-four men were grouped round a large, heavy oak table on which were spread out a number of maps. Neither Himmler nor Göring was present. The Führer himself was standing towards the middle of one of the long sides of the table, constantly leaning over the table to look at the maps, with Keitel and Jodl on his left. Stauffenberg took up a place near Hitler on his right. He placed his brief-case under the table, having started the fuse before he came in, and then left the room unobtrusively. He had been gone only a minute or two when, at 12.42 p.m., a loud explosion shattered the room, blowing out the walls and the roof, and setting fire to the debris which crashed down on those inside.

In the smoke and confusion, Hitler staggered out of the door on Keitel's arm. One of his trouser legs had been blown off; he was covered in dust, and he had sustained a number of injuries. His hair was scorched, his right arm hung stiff and useless, one of his legs had been burned, a falling beam had bruised his back, and both ear-drums were found to be damaged by the explosion. But he was alive. Those who had been at the end of the table where Stauffenberg placed the brief-case were either dead or badly wounded. Hitler had been protected, partly by the table-top over which he was leaning at the time, and partly by the heavy wooden support on which the table rested and against which Stauffenberg's brief-case had been pushed before the bomb exploded.

Although badly shaken Hitler was curiously calm, and in the early afternoon he appeared on the platform of the Headquarters station to receive Mussolini. Apart from a stiff right arm, he bore no traces of his experience and the account which he gave to Mussolini was marked by its restraint.

As soon as they reached Wolfsschanze Hitler took Mussolini to look at the wrecked conference room. Then, as he began to reenact the scene, his voice became more excited.

'After my miraculous escape from death today I am more than ever convinced that it is my fate to bring our common enterprise to a successful conclusion.' Nodding his head, Mussolini could only agree. 'After what I have seen here, I am absolutely of your opinion. This was a sign from Heaven.'

Some time passed before anyone at the Führer's headquarters realized what had happened—at first Hitler thought the bomb had been dropped from an aeroplane—and it was longer still before it was known that the attempted assassination had been followed by an attempted putsch in Berlin.

There, in the capital, a little group of the conspirators had gathered in General Olbricht's office at the General Staff Building in the Bendlerstrasse. Their plan was to announce that Hitler was dead and that an anti-Nazi government had been formed in Berlin, with General Beck as Head of State, Goerdeler as Chancellor, and Field-Marshal von Witzleben as Commander-in-Chief of the Armed Forces. Orders were to be issued in their name declaring a state of emergency and transferring all power to the Army in order to prevent the S.S. seizing control. It was hoped that—once Hitler himself had been removed—those officers who had hitherto refused to join the conspiracy would support the new Government. The smouldering hostility of the Army to the S.S. and the Party, the desperate position of Germany unless she could make a compromise peace, and, most important of all, the knowledge that the assassination had been successful, would, it was hoped, overcome all hesitations, and a number of sympathizers ready to act had already been secured in the different commands.

Everything depended upon two conditions, the successful assassination of Hitler and prompt, determined action in Berlin. Stauffenberg, in the confusion following the explosion, left the Führer's Headquarters convinced that no one in the conference could have survived. The first reports of the explosion to reach Olbricht, however, made it clear that Hitler was not dead, and he therefore decided not to issue the order for 'Valkyrie'. When Stauffenberg reached Rangsdorf airfield after a three-hour flight from East Prussia—still believing that Hitler had been killed—he persuaded Olbricht to start sending out orders for action. But three to four hours had been lost, and everything still remained to be done. Even in Berlin no move had been made to seize the radio station or the Gestapo headquarters, nor was any attempt made to arrest Goebbels.

Orders were hurriedly sent out to the chief army commands to carry out 'Operation Valkyrie', but shortly after 6.30 p.m. the German radio broadcast an announcement, telephoned by Goebbels, that an attempt had been made to kill Hitler but had failed. Once this became known, fear of Hitler's revenge became the dominant motive in the minds of that large number of officers who had hitherto waited to see if the putsch was successful before committing themselves.

Soon after 8 o'clock Keitel sent out a message by teleprinter to all commands directing all commanding officers to ignore orders not counter-signed by himself or by Himmler. An hour later the radio put out an announcement that Hitler would broadcast to the German people before midnight.

The plan to capture Berlin had totally miscarried and the situation of the little group of conspirators was now hopeless. In the course of the evening a group of officers loyal to Hitler, who had been placed under arrest in the Bendlerstrasse earlier in the day, broke out of custody, released General Fromm (whose office as Commander-in-Chief of the Home Army had been taken over by Höppner) and disarmed the conspirators. Fromm's own behaviour had been equivocal and he was now only too anxious to display his zealous devotion by getting rid of those who might incriminate him. When troops arrived to arrest the conspirators, Fromm ordered Stauffenberg, Olbricht, and two other officers to be shot in the courtyard. Beck was allowed the choice of suicide. Fromm was only prevented from executing the rest by the arrival of Kaltenbrunner, Himmler's chief lieutenant, who was far more interested in discovering what could be learned from the survivors than in shooting them out of hand, now that the putsch had failed. Himmler, reaching Berlin from East Prussia in the course of the evening, set up his headquarters at Goebbel's house, and the first examinations were carried out that night. The manhunt had begun.

Only in Paris were the conspirators successful. There they had been able to count on a number of staunch supporters, headed by General Heinrich von Stülpnagel, the Military Governor of France. As soon as he received the code word from Berlin Stülpnagel carried out the orders to arrest the 1,200 S.S. and S.D. men in Paris, and the Army was rapidly in complete command of the situation. But here, too, the conspirators were dogged by the same ill-luck that had pursued them throughout the day.

In the early months of 1944, Field-Marshal Rommel, then recently appointed to a command in the west, had been brought into contact with the group around Beck and Goerdeler. Rommel was a man of action, not much given to reflection, but he needed little convincing at this stage of the war that, if Germany was to be saved, Hitler must be got rid of. Rommel was opposed to an assassination of Hitler on grounds that they must avoid making a martyr of him. He proposed instead that Hitler should be seized and tried before a German court. He accepted the leadership of Beck and Goerdeler, however; he was willing to take over command of the Army or Armed Forces—his popularity would have been a considerable asset—and he proposed to initiate armistice negotiations with General Eisenhower on his own authority.

On 17 July, however, while Rommel was returning from the front, his car was attacked by British fighters and the Field-Marshal was severely injured. Thus, on 20 July, Rommel was lying unconscious in hospital, and his command was in the hands of Field-Marshal von Kluge, a horse of another colour.

Kluge had been approached by the conspirators as long ago as 1942, and he knew of Rommel's feelings, but when the attempt on Hitler's life failed, he refused to consider taking independent action in the west. Without the support of the commander in the field, Stülpnagel could do nothing: he had created an opportunity which there was no one to exploit. So, by dawn on the 21st, the putsch had collapsed in Paris as well as in Berlin, and Stülpnagel was summoned home to report. Now it was Hitler's turn to act, and his revenge was unsparing.

Half an hour after midnight on the night of 20–21 July all German radio stations relayed the shaken but still recognizable voice of the Führer speaking from East Prussia. 'If I speak to you today', he began, 'it is first in order that you should hear my voice and should know that I am unhurt and well, and secondly that you should know of a crime unparalleled in German history. A very small clique of ambitious, irresponsible, and at the same time senseless and stupid officers had formed a plot to eliminate me and the High Command of the Armed Forces. . . . This time we shall get even with them in the way to which we National Socialists are accustomed.'

Hitler's threats were rarely idle, and this time he was moved by a passion of personal vindictiveness. No complete figure can be given for the number of those executed after 20 July: a total of 4,980 has been accepted as the best estimate that can be made. Many thousands of others were sent to concentration camps. The investigations and executions of the Gestapo and S.D. went on without interruption until the last days of the war.

Hitler and Himmler also used the opportunity to imprison or kill many who had only the flimsiest connexion, or none at all, with the conspiracy, but who were suspected of a lack of enthusiasm for the régime. Few who had ever shown a trace of independence of mind could feel safe.

By the autumn sufficient evidence had been collected to rouse Hitler's suspicions of Rommel. After a slow recovery from his injuries, in October, Rommel received a brief message from the Führer offering him the choice between suicide and trial before the People's Court. For the sake of his family, Rommel chose the former. The cause of his death was announced as heart failure, due to the effects of his accident, and the Führer accorded him a State funeral. Hitler was not prepared to admit that the most popular general of the war had turned against him: 'His heart,' declared the funeral oration, which Rundstedt was called upon to read, 'belonged to the Führer.'

It was against the Officer Corps that Hitler's resentment was most sharply directed. To defeatism, cowardice, and conservatism the generals had now added the crime of treason. Had Hitler been free to give full rein to his anger, he would have made a clean sweep and imprisoned or shot every general within sight. But in the middle of a grave military crisis this was more than he could afford to do. Nor would his own prestige allow him to admit that the Army no longer had complete faith in his leadership. In public, therefore, elaborate measures were taken to conceal the split between the Army and its commander-in-chief. In his broadcast of 20–21 July Hitler insisted that only a small clique of officers was involved. Goebbels described the plot as a stab in the back aimed at the fighting front and crushed by the Army itself.

But in fact the humiliation of the Army was complete. The generals had now to accept the Waffen S.S. as equal partners with the Army, Navy, and Air Force. On 24 July the Nazi salute was made compulsory 'as a sign of the Army's unshakable allegiance to the Führer and of the closest unity between Army and Party'. On the 29th General Guderian issued a further order which insisted that henceforth every General Staff Officer must actively cooperate in the indoctrination of the Army with National Socialist beliefs, and National Socialist Political Officers were now appointed to all military headquarters.

Despite the measures taken to ensure loyalty, and despite the purge of the Officer Corps which followed the attempt, Hitler's distrust of the Army was henceforward unconcealed. This was bound to affect the desperate effort which had now to be made to hold the enemy outside the German frontiers. There was little enough hope of doing that in any case; there was less still when the Commander-in-Chief's attitude towards his own commanders was governed by invincible suspicion and vindictive spite.

CHAPTER FOURTEEN

THE EMPEROR WITHOUT HIS CLOTHES

By the end of July 1944 the Russian armies had cut off the German Army Group North by a thrust to the Baltic; had destroyed Army Group Centre and reached the Vistula; and had driven Army Group South (Ukraine) back into Rumania.

Hitler was forced to commit all his reserves in order to hold any line in the east, but he stubbornly refused to withdraw his troops from the Baltic States, where they were left to fight a local war which had no bearing on the main battle for the approaches to Germany. Hitler's reasons for this refusal were the possible effect of such a withdrawal on Sweden (with the all-important iron-ore supplies), and the loss of the Baltic training grounds for the new U-boats on which he set great store. He argued that Schörner was engaging a large number of Russian divisions which would otherwise be used on other and more vital fronts. The Russians, however, were not short of manpower, while the Germans were. Guderian protested strongly against the decision, but in vain. In fact, after the big German defeats of the summer in the east, Hitler was still trying to hold much-reduced forces a longer line than that through which the Russians had already broken. The man who had once proclaimed mobility as the key to success now rejected any suggestion of mobility in defence in favour of the utmost rigidity.

The Russian break-through in Poland was followed at the end of July by an American break-through in France. On 28 July the Americans captured Coutances, and two days later Avranches; by the 31st they were into Brittany. The German left flank collapsed, and the war of movement in the west began.

Patton's Third Army striking eastwards for Le Mans, and the threat of encirclement at Falaise, were the plainest possible indication that the time had come for an immediate German withdrawal behind the Seine. Hitler, remote from the battle in his East Prussian headquarters, and ignorant of the massive superiority of the Allied forces, especially in the air, refused to consider such a course. Kluge was ordered to counter-attack at once. The S.S. generals at the front were the loudest in their protests against the folly of gambling the few remaining armoured divisions on an attack which seemed certain to fail. Kluge's only reply was that these were Hitler's orders and that the Führer would tolerate no argument.

When the operation failed Hitler peremptorily ordered the attack to be renewed. On 15 August, when Kluge, up at the front, was out of touch with his headquarters for twelve hours, Hitler leaped to the conclusion that the Field-Marshal was trying to negotiate a surrender. The next day he summoned Model from the Eastern Front and ordered him to take over Kluge's command at once. On his way back to Germany Kluge committed suicide: he closed a long letter of self-defence to Hitler with the advice to end the war.

Model was one of the few generals whom Hitler trusted and whom he allowed to argue with him. A rough, aggressive character, who had nothing in common with the stiff caste conventions of the German military tradition, Model had identified his fortunes with those of Hitler's régime. But neither Model nor anyone else could prevent the collapse of the German front in France.

While Patton struck out boldly for the east, and Paris was liberated, the German Army in the west was streaming back across the Seine in headlong retreat. In the circumstances Model did well to preserve anything from the rout. On 29 August, he reported to Hitler that out of the sixteen infantry divisions which he had got back over the Seine he could raise sufficient men to form four, but was unable to equip them with more than small arms. Another seven infantry divisions had been totally destroyed, while of some 2,300 German tanks and assault guns committed in Normandy, only 100 to 120 were brought back.

France was lost. It was now a question of whether the Rhine could be held. In the first few days of September Patton's Third Army reached the Moselle, and the British Second Army liberated Brussels, Louvain, and Antwerp. On the evening of 11 September an American patrol crossed the German frontier: five years after the Polish campaign, the war had reached German soil.

In a conference with three of his generals on the afternoon of 31 August Hitler made it clear that, whatever happened and whatever the cost to Germany, he was determined to maintain the struggle. In a mood strangely compounded of inflexible determination and self-pity, he called on the German people for one more effort, and for the last time the German people responded. They no longer saw, or even heard, the man whose orders they obeyed, but the image of the Führer was still strong enough to carry conviction, and conviction was powerfully reinforced by fear.

It was to fear that Goebbels now openly appealed. The news of the Morgenthau Plan, which provided for the dismemberment of Germany, the destruction of her industrial resources, and her conversion into an agricultural and pastoral country, appeared to offer proof that Goebbels was right when he declared that the Allies intended the extermination of a considerable proportion of the German people and the enslavement of the rest. The grim picture which Goebbels had been drawing for months of the German people's fate under a Russian occupation was now supplemented by the prospect of an equally terrible revenge at the hands of the Western Allies. With the Red Army on the threshold of East Prussia, and the British and Americans on the edge of the Rhineland, the argument had an urgency it had never possessed before. To add point to it, Himmler announced on 10 September that the families of those deserting to the enemy would be summarily shot.

The Allies' plan was to burst into Germany before the winter came, and to strike at the basis of her war economy in the Ruhr and Rhineland. Bad luck, bad weather, difficulties of supply, and differences of opinion within the Allied High Command, combined to defeat their hopes. To these must be added the unexpected recovery of the German Army. By the end of September it had rallied along the line of the German frontier and succeeded in forming a continuous front again west of the Rhine. Field-Marshal von Rundstedt, whom Hitler had recalled to be Commander-in-Chief in the west at the beginning of September, had few illusions about the future, yet the measures taken by him and by Model, as the Commander of Army Group B, won for Hitler the breathing space of the winter before the Allies could bring their full weight to bear in the battle for Western Germany.

Hitler used this respite to build up as hurriedly as possible new forces with which to fill the gaps left by the summer's fighting. In the west alone 1944 had cost him a million men. On 24 August Goebbels announced a total mobilization which went far further than any previous measures. With this last reserve of manpower Hitler hoped not only to re-form the divisions which had been broken up on both the Western and Eastern Fronts, but to create twenty to twenty-five new Volksgrenadier divisions, eight to ten thousand men strong, under Himmler's direction. This was partly bluff, for units which had been reduced to the fighting value of no more than a battalion were retained as divisions in the German Order of Battle. Rather than use the men he had available to rebuild these to their full strength, or break them up completely, Hitler preferred to set up new divisions and retain the old formations in being at a half or a quarter of their strength. In this way he could keep up the illusion that he was still able to increase his forces to meet the crisis. As a final measure Hitler called up every able-bodied man between the ages of sixteen and sixty to form a Volkssturm, or Home Guard.

At the beginning of September 1944 the total paper strength of the German Armed Forces was still over ten million men. It was Hitler's own decision that kept these very considerable forces scattered over half the Continent, holding hopeless positions, instead of concentrating for the defence of the Reich itself. He refused to abandon hope of reversing the situation by a dramatic stroke. Thus Western Holland must be held to allow the V2s to be directed against London; Hungary and Croatia for the bauxite supplies necessary for the jet aircraft; the Baltic coast with its training-grounds; and the naval bases in Norway for the new U-boats on which he built so much.

Thanks to Speer, German armaments production had not yet been crippled by the bombing. The German aircraft factories and those producing other arms maintained, and in some cases even increased, the rate of production. The greatest material difficulty was the desperate shortage of oil and petrol, due to the systematic Allied bombing of the synthetic oil plants, refineries, and communications. By the end of September the Luftwaffe had only five weeks' supply of fuel left. Moreover, Speer maintained arms production only by drawing heavily on supplies of raw materials which could scarcely be replaced. Germany made a remarkable recovery in the last three months of 1944, but it was the last reserves of men, materials, and morale on which Hitler was now drawing; if he squandered these there was nothing left.

Everything turned upon the use which Hitler proposed to make of the forces he had scraped together. A momentary calm on all fronts encouraged his illusions. But the resumption of the Allied attacks was only a question of time, and the real weakness of the German position was shown by the success of the Red Army's autumn offensive in the Balkans.

For the Russians, having forced Hitler to throw in all his reserves on the Centre Front in the summer, now reaped their advantage in the South. On 20 August a new offensive opened with the invasion of Rumania, which capitulated in the first few days, and the Russians were able to occupy the oilfields. On 8 September the Red Army began the occupation of Bulgaria, and the loss of Germany's two Balkan satellites was accompanied by the withdrawal of Finland from the war. In October the British freed Athens, and the Russians reached Belgrade, where they joined Tito's partisan forces. By the beginning of December the Germans were besieged in Budapest.

Hitler did not ignore the danger from the southeast, and the Germans succeeded in prolonging the battle for the Hungarian capital into February 1945, but he had already made up his mind in the autumn that the new and re-formed divisions were to go to the west. At the same time the panzer and panzergrenadier divisions already stationed in the west were re-equipped, and over two-thirds of the Luftwaffe's planes deployed in their support.

In deciding for the west against the east, Hitler was not thinking in terms of defence of the German frontiers; he thought solely of an offensive which would take the Allies by surprise, enable him to recapture the initiative and so gain time for the development of the new weapons and of the split between the members of the Grand Alliance upon which he counted to win the war. If the basis of this calculation was slender, it was natural for Hitler to think along these lines. For him at least the only choice lay between victory or death. A defensive campaign could defer a decision, but would not alter the situation. The one chance of doing that was to stake what was left on the gamble of attack. He saw a greater possibility of success in the west, where distances were shorter, less fuel would be needed, and strategic objectives of importance were more within the compass of his forces. Nor did he believe the Americans and British were as tough opponents as the Russians.

Accordingly, Hitler and Jodl set to work in great secrecy to plan a counter-offensive in the west for the end of November. The object of the attack was the recapture of the principal Allied supply port of Antwerp by a drive through the Ardennes and across the Meuse, which would have the effect of cutting Eisenhower's forces in two and trapping the British Army in the angle formed by the Meuse and the Rhine as they turn westwards towards the sea.

The idea was excellent. The last thing the Allied commanders expected was a German attack. The Ardennes sector was the weakest point in their front, and the loss of Antwerp would have been a major blow at their supply lines. But the idea bore no relation to the stage of the war which had now been reached. The permanent disparity between the resources of Germany in 1944 and those of the three most powerful states in the world could not be redressed by a single blow with the forces which Hitler was able to concentrate in the west. The utmost Hitler could hope to inflict on the Allied armies was a set-back, not a defeat, and in the process he ran the heavy risk of throwing away his last reserves.

The attempt of the men in command to argue

with Hitler, and to persuade him to accept more limited objectives, proved as unsuccessful as all the other previous attempts. To have admitted that the generals were right would have meant admitting that the war was lost. The final plans were sent to Rundstedt with every detail cut and dried down to the times of the artillery bombardment, and with the warning in Hitler's own handwriting: 'Not to be altered.' In order to keep even tighter control over the handling of the battle Hitler moved his headquarters from East Prussia to Bad Nauheim, behind the Western Front.

Four days before the attack was due to begin, on 12 December, Hitler summoned all the commanders to a conference. After being stripped of their weapons and brief-cases and bundled into a bus, they were led between a double row of S.S. troops into a deep bunker. Hitler made a long, rambling speech which lasted for two hours, during which S.S. guards stood behind every chair and watched every movement that was made.

Much of what Hitler said was a justification of his career and of the war. He laid particular stress on the incongruity of the alliance with which Germany was faced and predicted the the collapse of the 'artificially bolstered common front' after a 'few more heavy blows' by the German Army.

At dawn on 16 December the attack was launched. Hitler at least achieved the satisfaction of taking his opponents by surprise, and in the first few days the German Army made considerable gains. Yet never for a moment were the Germans within sight of Antwerp. As soon as the Allies had recovered their balance the Germans found themselves thrown back on the defensive, and by Christmas it was evident that they would be well advised to break off the battle and withdraw.

Hitler furiously rejected any such suggestions. Twice Guderian, who was responsible for the defence of the Eastern Front, visited Hitler's headquarters and tried to persuade him to transfer troops to the east, where there were ominous signs of Russian preparations for a new offensive. Hitler impatiently rejected Guderian's reports, declaring that the Russians were bluffing. After reinforcements had been sent to Budapest, the reserves for a front of 750 miles in the east totalled no more than twelve and a half divisions. Yet Hitler refused to write off the Ardennes offensive. Not only was Model ordered to make another attempt to reach the Meuse, but a new attack was to be launched into northern Alsace.

Once again the German attack fell short of Hitler's objective—this time Strasbourg—while Model's second attempt to break through the Ardennes was no more successful than the first. On 8 January Hitler reluctantly agreed to the withdrawal of the German armour on the Ardennes front. It was a tacit admission that he had failed. He continued to claim that he had inflicted a heavy defeat on the enemy, but the figures do not bear him out. The First and Third U.S. Armies fighting in the Ardennes lost 8,400 killed, with 69,000 wounded and missing. The total German casualties were around 120,000, in addition to the loss of 600 tanks and assault guns and over 1,600 planes. Most important of all, while the Americans easily made good their losses, Hitler's were irreplaceable.

Only 75 divisions were now left to guard against the most dangerous threat of all, the possibility of a Red Army thrust across the northern plains directed at the industrial districts of Silesia, Saxony, and Berlin itself. The divisions and equipment so laboriously scraped together in the closing months of 1944 had been expended without strengthening the defences in the east, and there were no more reserves to replace them. When Guderian tried to point out the dangers to Hitler at a conference on 9 January he was met with an hysterical outburst of rage. 'He had,' says Guderian, 'a special picture of the world, and every fact had to be fitted into that fancied picture. As he believed, so the world must be: but, in fact, it was a picture of another world.'

Reality, however, was to prove stronger than fantasy. Hitler still insisted that priority must be given to the west and told Guderian he must make do with what he had in the east. But, on 12 January, the Red Army opened its offensive in Poland and the German defences went down like matchwood before the onslaught of Russian divisions attacking all along the line from the Baltic to the Carpathians. By the end of the month Marshal Zhukov was within less than a hundred miles of the German capital, and the Berlin Home Guard was being sent to hold the line of the Oder.

During the late summer and autumn Hitler's health became worse and he was frequently confined to his bed. His doctors urged him to go to the Obersalzberg for a rest cure, but he refused. So long as he remained in East Prussia, he declared, it would be held, but if he left it would fall to the Russians.

In the middle of September, however, he broke down completely and had to return to bed. Apart from continual headaches and an aggravation of his stomach cramps, he was troubled by his throat, and for a time his voice was scarcely recognizable, so weak had it become. Lying on a camp-bed between the naked concrete walls of the bunker, he appeared to have lost all desire to go on living.

By one more effort of will Hitler recovered sufficiently to get up and resume work, although those who saw him in the last six months of his life agree in their description of him as an old man, with an ashen complexion, shuffling gait, shaking hands and leg. Guderian, who was frequently in his company, writes: 'It was no longer simply his left hand, but the whole left side of his body that trembled. . . . He walked awkwardly, stooped more than ever, and his gestures were both jerky and slow. He had to have a chair pushed beneath him when he wished to sit down.' This was his state of health when he returned from the west in the middle of January, shortly after the beginning of the Russian offensive, and moved into the Reich Chancellory.

The vast pile which Hitler had built to overawe his tributaries was now surrounded by the ruins of a bombed city. Jagged holes had appeared in the Chancellery's walls; the windows were boarded up; the rich furnishings removed—except from Hitler's own quarters. During the frequent air-raids Hitler moved to the massive concrete shelter built in the Chancellery garden.

Hitler rarely moved out of the Chancellery building, and in the last month lived almost entirely in the deep shelter. One of the few visits he paid was in January, shortly after his return to Berlin, when he drove out to Goebbels's home and took tea with his wife and family. It was the first visit he had paid them for five years, an indication of Goebbels's return to favour in the latter part of the war. Hitler was accompanied by a bodyguard of six S.S. officers, his adjutant and his servant, the last carrying a brief-case in which were contained the Führer's own vacuum-flask and a bag of cakes. They spent the afternoon reviving memories and discussing the plans for rebuilding Berlin. When Hitler left, Frau Goebbels expressed her satisfaction with the remark: 'He wouldn't have gone to the Görings.'

From the period between September 1942 and the beginning of 1945 only a few scattered records of Hitler's table talk have survived. But there has recently come to light the transcript of seventeen conversations (or rather, monologues), which Bormann arranged to be recorded in February 1945. Their single theme is the war and Hitler's analysis of the mistakes which had brought Germany to the position in which she then found herself. It was one of the few times in his life when Hitler was prepared to admit that he had made any mistakes at all, and this alone would endow these talks with interest, quite apart from the fact that whenever Hitler discussed politics—as distinct from art or religion—he never failed to show the power of his twisted mind.

Had it been wrong to go to war? No, he had been jockeyed into war: 'It was in any case unavoidable; the enemies of German National Socialism forced it upon me as long ago as January 1933.' The same was true of the attack on Russia. 'I had always maintained that we ought at all costs to avoid waging war on two fronts, and you may rest assured that I pondered long and anxiously over Napolean and his experiences in Russia. Why, then, you may ask, this war against Russia, and why at the time I selected?'

Hitler gave several answers to this question. It was necessary to deprive Britain of her one hope of continuing the war; Russia was withholding the raw materials essential to Germany; Stalin was trying to blackmail him into concessions in Eastern Europe. But the reason to which he always returned, 'and my own personal nightmare, was the fear that Stalin might take the initiative before me. . . . If I felt compelled to settle my accounts with Bolshevism by force of arms . . . I have every right to believe that Stalin had come to the same decision even before he signed the pact [of 1939].'

'The disastrous thing about this war is the fact that for Germany it began both too soon and too late.' He needed, Hitler declared, twenty years in which to bring his new élite to maturity. But the war also came too late. From a military point of view, it would have been better to fight in 1938, not 1939. Czechoslovakia was a better issue than Poland; Britain and France would never have intervened, and Germany could have consolidated her position in Eastern Europe before facing world war several years later. 'At Munich we lost a unique opportunity of easily and swiftly winning a war that was in any case inevitable.' It was all Chamberlain's fault: he had already made up his mind to attack Germany, but was playing for time and by giving way all along the line robbed Hitler of the initiative.

Once the war had begun, 'I must admit that my unshakable friendship for Italy and Duce may well be held to be an error on my part.' The Italian alliance had been a disaster. Not only had the Italians lost every campaign on which they had embarked, but by stirring up the Balkans they cost Germany six vital weeks' delay in the attack on Russia. Arab contempt for the Italians and the Duce's 'ridiculous pretensions' of an Italian empire robbed Germany of her opportunity. A similar opportunity had been lost in the French empire, where it should have been German policy to rouse the Arabs and other colonial peoples to throw off the French yoke.

But the greatest mistakes, Hitler concluded, had been made by Britain and the United States. Britain ought to have seen that it was in her interests to ally with Germany, the rising continental power, in order to defend the imperial possessions which she was now certain to lose. Churchill, a 'Jew-ridden, half-American drunkard', failed to see the need to bring about a unification of Europe, in which Britain would 'still retain the chance of being able to play the part of arbiter in world affairs.'

The United States ought to have realized that she had no quarrel with the Third Reich and preserved her isolation, Hitler continued. 'This war against America is a tragedy. It is illogical and devoid of any foundation of reality.' Once again it was due to the same sinister influence, the Jewish world conspiracy against Nazi Germany. 'Well, we have lanced the Jewish abscess; and world of the future will be eternally grateful to us.'

Such was Hitler's reasoning as he faced the possibility of defeat. There was not even a passing thought for the millions of deaths, the untold suffering, and the destruction he had brought on Germany and Europe. If he admitted to errors of judgement, they sprang from insufficient hardness, from his own too great tolerance; and the blame for war and for defeat rested not on himself, but on others.

To this Hitler added a postscript. On 2 April, he delivered the last of his table talk monologues, in effect a political testament to the German nation. 'I have been Europe's last hope,' he had declared in February. If Germany was to suffer defeat after all, it would be utter and complete, and a tragedy for Europe as well as the German people. Then with a last burst of prophetic power he drew his picture of the future:

'With the defeat of the Reich and pending the emergence of the Asiatic, the African, and perhaps the South American nationalisms, there will remain in the world only two Great Powers capable of confronting each other—the United States and Soviet Russia. The laws of both history and geography will compel these two powers to a trial of strength, either military or in the fields of economics and ideology. These same laws make it inevitable that both Powers should become enemies of Europe. And it is equally certain that both these Powers will sooner or later find it desirable to seek the support of the sole surviving great nation in Europe, the German people.'

Those who dismiss Hitler's political gifts as negligible may well be asked how many in the spring of 1945, with the war not yet over, saw the future so clearly.

As the façade of power crumbled, Hitler rev
to his origins; there is a far closer resemblan

tween the early Hitler of the Vienna days and the Hitler of 1944–5 than between either and the dictator of Germany at the height of his power. The crude hatred, contempt, and resentment which were the deepest forces in his character appeared undisguised. They found expression in the increasing vulgarity of his language. It was the authentic voice of the gutter again.

The man who had made it his first principle never to trust anyone now complained bitterly that there was no one he could trust. Only Eva Braun and Blondi were faithful to him, he declared, quoting Frederick the Great's remark: 'Now I know men, I prefer dogs.'

Years before, Hermann Rauschning, describing Nazism as the St Vitus's Dance of the twentieth century, had diagnosed its essential element of nihilism. In his conversations with Hitler during the years 1932–4 he records many remarks that betray the underlying passion for destruction which was only cloaked during the period of his success.

In talking to Rauschning, Hitler frequently became intoxicated with the prospect of a revolutionary upheaval which would destroy the entire European social order. In 1934, when Rauschning asked him what would happen if Britain, France, and Russia made an alliance against Germany, Hitler replied: 'That would be the end. But even if we could not conquer them, we should drag half the world into destruction with us, and leave no one to triumph over Germany. There will not be another 1918. We shall not surrender.'

This was the stage Hitler had now reached, and he was as good as his word. Goebbels shared Hitler's mood, and Nazi propaganda in the final phase has a marked note of exultation in the climax of destruction with which the war in Europe ended. But Hitler's determination to drag Europe down with him was most clearly expressed in his insistence on continuing the war to the bitter end and in his demands for a 'scorched earth' policy in Germany. Speer did his best to dissuade Hitler on the grounds that the German people must still go on living even if the régime were to be overthrown. On 15 March Speer drew up a memorandum in which he set out his case. Within four to eight weeks, he wrote, Germany's final collapse was certain. A policy of destroying Germany's remaining resources in order to deny them to the enemy could not affect the result of the war. The overriding obligation of Germany's rulers, without regard to their own fate, was to ensure that the German people should be left with some possibility of reconstructing their lives in the future.

Hitler was adamant. On 19 March he issued categorical and detailed orders for the destruction of all communications, rolling-stock, lorries, bridges, dams, factories and supplies in the path of the enemy. Sending for Speer, he told him: 'If the war is to be lost, the nation also will perish. This fate is inevitable. There is no need to consider the basis even of a most primitive existence any longer. On the contrary, it is better to destroy even that, and to destroy it ourselves. The nation has proved itself weak, and the future belongs solely to the stronger eastern nation. Besides, those who remain after the battle are of little value; for the good have fallen.'

In these senseless orders to destroy everything and to shoot those who failed to comply with his directive he found some relief for the passion of frustrated anger which possessed him, and it was only thanks to the devotion of Speer that these orders were not fully carried out. But, as General Halder remarks, this mood was something more than the product of impotent rage. 'Even at the height of his power there was for him no Germany, there were no German troops for whom he felt himself responsible;

for him there was—at first subconsciously, but in his last years fully consciously—only one greatness, a greatness which dominated his life and to which his evil genius sacrificed everything—his own Ego.'

In order to keep alive the will to go on fighting, Hitler made desperate efforts to conceal the hopelessness of the situation. As soon as he came across the words: 'The war is lost,' in Speer's memorandum, he refused to read another line and locked it away in his safe.

Hitler turned for comfort to the example of Frederick the Great, who in 1757, when Prussia was invaded by half a dozen armies and all hope seemed gone, won his greatest victories and routed his foes. He kept Graff's portrait of Frederick hanging above his desk and told Guderian: 'When bad news threatens to crush my spirit I derive fresh courage from the contemplation of this picture. Look at those strong, blue eyes, that wide brow. What a head!'

His private conversation in the early hours of the morning, however, was increasingly pessimistic in tone. Before the war he had strongly condemned suicide, arguing that if only a man would hold on something would happen to justify his faith. Now he announced his conversion to Schopenhauer's view that life was not worth living if it brought only disillusionment. He was depressed by his own ill-health. 'If a man is no more than a living wreck, why prolong life? No one can halt the decay of his physical powers.'

His secretary, who had to endure many such outbursts, records that after his return to Berlin in January his conversation became entirely self-centred and was marked by the monotonous repetition of the same stories. His intellectual appetite for the discussion of such large subjects as the evolution of man, the course of world history, religion, and the future of science had gone; even his memory began to fail him. His talk was confined to anecdotes about his dog or his diet, interspersed with complaints about the stupidity and wickedness of the world.

Yet he still maintained his hold over those who were in daily contact with him: the sorcerer's magic was not yet exhausted. Goebbels, Göring, Himmler, Bormann, Ribbentrop—every one of them clung despairingly to the hope that the man to whom they owed everything would yet find a way out.

Himmler was unquestionably the second man in the rapidly dwindling Nazi empire and the most obvious heir to Hitler. But Himmler's position was not undisputed. In accepting the active command of an Army Group, Himmler made the mistake of removing himself from the Führer's court, while his failure to halt the Russian advance much reduced his standing with Hitler. In the last six months of the Third Reich it was Bormann, rather than Himmler, who was the rising power at the Führer's Headquarters.

For Bormann, content to keep in the background and appear solely as the devoted servant of the Führer, took care never to leave Hitler's side. He adjusted his way of life in order to go to bed and rise at the same time as Hitler, and he strengthened his control over access to him. Bormann was still not powerful enough to keep out Himmler, Speer, and Goebbels, but he soon made sure of Himmler's permanent representative with Hitler, Hermann Fegelein, took every opportunity to undermine Hitler's confidence in Speer, and concluded a tacit alliance with Goebbels.

In the middle of these rivalries Hitler's own position remained unchallenged, nor did anyone, except Speer, dare to question the wisdom of his decision to continue the war. The intrigues were aimed not at replacing him, but at securing his favour and a voice in the nomination of his successor. No more striking testimony to Hitler's hold over those around him can be imagined than the interest they still showed in the unreal question of who was to succeed him.

As day succeeded day in the isolated world of the Reich Chancellery and its garden shelter, the news grew steadily worse. Between 12 January, the day on which the Russians opened their offensive in Poland, and 12 April, the day on which the U.S. Ninth Army crossed the Elbe, the Allies inflicted a total defeat upon the German Army.

On the 13th the Russians captured Vienna, and on the 16th they broke the defence line on the Oder. The way to Berlin was open, and it was now only a question of time before the armies advancing from the west met those coming from the east and cut Germany in two.

Hitler had lost all control over events, and by April he had the greatest difficulty in discovering what was happening. The Germans went on fighting—in the east, with the courage of despair—but there was no longer any organized direction of the war. Records of the discussions of the military situation in the early months of 1945 are rambling, confused, and futile. Hours were wasted in discussion of questions of detail and local operations, interrupted by reminiscences and recriminations. Hitler's orders became wilder and more contradictory, his demands more impossible, his decisions more arbitrary. His one answer to every proposal was: No withdrawal.

Hitler had long scorned the belief that war can be waged without resort to terrorism. A succession of orders from his headquarters—such as the notorious 'Commissar' and 'Commando' orders—demanded deliberate brutality in dealing with the enemy. In February 1945 there were prolonged discussions of a proposal made by Goebbels and eagerly seized on by Hitler that the German High Command should denounce the Geneva and other international conventions, shoot all captured enemy airmen out of hand and make use of the new poison gases, Tabun and Sarin. Only with the greatest difficulty was he restrained from taking this desperate and irresponsible step.

Without bothering to investigate the facts he ordered the dismissal, degradation and even execution of officers who, after fighting against overwhelming forces, were forced to give ground. Even the Waffen S.S. was not exempt from his vicious temper. When Sepp Dietrich, once the leader of his personal bodyguard and now in command of the Sixth S.S. Panzer Army, was driven back into Vienna, Hitler radioed: 'The Führer believes that the troops have not fought as the situation demanded and orders that the S.S. Divisions Adolf Hitler, Das Reich Totenkopf, and Hohenstauffen, be stripped of their arm-bands.'

When Dietrich received this he summoned his divisional commanders and, throwing the message on the table, exclaimed: 'There's your reward for all that you've done these past five years.' Rather than carry out the order, he cabled back, he would shoot himself.

Hitler still tried to buoy himself up with the belief that the new weapons, of which he never ceased to talk, would work a miracle. But gradually these hopes too faded and his continued references to them became no more than the mechanical repetition of ritual phrases. The V1s and V2s had come and gone. The Ardennes offensive had been launched and failed. The jet fighters never took the air. The U-boat fleet, reinforced by the new types on which Hitler and Dönitz had built the most extravagant expectations, put to sea but were routed.

The last hope of all was a split in the Grand Alliance, but Churchill, Roosevelt, and Stalin, meeting at Yalta in February, patched up their differences and contrived an agreement which, however impermanent, outlasted Hitler. The demand for unconditional surrender was reaffirmed, and the Allied armies never paused in their advance.

In the middle of April the Nazi Empire which had once stretched to the Caucasus and the Atlantic was reduced to a narrow corridor in the heart of Germany little more than a hundred miles wide. Hitler had reached the end of the road.

In April Eva Braun arrived unexpectedly in Berlin and, defying Hitler's orders, announced her intention of staying with him to the end. For some time Goebbels had been urging Hitler to remain in Berlin and make an ending in the besieged city worthy of an admirer of Wagner's *Götterdämmerung*. Goebbels scorned any suggestion that by leaving the capital he might allow the two million people still living there to escape the horrors of a pitched battle fought in the streets of the city. 'If a single white flag is hoisted in Berlin,' he declared, 'I shall not hesitate to have the whole street and all its inhabitants blown up. This has the full authority of the Führer.'

None the less Hitler's mind was not yet made up. Preparations were in train for the Government to leave Berlin and move to the 'National Redoubt' in the heart of the Bavarian Alps, round Berchtesgaden, the homeland of the Nazi movement, where the Führer was expected to make his last stand. Various ministries and commands had already been transferred to the Redoubt area, and the time had come when Hitler himself must follow if he was still to get through the narrow corridor left between the Russian and American armies.

Hitler's original plan was to leave for the south on 20 April, his fifty-sixth birthday, but at the conference on the 20th, following the reception and congratulations, he still hesitated. For the last time, all the Nazi hierarchs were present—Göring, Himmler, Goebbels, Ribbentrop, Bormann, Speer—together with the chiefs of the three Services. They advised his leaving Berlin. The most Hitler would agree to, however, was the establishment of Northern and Southern Commands, in case Germany should be cut in two by the Allied advance. There and then he appointed Admiral Dönitz to assume the full responsibility in the north, but, although Kesselring was nominated for the Southern Command, Hitler left open the possibility that he might move to the south and take the direction of the war there into his own hands.

On the 21st Hitler ordered an all-out attack on the Russians besieging Berlin, and he built the most exaggerated hopes on the success which he anticipated from the operation. It was the disappointment of these hopes which led him finally to make up his mind and refuse to leave the capital.

For the attack was never launched. The withdrawal of troops to provide the forces necessary allowed the Russians to break through the city's outer defences in the north, and Hitler's plan foundered in confusion. Throughout the morning of the 22nd a series of telephone calls from the bunker failed to elicit any news of what was happening. By the time the conference met at three o'clock in the afternoon Hitler was on the verge of one of his worst outbursts.

The storm burst during the conference, which lasted for three hours and left everyone who took part in it shaken and exhausted. In a universal gesture of denunciation Hitler cursed them all for their cowardice, treachery, and incompetence. The end had come, he declared. There was nothing left but to die. He would meet his end there, in Berlin; those who wished could go to the south, but he would never move. From this resolution he was not to be swayed.

Now that he was forced to admit the fact of defeat, the man who had insisted on prolonging the war against the advice of his generals refused to take any further responsibility. All the grandiloquent talk of dying in Berlin cannot disguise the fact that this petulant decision was a gross dereliction of his duty to the troops still fighting under his command and an action wholly at variance with the most elementary military tradition.

The setting in which Hitler played out the last scene of all was well suited to the end of so strange a history. The Chancellery air-raid shelter, in which the events of 22 April had taken place, was buried fifty feet beneath the ground, and built in two storeys covered with a massive canopy of reinforced concrete. The lower of the storeys formed the Führerbunker. It was divided into eighteen small rooms grouped on either side of a central passageway. Half of this passage was closed by a partition and used for the daily conferences. Eva Braun had a bed-sitting-room, a bathroom, and a dressing-room; Hitler a bedroom and a study, the sole decoration in which was the portrait of Frederick the Great. A map-room used for small conferences, a telephone exchange, a power-house, and guard rooms took up most of the rest of the space, and there were two rooms for Goebbels, whom Hitler had invited to join him in the Führerbunker, and two for Stumpfegger, now Hitler's surgeon. Frau Goebbels, who insisted on remaining with her husband, together with her six children, occupied four rooms on the floor above, where the kitchen, servants' quarters and dining-hall were also to be found. Other shelters near-by housed Bormann, his staff and the various Service officers; and Mohnke, the S.S. commandant of the Chancellery, and his staff.

The physical atmosphere of the bunker was oppressive, but this was nothing compared to the pressure of the psychological atmosphere. The incessant air-raids, the knowledge that the Russians were now in the city, nervous exhaustion, fear, and despair produced a tension bordering on hysteria, which was heightened by propinquity to a man whose changes of mood were not only unpredictable but affected the lives of all those in the shelter.

Hitler had been living in the bunker for some time. Such sleep as he got in the last month appears to have been between eight and eleven o'clock in the morning. As soon as the mid-morning air attacks began, Hitler got up and dressed. He had a horror of being caught either lying down or undressed.

Much of the time was still taken up with conferences. The midday or afternoon conference was matched by a second after midnight which sometimes lasted till dawn. The evening meal was served between 9 and 10 p.m., and Hitler liked to drag it out in order not to be left alone during a night air-raid. Sometimes he would receive his secretaries at six in the morning, after a late-night conference. He would make an effort to stand up and greet them, but rapidly sank back exhausted on to the sofa. The early morning meal was the one he most enjoyed, and he would eat greedily of chocolate and cakes, playing with Blondi and the puppies which she produced in March. To one of these puppies Hitler gave his own old nickname, Wolf, and brought it up without anyone's help. He would lie with it on his lap, stroking it and repeating its name until the meal was over and he tried to get some sleep.

Between 20 and 24 April a considerable number of Hitler's entourage—including Göring and Morell—left for the south. In the last week of his life Hitler shared the cramped accommodation of the Führerbunker with Eva Braun, the Goebbels and their children, his surgeon, his valet, his S.S. adjutant, his two remaining secretaries, his vegetarian cook, and Goebbels's adjutant. Frequent visitors to the Führerbunker from the neighbouring shelters were Bormann; General Krebs, who had succeeded Guderian as the Army's Chief of Staff; General Burgdorf, Hitler's chief military adjutant; Artur Axmann, the leader of the Hitler Youth (a thousand of whom took part in the defence of Berlin), and a crowd of aides-de-camp, adjutants, liaison officers, and S.S. guards.

On Monday 23, April, having at last come to a decision, Hitler was in a calmer frame of mind. Speer, who flew back from Hamburg to say farewell, made a full confession of the steps he had taken to thwart Hitler's orders for scorching the German earth. Hitler undoubtedly had a genuine affection for Speer, but it is surprising that he was moved, rather than incensed, by his frankness. Speer was allowed to go free, and like everyone else who saw Hitler that day he was impressed by the change in him, the serenity which he appeared to have reached after months of desperate effort to maintain his conviction, in the face of all the facts, that the war could still be won. Now that he had abandoned the attempt to flog himself and those around him into keeping up the pretence, he was resigned to facing death as a release from the difficulties which overwhelmed him. He repeated to Speer what he had told Jodl and Keitel the day before, that he would shoot himself in the bunker and have his body burned to avoid its falling into the hands of the enemy. This was stated quietly and firmly, as a matter no longer open to discussion.

While it is true, however, that Hitler never varied this decision, his moods remained as unstable as ever, anger rapidly succeeding to resignation, and in turn yielding to the brief revival of hope. This is well illustrated by the incident of Göring's dismissal.

When Göring flew to the south he left behind as his representative General Koller, the Chief of Staff of the Air Force. On 23 April Koller appeared at the Obersalzberg and reported the decisions of the fateful conference in the bunker the day before. Hitler's intentions appeared to be clear enough: 'if it comes to negotiating the Reichsmarshal can do better than

I can.' But Göring was afraid of the responsibility, afraid in particular of Bormann. He anxiously fetched from the safe the decree of June 1941 which named him as the Führer's successor, and finally decided to wireless Hitler for confirmation: 'In view of your decision to remain at your post in the fortress of Berlin, do you agree that I take over, at once, the total leadership of the Reich, with full freedom of action at home and abroad, as your deputy, in accordance with your decree of 29 June 1941? If no reply is received by ten o'clock tonight I shall take it for granted that you have lost your freedom of action, and shall consider the conditions of your decree as fulfilled, and shall act for the best interests of our country and our people. . . .'

When Göring's message reached the bunker it did not take long for Bormann, Göring's sworn enemy, to represent it as an ultimatum. Speer, who was present, reports that Hitler became unusually excited, denouncing Göring as corrupt, a failure, and a drug addict, but adding: 'He can negotiate the capitulation all the same. It does not matter anyway who does it.'

The addition is revealing. Hitler was clearly angry at Göring's presumption—the habits of tyranny are not easily broken—he agreed to Bormann's suggestion that Göring should be arrested for high treason, and he authorized his dismissal from all his offices, including the succession—yet 'it does not matter anyway'. As Speer pointed out at Nuremberg, all Hitler's contempt for the German people was contained in the off-hand way in which he made this remark.

To try to make too much sense out of what Hitler said or ordered in those final days would be wholly to misread both the extraordinary circumstances and his state of mind. Those who saw him at this time and who were not so infected by the atmosphere of the bunker as to share his mood regarded him as closer than ever to that shadowy line which divides the world of the sane from that of the insane. He spoke entirely on the impulse of the moment, and moods of comparative lucidity were interspersed with wild accusations, wilder hopes and half-crazed ramblings.

Hitler found it more difficult than ever to realize the situation outside the shelter, or to grasp that this was the end. Conferences continued until the morning of the day on which he committed suicide. On the 24th he sent an urgent summons for Colonel General Ritter von Greim, in command of Air Fleet 6, to fly from Munich to Berlin. Greim made the hazardous journey into the heart of the capital, with the help of a young woman test-pilot, Hanna Reitsch, at the cost of a severe wound in his foot. When Greim arrived it was to find that Hitler had insisted on this simply in order to inform him personally that he was promoting him to be Commander-in-Chief of the Luftwaffe in succession to Göring, an appointment that he could perfectly well have made by telegram.

Later that night Hitler gave Hanna Reitsch a vial of poison. 'Hanna, you belong to those who will die with me. Each of us has a vial of poison such as this. I do not wish that one of us falls into the hands of the Russians alive, nor do I wish our bodies to be found by them.' At the end of a highly emotional interview Hitler reassured her: 'But, my Hanna, I still have hope. The Army of General Wenck is moving up from the south. He must and will drive the Russians back long enough to save our people. Then he will fall back to hold again.'

Hitler's resentment found expression in constant accusations of treachery, which were echoed by Goebbels and the others. Hanna Reitsch describes Eva Braun as 'raving about all the ungrateful swine who had deserted their Führer and should be destroyed. It appeared that the only good Germans were those who were caught in the bunker and that all the others were traitors because they were not there to die with him.' Eva regarded her own fate with equanimity. She had no desire to survive Hitler, and spent much of her time changing her clothes and caring for her appearance in order to keep up his spirits. Her perpetual complaint was: 'Poor, poor Adolf, deserted by everyone, betrayed by all. Better that ten thousand others die than that he should be lost to Germany.'

On the night of the 26th the Russians began to shell the Chancellery, and the bunker shook as the massive masonry split and crashed into the courtyard and garden. The Russians were now less than a mile away.

The climax came on the night of Saturday–Sunday, 28–9 April. Between nine and ten o'clock on the Saturday evening Hitler was talking to Ritter von Greim when a message was sent to him which determined him to end at last the career which had begun twenty-seven years before, at the end of another lost war. It consisted of a brief Reuter report that Himmler had been in touch with the Swedish Count Bernadotte for the purpose of negotiating peace terms.

Since the beginning of 1945 Himmler had been secretly urged by Walter Schellenberg, the youngest of his S.S. generals, to open negotiations with the Western Powers on his own initiative, and when Count Bernadotte visited Berlin in February Schellenberg arranged for Himmler to meet him. At that stage the reluctant Reichsführer S.S., much troubled by his loyalty to Hitler, had been unwilling to commit himself. Even when Bernadotte paid a second visit to Berlin in April Himmler could not make up his mind to speak out. But reports of the dramatic scene at the conference of 22 April and Hitler's declaration that the war was lost, and that he would seek death in the ruins of Berlin, made much the same impression on Himmler that it had made on Göring. Both men concluded that loyalty to Hitler was no longer inconsistent with steps to end the war, but while Göring telegraphed to Hitler for confirmation of his view, Himmler more wisely acted in secret. On the night of 23–4 April, Himmler accompanied Schellenberg to Lübeck for another meeting with Count Bernadotte at the Swedish Consulate. This time Himmler was prepared to put his cards on the table. Hitler, he told Bernadotte, was quite possibly dead; if not, he certainly would be in the next few days. 'In the situation that has now arisen,' Himmler continued, 'I consider my hands free. I admit that Germany is defeated. In order to save as great a part of Germany as possible from a Russian invasion I am willing to capitulate on the Western Front in order to enable the Western Allies to advance rapidly towards the east. But I am not prepared to capitulate on the Eastern Front.'

Bernadotte agreed to forward a proposal, although he warned the two Germans that he did not believe there was the least chance that Britain and the U.S.A. would agree to a separate peace.

On 27 April, Bernadotte returned with the news that the Western Allies insisted on unconditional surrender. This was a heavy blow, especially to Schellenberg. But worse was to follow: on the 28th the fact that Himmler had been taking part in such negotiations was reported from London and New York. Himmler was now to discover, as Göring had before him, that it was unwise to discount Hitler before he was really dead.

Hitler was beside himself at the news. Göring had at least asked permission first before beginning negotiations; Himmler, in whose loyalty he had placed unlimited faith, had said nothing. That Himmler should betray him was the bitterest blow of all, and it served to crystallize the decision to commit suicide which Hitler had threatened on the 22nd, but which he had not yet made up his mind to put into effect. This final decision followed the pattern of all the others: a period of hesitation, then a sudden resolution from which he was not to be moved. Throughout the week Hitler spoke constantly of taking his own life, and on the night of the 27th—if Hanna Reitsch's report is to be believed—he held a conference at which the plans for a mass suicide were carefully rehearsed and everyone made little speeches swearing allegiance to the Führer and Germany. But still he waited and hoped—until the night of the 28th, the night of decisions.

Shortly after he received the news of Himmler, Hitler disappeared behind closed doors with Goebbels and Bormann, the only two Nazi leaders in whom he now felt any confidence. Hitler's first thought was revenge, and Bormann had at least the satisfaction of removing Himmler as well as Göring before the Third Reich crumbled into dust.

Himmler's representative with the Führer, Fegelein, had already been arrested after it had been discovered that he had slipped quietly out of the bunker with the apparent intention of making a discreet escape before the end. The fact that he was married to Eva Braun's sister was no protection. He was taken into the courtyard of the Chancellery to be shot. Himmler was more difficult to reach, but Hitler ordered Greim and Hanna Reitsch to make an attempt to get out of Berlin by plane and entrusted them with the order to arrest Himmler at all costs.

Greim and Hanna Reitsch left between midnight and 1 a.m. on the morning of Sunday 29 April, and Hitler now turned to more personal matters. One human being at least had remained true and she should have her reward. Now that he had decided to end his life, the argument he had always used against marriage—that it would interfere with his career—no longer carried weight. So, between 1 a.m. and 3 a.m. on the 29th, Hitler married Eva Braun. The ceremony was hurriedly performed by a municipal councillor, Walter Wagner, then serving in the Volkssturm and called in by Goebbels. Both bride and bridegroom swore that they were 'of pure Aryan descent.' Goebbels and Bormann were the witnesses. Afterwards the bridal party returned to their private suite to drink champagne and talk nostalgically of the old days.

The celebration went on while Hitler retired to the adjoining room with his secretary, Frau Junge. There, in the early hours of 29 April, he dictated his will and his political testament.

Facing death and the destruction of the régime he had created, this man who had exacted the sacrifice of millions of lives rather than admit defeat was still recognizably the old Hitler. From first to last there is not a word of regret, nor a suggestion of remorse. The fault is that of others, above all that of the Jews, for even now the old hatred is unappeased. Word for word, Hitler's final address to the German nation could be taken from almost any of his early speeches of the 1920s or from the pages of *Mein Kampf*. Twenty-odd years had changed and taught him nothing.

The first part of the Political Testament consists of a general defence of his career. Hitler then turned to defend his decision to stay in Berlin and to speak of the future. 'After six years of war, which in spite of all set-backs will go down one day in history as the most glorious and valiant demonstration of a nation's life-purpose, I cannot forsake the city which is the capital of the Reich. . . .

'. . . I beg the heads of the Army, Navy, and Air Force to strengthen by all possible means the spirit of resistance of our soldiers in the National Socialist sense, with special reference to the fact that I myself, as founder and creator of this movement, have preferred death to cowardly abdication or even capitulation.'

The second part of the Testament contains Hitler's provisions for the succession. He began by expelling Göring and Himmler from the Party and from all offices of State, and accused them of causing immeasurable harm to Germany. As his successor he appointed Admiral Dönitz President of the Reich, Minister of War, and Supreme Commander of the Armed Forces—and promptly proceeded to nominate his Government for him. Goebbels and Bormann had their reward, the first as the new Chancellor, the second as Party Minister.

After naming other officials to be appointed, Hitler in the last paragraph returned once more to the earliest of his obsessions: 'Above all I charge the leaders of the nation and those under them to scrupulous observance of the laws of race and to merciless opposition to the universal poisoner of all peoples, international Jewry.'

The Testament was signed at four o'clock in the morning of Sunday 29 April, and witnessed by Goebbels and Bormann for the Party, by Burgdorf and Krebs, as representatives of the Army. At the same time Hitler signed his will, a shorter and more personal document:

'Although I did not consider that I could take the responsibility during the years of struggle of contracting a marriage, I have now decided, before the end of my life, to take as my wife the woman who, after many years of faithful friendship, of her own free will entered this town, when it was already besieged, in order to share my fate. At her own desire she goes to death with me as my wife. This will compensate us for what we have both lost through my work in the service of my people.

'What I possess belongs—in so far as it has any

value—to the Party, or, if this no longer exists, to the State. Should the State too be destroyed, no further decision on my part is necessary.

'My pictures, in the collection which I have bought in the course of years, have never been collected for private purposes, but only for the establishment of a gallery in my home-town of Linz on the Danube.

'It is my heartfelt wish that this bequest should be duly executed.

'As my executor I nominate my most faithful Party comrade, Martin Bormann. He is given full legal authority to make all decisions. He is permitted to hand to my relatives anything which has a sentimental value or is necessary for the maintenance of a modest standard of life; especially for my wife's mother and my faithful fellow-workers who are well known to him. The chief of these are my former secretaries, Frau Winter, etc., who have for many years helped me by their work.

'I myself and my wife choose to die in order to escape the disgrace of deposition or capitulation. It is our wish to be burned immediately in the place where I have carried out the greater part of my daily work in the course of my twelve years' service to my people.'

Hitler's choice of Dönitz as his successor is surprising, and to no one did it come as more of a surprise than to Dönitz himself. Hitler, who had come to look upon the Navy with different eyes, attached the greatest importance to the U-boat campaign, and contrasted the 'National Socialist spirit' of the Navy under Dönitz with what he regarded as the treachery and disaffection of the Army and Air Force. With Göring and Himmler excluded, Goebbels was the obvious choice as Hitler's successor, but Goebbels would never have been accepted by the soldiers. To command the Armed Forces—which, in effect, meant to negotiate a surrender—someone else, preferably a serving officer, must become Head of the State and

Minister for War. Goebbels was thus to succeed Hitler as Chancellor, but Dönitz was to become Head of the State and Supreme Commander. By choosing an officer from the Navy, rather than from the Army, Hitler offered a last deliberate insult to the military caste on whom he laid the blame for losing the war.

Characteristically, Hitler's last message to the German people contained at least one striking lie. His death was anything but a hero's end; by committing suicide he deliberately abandoned his responsibilities and took a way out which in earlier years he had strongly condemned as a coward's. The words in the Testament are carefully chosen to conceal this; he speaks of his 'unity with our soldiers unto death', and again of fulfilling his duty unto death. The fiction was maintained in the official announcement, and Dönitz, in his broadcast of 1 May, declared that the Führer had died fighting at the head of his troops.

During the 29th, while messengers were setting out from the bunker to deliver copies of the Führer's Political Testament, the news arrived of Mussolini's end. The Duce, too, had shared his fate with his mistress; together with Clara Petacci, he had been shot by the Partisans on the shore of Lake Como on 28 April. Their bodies were taken to Milan and hung in the Piazzale Loreto. The news can only have confirmed Hitler in the decision he had taken about his own end. Even when dead he was determined not to be put on show.

He now began to make systematic preparations for taking his life. He had his Alsatian bitch, Blondi, destroyed, and in the early hours of Monday 30 April assembled his staff in the passage and silently said farewell by shaking each by the hand. Shortly afterwards Bormann sent out a telegram to Dönitz, instructing him to proceed 'at once and mercilessly' against all traitors.

On the morning of the 30th Hitler was given the

latest reports on the situation in Berlin at the usual conference. The Russians had reached the Potsdamer Platz, only a block or two away from the Chancellery. Hitler received the news without excitement, and took lunch at two o'clock in the afternoon in the company of his two secretaries and his cook. Eva Hitler remained in her room and Hitler behaved as if nothing unusual were happening.

In the course of the early afternoon Erich Kempka, Hitler's chauffeur, was ordered to send two hundred litres of petrol to the Chancellery Garden.

Meanwhile, Hitler went to fetch his wife from her room, and for the second time they said farewell to those who remained in the bunker. Hitler then returned to the Führer's suite with Eva and closed the door. A few minutes passed while those outside stood waiting in the passage. Then a single shot rang out.

After a brief pause the little group outside opened the door. Hitler was lying on the sofa, which was soaked in blood: he had shot himself through the mouth. On his right-hand side lay Eva Braun, also dead: she had swallowed poison. The time was half past three on the afternoon on Monday 30 April 1945, ten days after Hitler's fifty-sixth birthday.

Hitler's instructions for the disposal of their bodies had been explicit, and they were carried out to the letter. Hitler's own body, wrapped in a blanket, was carried out and up to the garden by two S.S. men. Eva's body was picked up by Bormann, who handed it to Kempka. They made their way up the stairs and out into the open air, accompanied by Goebbels, Günsche, and Burgdorf. The doors leading into the garden had been locked and the bodies were laid in a shallow depression of sandy soil close to the porch. Picking up the five cans of petrol, one after another, Günsche, Hitler's S.S. adjutant, poured the contents over the two corpses and set fire to them with a lighted rag.

A sheet of flame leapt up, and the watchers withdrew to the shelter of the porch. A heavy Russian bombardment was in progress and shells continually burst on the Chancellery. Silently they stood to attention, and for the last time gave the Hitler salute; then turned and disappeared into the shelter.

The rest of the story is briefly told. Bormann at once informed Dönitz by radio that Hitler had nominated him as his successor, but he concealed the fact of Hitler's death for another twenty-four hours. During the interval, on the night of 30 April, Goebbels and Bormann made an unsuccessful effort to negotiate with the Russians. The Russian reply was 'unconditional surrender'. Then, but only then, Goebbels sent a further cable to Dönitz, reporting Hitler's death. The news was broadcast on the evening of 1 May to the solemn setting of music from Wagner and Bruckner's Seventh Symphony: the impression left was that of a hero's death, fighting to the last against Bolshevism.

An attempt at a mass escape by the men and women crowded into the network of bunkers round the Chancellery was made on the night of 1–2 May, and a considerable number succeeded in making their way out of Berlin. Among them was Martin Bormann: whether he was killed at the time or got away has never been established. Goebbels did not join them. On the evening of 1 May, after giving poison to his children, Goebbels shot his wife and himself in the Chancellery Garden. The bodies were set fire to by Goebbels's adjutant, but the job was badly done, and the charred remains were found next day by the Russians. After Goebbels's death the Führerbunker was set on fire.

In the following week Dönitz attempted to negotiate terms of surrender with the Western Allies, but their reply was uncompromising. The German Army in Italy had already capitulated and the British and Americans refused to be drawn by Dönitz's clumsy efforts to secure a separate peace and split the Grand Alliance. On 4 May Admiral von Friedeburg signed an armistice providing for the surrender of the German forces in north-west Europe, and early on the morning of the 7th General Jodl and Friedeburg put their signatures to an unconditional surrender of all the German forces presented to them jointly by the representatives of the U.S.A., Great Britain, the U.S.S.R., and France at Rheims.

The Third Reich had outlasted its founder by just one week.

EPILOGUE

Many attempts have been made to explain away the importance of Hitler, from Chaplin's brilliant caricature in *The Great Dictator* to the much less convincing picture of Hitler the pawn, a front man for German capitalism. Others have argued that Hitler was nothing in himself, only a symbol of the restless ambition of the German nation to dominate Europe; a creature flung to the top by the tides of revolutionary change, or the embodiment of the collective unconscious of a people obsessed with violence and death.

These arguments seem to me to be based upon a confusion of two different questions. Obviously, Nazism was a complex phenomenon to which many factors—social, economic, historical, psychological—contributed. But whatever the explanation of this episode in European history—and it can be no simple one—that does not answer the question with which this book has been concerned, what was the part played by Hitler. It may be true that a mass movement, strongly nationalist, anti-Semitic, and radical, would have sprung up in Germany without Hitler. But so far as what actually happened is concerned—not what might have happened—the evidence seems to me to leave no doubt that no other man played a role in the Nazi revolution or in the history of the Third Reich remotely comparable with that of Adolf Hitler.

The conception of the Nazi Party, the propaganda with which it must appeal to the German people, and the tactics by which it would come to power—these were unquestionably Hitler's. After 1934 there were no rivals left and by 1938 he had removed the last checks on his freedom of action. Thereafter, he exercised an arbitrary rule in Germany to a degree rarely, if ever, equalled in a modern industrialized state.

At the same time, from the re-militarization of the Rhineland to the invasion of Russia, he won a series of successes in diplomacy and war which established an hegemony over the continent of Europe comparable with that of Napoleon at the height of his fame. While these could not have been won without a people and an Army willing to serve him, it was Hitler who provided the indispensable leadership, the flair for grasping opportunities, the boldness in using them. In retrospect his mistakes appear obvious, and it is easy to be complacent about the inevitability of his defeat; but it took the combined efforts of the three most powerful nations in the world to break his hold on Europe.

Luck and the disunity of his opponents will account for much of Hitler's success—as it will of Napoleon's—but not for all. He began with few advantages, a man without a name and without support other than that which he acquired for himself, not even a citizen of the country he aspired to rule. To achieve what he did Hitler needed—and possessed—talents out of the ordinary which in sum amounted to political genius, however evil its fruits.

His abilities have been sufficiently described in the preceding pages: his mastery of the irrational factors in politics, his insight into the weaknesses of his opponents, his gift for simplification, his sense of timing, his willingness to take risks. An opportunist entirely without principle, he showed both consistency and an astonishing power of will in pursuing his aims. Cynical and calculating in the exploitation of his histrionic gifts, he retained an unshaken belief in his historic role and in himself as a creature of destiny.

The fact that his career ended in failure, and that his defeat was pre-eminently due to his own mistakes, does not by itself detract from Hitler's claim to greatness. The flaw lies deeper. For these remarkable powers were combined with an ugly and strident egotism, a moral and intellectual cretinism. The passions which ruled Hitler's mind were ignoble: hatred, resentment, the lust to dominate, and, where he could not dominate, to destroy. His career did not exalt but debased the human condition, and his twelve years' dictatorship was barren of all ideas save one—the further extension of his own power and that of the nation with which he had identified himself. Even power he conceived of in the crudest terms: an endless vista of military roads, S.S. garrisons, and concentration camps to sustain the rule of the Aryan 'master race' over the degraded subject peoples of his new empire in the east.

The great revolutions of the past, whatever their ultimate fate, have been identified with the release of certain powerful ideas: individual conscience, liberty, equality, national freedom, social justice. National Socialism produced nothing. Hitler constantly exalted force over the power of ideas and delighted to prove that men were governed by cupidity, fear, and their baser passions. The sole theme of the Nazi revolution was domination, dressed up as the doctrine of race, and, failing that, a vindictive destructiveness. It is this emptiness, this lack of anything to justify the suffering he caused rather than his own monstrous and ungovernable will which makes Hitler both so repellent and so barren a figure.

The view has often been expressed that Hitler could only have come to power in Germany, and it is true—without falling into the same error of racialism as the Nazis—that there were certain features of German historical development, quite apart from the effects of the Defeat and the Depression, which favoured the rise of such a movement.

This is not to accuse the Germans of Original Sin, or to ignore the other sides of German life which were only grossly caricatured by the Nazis. But Nazism was not some terrible accident which fell upon the German people out of a blue sky. It was rooted in their history, and while it is true that a majority of the German people never voted for Hitler, it is also true that thirteen millions did. Both facts need to be remembered.

From this point of view Hitler's career may be described as a *reductio ad absurdum* of the most powerful political tradition in Germany since the Unification. This is what nationalism, militarism, authoritarianism, the worship of success and force, the exaltation of the State, and *Realpolitik* lead to, if they are projected to their logical conclusion.

There are Germans who reject such a view. They argue that what was wrong with Hitler was that he lacked the necessary skill, that he was a bungler. If only he had listened to the generals—or Schacht—or the career diplomats—if only he had not attacked Russia, and so on. There is some point, they feel, at which he went wrong. They refuse to see that it was the ends themselves, not simply the means, which were wrong: the pursuit of unlimited power, the scorn for justice or any restraint on power; the exaltation of will over reason and conscience; the assertion of an arrogant supremacy, the contempt for others' rights.

The Germans, however, were not the only people who preferred in the 1930s not to know what was happening and refused to call evil things by their true names. The British and French at Munich; the Italians, Germany's partners in the Pact of Steel; the Poles, who stabbed the Czechs in the back over Teschen; the Russians, who signed the Nazi–Soviet Pact to partition Poland, all thought they could buy Hitler off, or use him to their own selfish advantage. They did not succeed, any more than the German Right or the German Army. In the bitterness of war and occupation they were forced to learn the truth of the words of John Donne which Ernest Hemingway set at the beginning of his novel of the Spanish Civil War: 'No man is an Iland, intire of it selfe; every man is a peece of the Continent, a part of the maine; If a clod bee washed away by the Sea, Europe is the lesse, as well as if a Promontorie were, as well as if a Mannor of thy friends or of thine own were; Any man's death diminishes me, because I am involved in Mankinde; And therefore never send to know for whom the bell tolls; It tolls for thee.'

Hitler, indeed, was a European, no less than a German phenomenon. The conditions and the state of mind which he exploited, the *malaise* of which he was the symptom, were not confined to one country, although they were more strongly marked in Germany than anywhere else. Hitler's idiom was German, but the thoughts and emotions to which he gave expression have a more universal currency.

Hitler recognized this relationship with Europe perfectly clearly. He was in revolt against 'the System' not just in Germany but in Europe, against the liberal bourgeois order, symbolized for him in the Vienna which had once rejected him. To destroy this was his mission, the mission in which he never ceased to believe; and in this, the most deeply felt of his purposes, he did not fail. Europe may rise again, but the old Europe of the years between 1789, the year of the French Revolution, and 1939, the year of Hitler's War, has gone for ever—and the last figure in its history is that of Adolf Hitler, the architect of its ruin. '*Si monumentum requiris, circumspice*'—'If you seek his monument, look around.'